Laboratory Guide to Insect Pathogens and Parasites

Laboratory Guide to Insect Pathogens and Parasites

George O. Poinar, Jr.
and Gerard M. Thomas

College of Natural Resources
University of California
Berkeley, California

PLENUM PRESS • NEW YORK AND LONDON

Library of Congress Cataloging in Publication Data

Poinar, George O.
Laboratory guide to insect pathogens and parasites.

Includes bibliographical references and index.
1. Insects—Diseases—Diagnosis—Laboratory manuals. 2. Insects—Parasites—Identification—Laboratory manuals. 3. Micro-organisms, Pathogenic—Identification—Laboratory manuals. I. Thomas, Gerard M. II. Title.
SB942.P65 1984 595.7′02 84-9875
ISBN 0-306-41680-8

This book is a revised and expanded edition of *Diagnostic Manual for the Identification of Insect Pathogens,* published in 1978

A Division of Plenum Publishing Corporation
233 Spring Street, New York, N.Y. 10013

Printed in the United States of America

Dedicated to the memory of
ROBERT VAN DEN BOSCH,
for his interest, encouragement, and tenacity
in promoting biological pest control

A. Dark red color of a larva of the wax moth, *Galleria mellonella,* infected with *Heterorhabditis* nematodes. The color is produced by the bacterium *Xenorhabdus luminescens,* which is symbiotically associated with nematodes of the genus *Heterorhabditis.* Another insect pathogen that produces a red color in the host is the bacterium *Serratia marcescens.*

B. Microsporidans, such as this species of *Thelohania* in the larva of the mosquito *Culiseta incidens,* often turn the host a whitish color as the spores accumulate in the infected tissues. A noninfected host is on the right. (Photo by H. Chapman, courtesy of R. Goodwin.)

C. Many granulosis viruses, such as this one infecting the lepidopteran *Trichoplusia ni,* turn the insect host white or cream color owing to an accumulation of inclusion bodies. A similar color can also be produced in some nuclear polyhedrosis virus infections.

D. Insects covered with green, powdery spores are likely to be infected by the fungus *Metarrhizium anisopliae,* as shown here. Other fungi which produce green spores during their development are *Nomuraea rileyi* and some species of *Aspergillus.* Some of the chloriridoviruses also turn their hosts green, and strains of the bacterium *Pseudomonas aeruginosa* may produce a green pigment inside their hosts.

E. Blue is an unusual color for diseased insects, but it is imparted to some hosts infected with iridoviruses, such as this scarabaeid larva, *Sericesthis,* infected with *Sericesthis* iridovirus. (Photo by R. Goodwin.)

F. Orange color is imparted to many hosts infected with fungi of the order Entomophthorales. Here are aphids attacked by *Erynia aphidis.* Note the circles of white spores (conidia) that are forcibly discharged from the hosts.

G. Insects covered with a white, cottony growth may be infected with fungi of the genus *Hirsutella,* as in the case of this scarabaeid grub, *Rhopoea morbellosa.* Similar white mats of mycelium are produced on insects infected with *Beauveria bassiana* and other less commonly encountered species of fungi. (Photo by R. Goodwin.)

H. White-ringed orange-red spots on citrus leaves are probably the result of scale insects or whiteflies infected with the fungus *Aschersonia* sp., as shown here.

PREFACE

After the publication of the *Diagnostic Manual for the Identification of Insect Pathogens,* the authors received many queries asking why they had not included the larger metazoan parasites as well as the microbial forms. An examination of the literature indicated that pictorial guides to the identification of nematodes and the immature stages of insect parasites were unavailable. Consequently we decided to rewrite the sections covering insect pathogens and combine these with new sections on entomogenous nematodes and the immature stages of insect parasites. The result is the present laboratory guide, which is unique in covering all types of biotic agents which are found inside insects and cause them injury or disease. Included as parasites are insects and nematodes. Among the pathogens included are viruses, rickettsias, bacteria, fungi, and protozoans.

Emphasis is placed on identification with an attempt to use the most easily recognizable characters. Use of a certain number of technical terms is unavoidable, and explanations of these can be found in most biological dictionaries or the glossary of invertebrate pathology prepared by Steinhaus and Martignoni (1970).

For each of the separate biotic categories included in the guide, a brief discussion of the following topics is included: (1) types of associations with insects, (2) taxonomic status, (3) life cycle, (4) characteristics of infected insects, (5) methods of examination, (6) isolation and cultivation, (7) characters for identification, (8) testing for infectivity, (9) storage, (10) literature, and (11) an illustrated key to the groups (families) or genera.

An extra section on the availability and sources of biological pest control agents is also included.

We are deeply indebted to a number of scientists, mentioned in the acknowledgments, for sending us photographs or living material to be photographed. Their support and enthusiasm helped make this work possible. We hope the guide will serve its intended purpose as an aid to the identification of damaging biotic agents occurring inside insects. We welcome any suggestions which would make it more complete.

George O. Poinar, Jr.
Gerard M. Thomas

Berkeley

ACKNOWLEDGMENTS

We would like to thank the following persons who supplied us with pathogens or photographs of pathogens unavailable to us: R. Hess, L. Etzel, L. Caltagirone, J. Wright, K. Lakin, R. M. Bohart, J. K. Clark, J. Huber, A. Slater, B. Villegas, J. Evans, J. Harper, S. G. Moore, R. Akhurst, R. Zarate, B. A. Keddie, E. Kurstak, R. Granados, T. Hukuhara, Y. Iwashita, C. Payne, B. Hillman, L. Bailey, J. Hurliman, J. Weiser, C. Splittstoesser, D. Sanders, B. Nelson, E. M. McCray, Jr., M. Martignoni, W. R. Kellen, A. Kaplan, M. Laird, A. M. Huger, D. Hoffmann, E. Hazard, E. Greiner, R. Goodwin, T. Fukuda, D. Forgach, B. F. Eldridge, J. Couch, H. Chapman, W. Burgdorfer, L. P. Brinton, and K. S. S. Nair.

The authors would especially like to thank Roberta Hess for her assistance in preparing the chapter on viruses and L. Caltagirone for his help in preparing the section on insect endoparasites and for supplying the photograph of Dr. R. van den Bosch.

CONTENTS

IDENTIFICATION OF THE CATEGORIES OF INSECT PATHOGENS AND PARASITES

It will often be obvious to what group a particular pathogen or parasite belongs. However, it should be remembered that simply finding one of the biotic agents mentioned in this guide inside an insect does not necessarily mean that it is the cause of mortality. Many insects, nematodes, fungi, and bacteria are saprophytic and quickly invade an insect cadaver. Unless the agent is a known pathogen, Koch's postulates (given in Techniques) should be followed to determine if the agent is indeed infectious.

The beginning of any study of insect mortality or disease is the identification of the causal agent, and that is the purpose of this guide. The following key will provide assistance in determining which category of biotic agent is involved.

KEY TO THE CATEGORIES OF INSECT PATHOGENS AND PARASITES

1. Insects covered with mycelium, often bearing spores, or with string- or hornlike structures emerging from their bodies; cadaver often mummified, hard or cheeselike in consistency; tissues containing mycelium and/or spores—Fungi

1. Insects not covered with mycelium or elongate structures—**2**

2. Insects containing metazoan (many-celled) parasites in their body cavity—**3**

2. Insects containing unicellular (single-celled) particles in their body cavity—**4**

3. Parasites mostly elongate and wormlike; lacking segments, appendages, and head capsules—Phylum Nematoda

3. Parasites mostly cigar- or pear-shaped; containing segments; head region often containing a distinct capsule with simple antennal projections—Class Insecta

4. Particles spherical, stain reddish with Sudan III (Fig. 1) (see Techniques)—Fat globules

4. Particles variable in shape, do not stain reddish with Sudan III—**5**

5. Particles showing birefringence under polarized light (Fig. 2)—Urate or other types of crystals

5. Particles not showing birefringence under polarized light—**6**

6. Particles generally motile and rod-shaped (occasionally spherical and nonmotile); develop in the hemocoel (rarely only in the intestine); spores may be present—Bacteria

6. Particles generally nonmotile and not rod-shaped (except Rickettsia and some Protozoa); develop in host cells or tissues (some ciliate protozoans are motile and multiply in the hemocoel of their hosts); infectious particles (spores, sporozoites, etc.) or structures containing infectious particles (oocysts, etc.) variable in size and shape—**7**

7. Infectious particles rod-shaped; just visible under the light microscope—Rickettsia

7. Infectious particles (spores, sporozoites, etc.) or structures containing infectious particles (cysts, etc.) ellipsoidal, spherical, or rarely rod-shaped; generally easily visible under the light microscope—**8**

8. Small to minute particles (greatest diameter generally less than 10.0 μm) consisting of polyhedra and capsules which usually dissolve in a weak solution of NaOH—Inclusion viruses

8. Particles of various shapes (diameter generally ranges from 2.0 to 20.0 μm) which do not dissolve in a weak solution of NaOH—**9**

9. Infectious particles (spores, sporozoites, etc.) or structures containing infectious particles (cysts, etc.) generally formed within host tissues; some ciliated forms multiply in the host's hemolymph—Protozoa

9. Infectious particles (spores) or structures containing infectious particles (cysts) generally formed in the host's hemolymph; cilia absent—Fungi

VIRUSES

INTRODUCTION

Insect virology is a very dynamic field owing to the continuous discovery of new viruses and new host–pathogen relationships. While viruses originally were thought to be restricted to a few specific insect groups, it is now apparent that they infect representatives of many insect orders. Viruses very similar to those attacking insects have also been reported in mites, shrimp, and even some vertebrates. Widespread use of these viruses is now being discussed, and safety considerations for the baculoviruses have already been examined (Summers *et al.*, 1975; Mittenburger, 1980).

Viruses which cause diseases in plants and vertebrates may be transmitted by insects. These vectors may carry the virus mechanically or actually serve as a second host. In the latter case, the virus multiplies in the vector and may act as a pathogen by, for example, reducing longevity or damaging tissues. Thus, Mims *et al.* (1966) discovered that salivary glands of *Aedes aegypti* were destroyed by an arbovirus. Likewise, plant viruses may produce cytopathic changes, a shortened life span, reduction in reproductive ability, or even death in their insect vectors (Maramorosch, 1968a). Even though most of the viruses which are vectored by insects are not pathogenic to their host, they may be encountered when working with these insects.

TAXONOMIC STATUS

Viruses can be defined as submicroscopic, obligate, intracellular, pathogenic entities. Experts claim that about 450 different pathogenic viruses have been isolated from insects and mites and that about 90% of these have inclusion bodies (=occluded) (Ignoffo, 1974).

Insect viruses were originally classified based on the morphology of their inclusion bodies and virions or virus particles. Host group and tissue affinity were also characteristic features. At one stage, virus groups were given generic names such as *Borrelinavirus, Smithiavirus,* and *Bergoldiavirus,* which corresponded to the nuclear polyhedrosis (NPV), cytoplas-

Table 1. Types of Viruses Found in Insects

	Occluded	Nonoccluded
Enveloped, dsDNA	Baculoviridae, nuclear polyhedrosis viruses	Baculoviridae, nonoccluded rod-shaped nuclear viruses
	Baculoviridae, granulosis viruses	Baculoviridae, [1]nonoccluded nuclear viruses with a polydisperse DNA genome
	Poxviridae, entomopoxviruses	Baculoviridae, [2]filamentous bee virus group
Enveloped, ssRNA		Rhabdoviridae Bunyaviridae Togaviridae
Nonenveloped, dsDNA		[3]Iridoviridae, Iridovirus [3]Iridoviridae, Chloriridovirus
Nonenveloped, ssDNA		Paroviridae, Densovirus
Nonenveloped, dsRNA	Reoviridae, cytoplasmic polyhedrosis viruses	[1]Bisegmented dsRNA viruses Reoviridae, Orbivirus
Nonenveloped, ssRNA		Nodaviridae Picornaviridae Nudaurelia β virus group [4]Chronic bee paralysis virus group Caliciviridae

[1]Proposed status. [3]May possess cell-derived envelope.
[2]Possible member. [4]Unclassified.

mic polyhedrosis (CPV), and granulosis (GV) viruses, respectively. However, as more basic biochemical and structural data were obtained on insect and other viruses, it became possible to compare all known viruses, and new sets of families, genera, and species were proposed (see Table 1). Thus, the NPV and GV were placed together in the family Baculoviridae (Matthews, 1982).

From the practical standpoint of identification, insect viruses fall into two groups, those possessing inclusion bodies (usually just visible under the light microscope) and those without inclusion bodies (visible only with the electron microscope).

Those without inclusion bodies consist of virus particles (virions) which may be rod-, bullet-, oval-, anisometrically, or isometrically shaped. When inclusion bodies occur, the virions usually are embedded within them.

LIFE CYCLE

Except for specific cases of transovarial transmission, insect viruses generally enter a host through the mouth and digestive tract. The virus particles, which are ingested directly or released from inclusion bodies, infect or pass through the gut epithelium to enter susceptible host tissues. Some viruses appear to infect specific tissues, while others are polytrophic, capable of infecting most, if not all, tissues of the host. While most viruses produce acute infections, others are occult and produce an inapparent disease which cannot be detected by external examination of the host. Occult viruses may be induced to the acute pathogenic stage by subjecting the insects to stress conditions such as crowding or starvation. These conditions are probably responsible for the sudden and dramatic outbursts of virus diseases by activating latent infections. The infection process of various insect viruses has been summarized by David (1975), Harrap (1973), Delgarno and Davey (1973), and Vaughn (1974).

CHARACTERISTICS OF INFECTED INSECTS

Insects suffering from virus infections may exhibit morphological, physiological, and behavioral symptoms. The extent and type of symptoms depend on the virus and host involved. Lepidopteran larvae infected with NPV may show behavioral abnormalities such as moving toward the tops of plants, where they cease feeding and become flaccid. However, symptoms usually occur late in infection. The integument frequently changes color (often whitish at first, then dark) and becomes distended. Infected insects may hang by their prolegs from foliage. Death quickly follows, with disintegration of internal tissues and release of inclusion bodies, which may cloud the hemolymph. The integument may rupture if the hypodermis was infected. Pupae may show similar symptoms when late larval instars become infected. NPV in Hymenoptera infect only the midgut epithelium. Sawfly larvae infected with NPV may exhibit a faint, yellow discoloration (especially on the third to fifth abdominal segments), lose their appetite, and become inactive. A brown or milky fluid is often exuded from the anus. Larvae of *Tipula paludosa* (Meigen) infected with NPV become lighter as the disease progresses and finally turn chalky white. Unusual crescent-shaped inclusion bodies are produced in this infection. In the mosquito, NPV infections appear to occur only in the midgut.

Whereas NPV infect Lepidoptera, Hymenoptera, Coleoptera, Trichoptera, Neuroptera, and some Diptera, GV have been observed only in Lepidoptera. Symptoms associated with infection with GV are nonspecific and vary considerably from one insect to another. The first symptom is usually a paling in color, followed by loss of appetite. There may be a mottling of the integument, and very often the ventral surface will become progressively pale whitish or milky yellow owing to the infected fat bodies (color plate, p. vii, C). As infected tissues disintegrate, large numbers of capsules are released, and the hemolymph becomes turbid and milky. In some species, there may be a liquification of internal tissues after death, and when the epidermis is infected, the integument becomes very fragile, similar to cases of NPV infection. However, if the epidermis is not affected, the integument remains relatively firm.

Lepidopteran larvae infected with CPV usually are retarded in growth. There is a loss of appetite and feeding is reduced or may cease since the virus infects only the midgut cells. Larvae become sluggish and inactive. As the disease progresses, the infected midgut may be visible through the integument as a pale yellow or whitish area. In later stages, polyhedra are often regurgitated or passed out with the feces. The skin of diseased larvae usually does not rupture as with NPV infections.

As with other occlusion-body-producing viruses, the entomopoxviruses (EPV) usually produce a whitish change in the coloration of the insect, with the accumulation of occlusion bodies in the target tissues. In later stages of infection, sluggishness and slow development sometimes occur. Occlusion bodies may be regurgitated or defecated.

Larvae infected with iridoviruses can often be recognized by the opalescent, iridescent, blue, green, or brown color of the infected tissues (color plate, E). The fat body and hemocytes are the primary tissues infected, although the virus systemically infects the insect. The fat body generally becomes extensively infected, resulting in the death of the larvae.

The symptoms of other nonoccluded viruses seem to be peculiar to their host and the type of disease produced. For example, the inherited "sigma virus" of *Drosophila* sp. leaves the flies sensitive to carbon dioxide. Honeybee larvae which die from the sacbrood virus are extended lengthwise along the floor of the cell with the head darker than the rest of the body. Most small RNA viruses infect the midgut initially, and this often causes vomiting and diarrhea. Cricket paralysis virus causes paralysis in the adult hosts, and the viruses in the Nodaviridae also induce paralysis in the host. However, many of the nonoccluded viruses may be present in inapparent infections causing no overt symptoms in the host insect.

METHODS OF EXAMINATION

If a virus disease is suspected, an examination of the host's tissues should be made with the light microscope. Since the majority of reported insect viruses belong to the occluded type, their variously shaped inclusion bodies are usually visible under the light microscope. The inclusion bodies of NPV and CPV appear refringent (shining white) in bright field and phase contrast, whereas the capsules of GV appear white in bright field and gray in phase contrast. Small inclusion bodies may exhibit Brownian movement in wet mounts.

Uric acid crystals (Fig. 2) often occur in diseased insects and may resemble inclusion bodies. However, urate crystals are birefringent and often give a characteristic cross appearance under polarized light, while inclusion bodies are monorefringent. The latter can be distinguished from spherical fat droplets (Fig. 1), which turn red in the presence of aqueous Sudan III (10–15 min). The inclusion bodies do not stain with Sudan; however, their ability to dissolve in strong alkali (1 N NaOH) is characteristic (Fig. 19).

Inclusion bodies can be further demonstrated in diseased tissues by special staining methods, such as the Feulgen–Schiff reaction, slow Giemsa staining with acid hydrolysis, and the iron hematoxylin methods described by Huger (1961). Sikorowski *et al.* (1971) described a method for detecting polyhedra of a CPV in *Heliothis*. These methods are described in Techniques.

Noninclusion viruses can only be detected with the electron microscope. However, noninclusion virus infections sometimes can be determined from various signs and symptoms of infected hosts, along with the absence of other pathogens.

ISOLATION

There are several methods of isolating viruses for further studies. The simplest of these involves placing the diseased insects in a culture tube with water. After two or three days, the inclusion bodies will accumulate as a white layer on the bottom of the tube. Cell and tissue remnants, bacterial cells, and other breakdown products can be separated from the inclusion bodies by repeated washing and differential centrifugation. If the virus is to be used for infection studies, the pellet can be treated with antibiotics for 24 hr and washed several times with sterile distilled water, although differential certrifugation generally removes most bacteria. Inocula used for injection studies should be assayed for viable bacteria or fungi by culturing an aliquot on an enriched agar medium such as brain–heart infusion agar or AC medium (see Techniques). A variation which is particularly helpful in purifying very small inclusion bodies, such as granulosis capsules and small cytoplasmic polyhedra, is to mix the triturated diseased insect with an equal volume of carbon tetrachloride, shake vigorously, and centrifuge at 3000 rpm for about 1 hr. A layered plug of inclusion bodies forms between the CCl_4 (bottom) and the water phase, while fat accumulates on top of the water phase. The water and fat can be decanted off without disturbing the plug, and the inclusion bodies, which form the top layer of the plug, can be carefully washed or scraped off with a small spatula and suspended in water. Further purification may be accomplished by repeated washing and differential centrifugation. If convenient, a sucrose-gradient centrifugation will give a purer preparation of capsules and polyhedra than the CCl_4 method.

With the nonoccluded iridoviruses, the first step in the isolation process is to dissect out the infected tissues (identified by their iridescent color), triturate them in buffer, and leave them for 24 hr. The action of autolytic enzymes will further disintegrate the tissues and free the virus particles. The suspension should then be centrifuged at 3000 rpm for about 30 min and the virus-containing supernatant saved. The supernatant can be freed of most bacteria by differential and sucrose-gradient centrifugation or by passage through a 0.45-μm Millipore filter. Any fat will settle on top of the supernatant and must be removed before filtration. Centrifugation of the above supernatant at 10,000 rpm will concentrate the virus in a pellet which can then be washed and purified by differential centrifugation.

Another method of obtaining almost pure iridescent virus particles is to infect lepidopteran larvae containing large silk glands (e.g., *Bombyx mori* L., *Pseudaletia* sp., or *Galleria mellonella* L.). After the infection is well advanced, the heavily infected blue silk glands can be dissected out and the virus particles removed.

Noniridescent, noninclusion viruses are much more difficult to detect and purify. A general method for unenveloped viruses which can be followed is that used by Bailey *et al.* (1964) for the isolation of adult bee paralysis and sacbrood viruses of honeybees. Whole insects, or washed parts of them, are ground in tap water and a quarter volume of CCl_4. The resulting emulsion is coarsely filtered (e.g., through cheese cloth) and the filtrate purified by centrifugation at 8000*g* for 10 min. The water phase containing the virus is at the top and can be used as is or purified and concentrated by high-speed centrifugation. Viruses in the resulting supernatant can be examined in the electron microscope.

IDENTIFICATION

Most inclusion viruses can still be identified as to group on the basis of the morphology of the inclusion body and enclosed virus particles and the site of replication. The morphology of the virus particles is determined with the electron miscrocope (Figs. 3 and 4). Virions can be initially classified based on the presence or absence of an envelope, the shape of the virus (rod, bullet, oval, filamentous, aniosometric, or isometric), and its size. Additional diagnostic characteristics observed with the electron microscope are surface modifications on the virion (such as projections, spikes, peplomers, or knobs) and the presence of an envelope or capsid with particular staining properties (such as cup-shaped depressions, capsomer structure, and the presence of a core). The key provided can be used to identify the family of nonoccluded viruses based on morphology and size. However, size and shape are no longer sufficient characters for all of the nonoccluded viruses, and knowledge of the type and strandedness of the nucleic acid plus analysis of the viral structural proteins and other physicochemical characteristics are necessary. Serological techniques such as the enzyme-linked immunosorbent assay and the radioimmunoassay are also currently useful in the identification and characterization of viruses. The technical aspects of virus purification and characterization are presented in books edited by Howard (1982) and Fraenkel-Conrat and Wagner (1981) for viruses in general. Specific information on insect virus groups can be obtained by consulting individual papers listed in the section on literature. Payne and Kelly (1981) list references for guidance on purification and antiserum production for individual insect virus groups.

Insect viruses have been placed in eleven families (Matthews, 1982). In addition about 30 small RNA viruses of unknown affinities have yet to be classified, along with some other DNA and RNA viruses. The classification of insect viruses will certainly change in the future; however, a wise move has been made in treating the invertebrate viruses in the same general classification as the vertebrate and plant viruses.

In identifying insect viruses, characteristics of the disease are also important. The type of tissue infected, as well as any abnormal symptoms and the hosts involved, are all useful aids. Aside from specialized instances, however, host groups cannot be used for virus identification. There are now too many instances of similar viruses appearing in unrelated hosts.

TESTING FOR PATHOGENICITY

Since viruses normally enter their hosts per os, the simplest method for obtaining experimental infections is to introduce the virus particles into the mouth of a test insect. A calibrated hypodermic syringe can be used for introducing a measured amount of virus suspension into the mouth of test insects. An indirect approach is to contaminate the insect's food; for example, foliage can be immersed in virus suspensions and fed to insect larvae. For intrahemocoelic injections (for nonoccluded viruses), the virus suspension must be purified or freed of bacteria that could multiply and destroy any potential host. This can be done by first passing the suspension through a 0.45-μm Millipore filter or treating it with a 5000 unit/ml mixture of penicillin–streptomycin.

STORAGE

Inclusion bodies can be stored either in a purified condition, within the host tissues under refrigeration (5°C), or frozen. All occluded insect viruses apparently can survive lyophilization (freeze drying), and this may prove to be the most lasting method of storage.

LITERATURE

Basic coverage of insect viruses is presented by Smith (1967), who later reviewed the polyhedroses and granuloses of insects (Smith, 1971). Vaughn (1974) presented a general review of the virus diseases of insects. Other groups of insect viruses are discussed in works edited by Steinhaus (1963) and Maramorosch (1968a,b).

Other discussions of insect viruses include those of the replication of baculoviruses, CPV, and iridoviruses by Delgarno and Davey (1973); a general discussion of virus infections in invertebrates by Harrap (1973); and descriptions of associations between viruses and Diptera (Marshall, 1973), viruses and Lepidoptera (Longworth, 1973), viruses and Hymenoptera (Bailey, 1973b), and viruses and leafhoppers (Sinka, 1973).

General reviews of insect pathogenic viruses are presented in chapters by Tinsley and Harrap (1978), Harrap and Payne (1979), and Payne and Kelly (1981). In these works the morphology, biology, replication, and physiochemical characteristics of major insect virus groups are discussed.

The occluded viruses have been extensively studied. Of these the baculoviruses are discussed by Summers (1977), Faulkner (1981), and Tanada and Hess (1984). Bergoin and Dales (1971) presented a detailed account of poxviruses of invertebrates and vertebrates, and the biology and morphology of EPV are reviewed by Kurstak and Garzon (1977) and Granados (1981). Aruga and Tanada (1971) edited a book on the CPV of the silkworm. Additional comprehensive treatments of CPV include those by Payne and Harrap (1977) and Payne (1981).

Two recent works deal generally with nonoccluded insect viruses. Longworth (1978) discusses small isometric viruses of invertebrates, focusing on parvoviruses and small RNA viruses. Kelly (1981) covers the nonoccluded baculoviruses, bisegmented dsRNA viruses, reoviruses, retroviruses, and rhabdoviruses.

More specific works which deal with nonoccluded viruses include those on bee viruses by Bailey (1976) and Bailey and Woods (1977). Nonoccluded nuclear viruses with a polydisperse DNA genome found associated with the calyx of parasitic female Hymenoptera are covered by Stoltz and Vinson (1979). The Parvoviridae have been reviewed by Kurstak (1972) and Kurstak *et al.* (1977). Iridoviruses, particularly of *Tipula,* are dealt with in a chapter by Lee (1977), and Kelly and Robertson (1973) looked at the general properties of insect iridoviruses. Discussions on the sigma virus of *Drosophila,* a rhabdovirus, were presented by Sylvester (1977) and Teninges *et al.* (1980). The bisegmented dsRNA viruses, including *Drosophila* X virus, were treated by Dobos *et al.* (1979).

Of the other insect viruses not mentioned, many individual papers discuss their properties and morphology. Some references are given in the key while others will be found among the general reviews in the References. Discussions on the use of viruses for insect control have been presented by Stairs (1971), Ignoffo (1968, 1974), Bailey (1973a), and Tinsley (1978, 1979). Martignoni and Iwai (1975) and Martignoni (1981) have compiled a useful computer-based catalog of viral diseases of insects an mites.

KEY TO THE FAMILIES, SUBFAMILIES, AND GROUPS OF VIRUSES

1. Inclusion bodies (protein crystals enclosing virus particles) present (visible with the light microscope) (Figs. 3, 5–22)—**2**

1. Inclusion bodies absent; only virions or virus particles formed in host cells (visible only with the electron miscroscope) (Figs. 4, 23–37)—**5**

2. Inclusion bodies (capsules) ovoid, ellipsoidal, small; 0.2–0.5 μm in length, appear dark under phase contrast; virions bacilliform, single nucleocapsid per envelope, 40–60 nm × 200–400 nm, replicate in nucleus and cytoplasm, contain dsDNA, thus far found only in larvae of Lepidoptera—Baculoviridae, granulosis viruses. Example: *Trichoplusia ni* granulosis virus. See Summers (1971) (Figs. 3, 5–9).

2. Inclusion bodies irregular, polyhedral (0.2–20.0 μm in diameter); appear light under phase contrast; occur in representatives of many orders—**3**

3. Inclusion bodies ovoid, 2.0–20.0 μm in diameter; noninfective spindle-shaped bodies may be also present; virions ovoid- or cuboid-shaped, 170–250 nm × 300–400 nm, enveloped with globular surface units which give virions mulberrylike appearance, replicate in cytoplasm containing dsDNA—Poxviridae, Entomopoxvirinae. Example: *Amsacta moorei* entomopoxvirus. See Granados (1973). Three probable genera exist in entomopoxviruses based on morphology of the virions, host range, and molecular weight of the genome: A, with one lateral body and a unilateral concave core, surface globular units 22 nm, found in Coleoptera; B, with a sleeve-shaped lateral body and a cylindrical core, surface globular units 40 nm, found in Lepidoptera; C, with two lateral bodies and a biconcave core, found in Diptera (Figs. 3, 10–13).

3. Inclusion bodies irregular, polyhedral; 0.2–15.0 μm in length; virions rod-shaped or isometric, contain DNA or RNA—**4**

4. Inclusion bodies usually 0.2–2.5 μm in diameter, formed in the cytoplasm of midgut epithelial cells; virions isometric, nonenveloped, 50–65 nm in diameter with twelve hollow spikes at vertices of the icosahedral, containing dsRNA—Reoviridae, cytoplasmic polyhedrosis viruses. Example: *Bombyx mori* cytoplasmic polyhedrosis virus. See Kobayashi (1971) (Figs. 3, 14–17).

4. Inclusion bodies usually 0.5–15.0 μm in diameter, formed in the nuclei of various tissues; virions rod-shaped, with a single nucleocapsid (SNPV) or multiple nucleocapsids (MNPV) per envelope, containing dsDNA—Baculoviridae, nuclear polyhedrosis viruses. Example: *Autographa californica* nuclear polyhedrosis virus (AcMNPV). See Falcon and Hess (1977) (Figs. 3, 18–22).

5. Infected tissues showing iridescence with reflected light—**6**

5. Infected tissues noniridescent with reflected light—**7**

6. Diseased insects and/or infected tissue showing blue iridescence with reflected light; virions icosahedral, nonenveloped (may have host-derived membrane) with dense core, 120 nm in diameter; replicate in cytoplasm, containing dsDNA—Iridoviridae, Iridovirus. Example: *Tipula* iridescent virus. See Lee (1977) (Figs. 4, 23, 24, and color plate, E).

6. Diseased insects and/or infected tissues showing a yellow-green iridescence with reflected light; virions icosahedral, 180 nm in diameter, nonenveloped (may have host-derived membrane) with dense core, replicate in cytoplasm, containing dsDNA—Iridoviridae, Chloriridovirus. Example: Mosquito iridescent virus, type 3, regular strain from *Aedes taeniorrhynchus*. See Clark *et al.* (1965).

7. Virions bacilliform or bullet-shaped, enveloped—**8**

7. Virions isometric, enveloped or unenveloped—**10**

8. Virions 130–380 nm × 50–95 nm with surface projections on envelope, virions bud from cell membranes, containing ssRNA—Rhabdoviridae. Example: Sigma virus of *Drosophila*. See Teninges *et al.* (1980) (Fig. 4). CAUTION: Arboviruses are in this group.

8. Envelope smooth, without projections, contain dsDNA—**9**

9. Virions 80 nm × 220 nm, replicate in nucleus; proposed subgroup of Baculoviridae, nonoccluded rod-shaped nuclear viruses. Example: *Oryctes rhinoceros* virus. See Huger (1966) (Fig. 25).

9. Virions of uniform size or may be variable, replicate in nucleus; located in the calyx of parasitic female Hymenoptera and transferred to the host at time of oviposition (therefore found in parasitized hosts)—Proposed subgroup of Baculoviridae; nonoccluded nuclear viruses with a polydisperse DNA genome. Example: *Phanerotoma flavitestacae* calyx virus. See Poinar *et al.* (1976) (Fig. 26).

10. Virions anisometric—**11**

10. Virions isometric spherical or oval—**12**

11. Virions mainly ellipsoidal, varying lengths with modal width of 22 nm, unenveloped, develop in cytoplasm, containing ssRNA—Unclassified virus, chronic bee paralysis group. Example: Chronic bee paralysis virus. See Bailey (1976) (Fig. 27).

11. Virions anisometric in sectioned material, oval when isolated, enveloped, nucleocapsid filamentous, 20–22 nm in diameter and up to 3 μm in length, replicate in nucleus, probably containing DNA—Possible Baculoviridae; filamentous virus. Example: filamentous virus of *Apis mellifera*. See Clark (1978) (Figs. 28, 29).

12. Virions 18 nm or greater in diameter; enveloped or unenveloped; occur alone—**13**

12. Virions 13–17 nm in diameter; unenveloped; may occur with other viruses; containing RNA—Unclassified satellite or mini viruses. Example: Chronic bee paralysis virus associate. See Bailey (1976).

13. Virions large, 35–100 nm in diameter; enveloped or unenveloped—**17**

13. Virions small, isometric, 18–30 nm in diameter; unenveloped—**14**

14. Virions 18–26 nm in diameter, with a core 14–17 nm in diameter, replicate in nucleus; contain ssDNA—Parvoviridae, Densovirus. Example: Densovirus of *Galleria mellonella*. See Kurstak *et al.* (1977) (Figs. 4, 30–32).

14. Virions containing ssRNA—**15**

15. Virions 29 nm in diameter; with hollow surface projections—Unclassified Kelp virus group. Example: Kelp fly virus. See Scotti *et al.* (1976).

15. Virions without hollow surface projections—**16**

16. Virions 22–30 nm in diameter, replicate in cytoplasm, monopartite genome—Picornaviridae. Example: Cricket paralysis virus. See Reinganum *et al.* (1970) (Figs. 4, 33, 34)

16. Virions 29 nm in diameter, replicate in cytoplasm, bipartite genome—Nodaviridae, Nodavirus. Example: Nodamura virus. See Bailey *et al.* (1975). Caution: This virus will grow in suckling mice and can cause fatal infections (Scherer *et al.*, 1968). It has a wide host range and can be grown in vertebrate and invertebrate tissue culture lines (Fig. 4).

17. Virons enveloped; spherical to oval—**18**

17. Virons unenveloped—**19**

18. Virions 40–70 nm in diameter, core 25–36 nm in diameter, with surface projections; multiply in cytoplasm; contain ssRNA (positive-sense genome)—Togaviridae; Alphavirus, Flavivirus. Examples: Sindbis virus, yellow fever virus. See Bishop and Shope (1979). Caution: These viruses multiply in arthropods as well as in vertebrates (Arboviruses) (Fig. 4).

18. Virions 90–100 nm in diameter, with surface projections; replicate in cytoplasm; contain ssRNA (negative-sense genome)—Bunyaviridae: Bunyavirus, Phelobovirus, Nairovirus, Uukuvirus. Examples: Bunyamwera virus, Sandfly fever Sicilian virus, Nairobi sheep disease. See Murphy (1980). Caution: The host range for these viruses includes warm- and cold-blooded vertebrates as well as arthropods (Arboviruses) (Fig. 4).

19. Virons 30–45 nm in diameter, with ssRNA—**20**
19. Virions 55–85 nm in diameter, with dsRNA—**21**

20. Virions 35 nm in diameter; replicate primarily in cytoplasm, with monopartite genome—Nudaurelia B virus group. Examples: *Nudaurelia cytherea capenis* virus, Trichoplusia RNA virus. see Morris *et al.* (1979) (Figs. 4, 35, 36).
20. Virions 35–39 nm in diameter, with 32 cup-shaped surface depressions; mature in cytoplasm, with monopartite genome—Caliciviridae, Calicivirus. Example: Calicilike virus of navel orange worm. See Hillman *et al.* (1982) (Figs. 4, 37).

21. Virions 60–64 nm in diameter, cores 45 nm in diameter in sectioned material—Bisegmented dsRNA viruses. Example: *Drosophila* X virus. See Teninges *et al.* (1979).
21. Virions 65–80 nm in diameter, with two protein coats; outer capsid layer diffuse and structureless under electron microscope; cores with twelve spikes; replicate in cytoplasm—Reoviridae, Orbivirus. Example: Bluetongue virus. See Verwoerd *et al.* (1979). CAUTION: The host range for these viruses includes insects and other arthropods as well as vertebrates including man (Arbovirus).

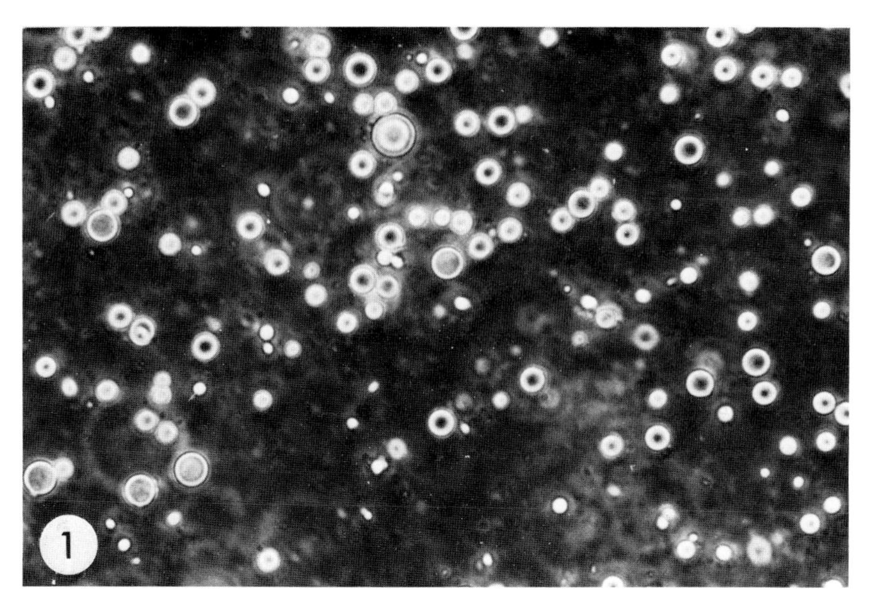

FIGURE 1. Fat globules from the hemocoel of a healthy insect. ×640.

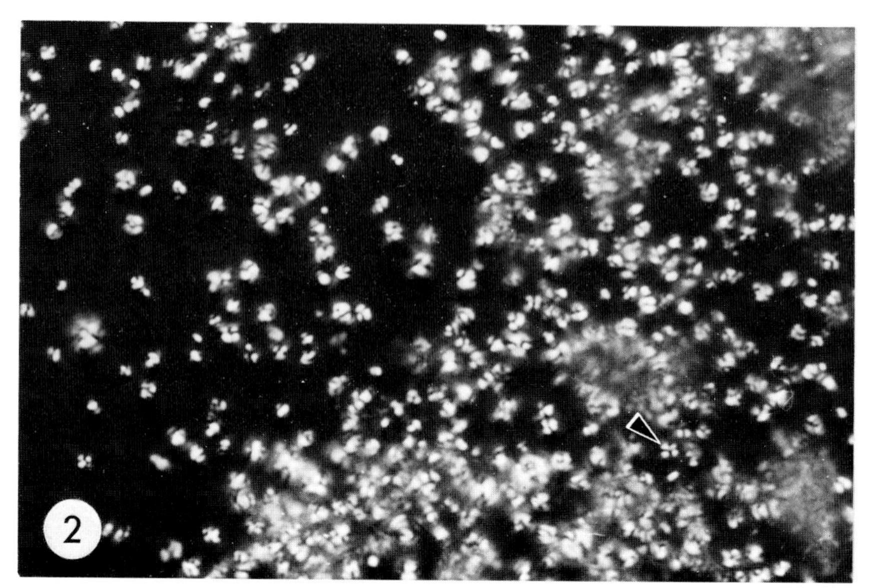

FIGURE 2. Urate crystals under polarized light. Note cross (arrowhead). ×640.

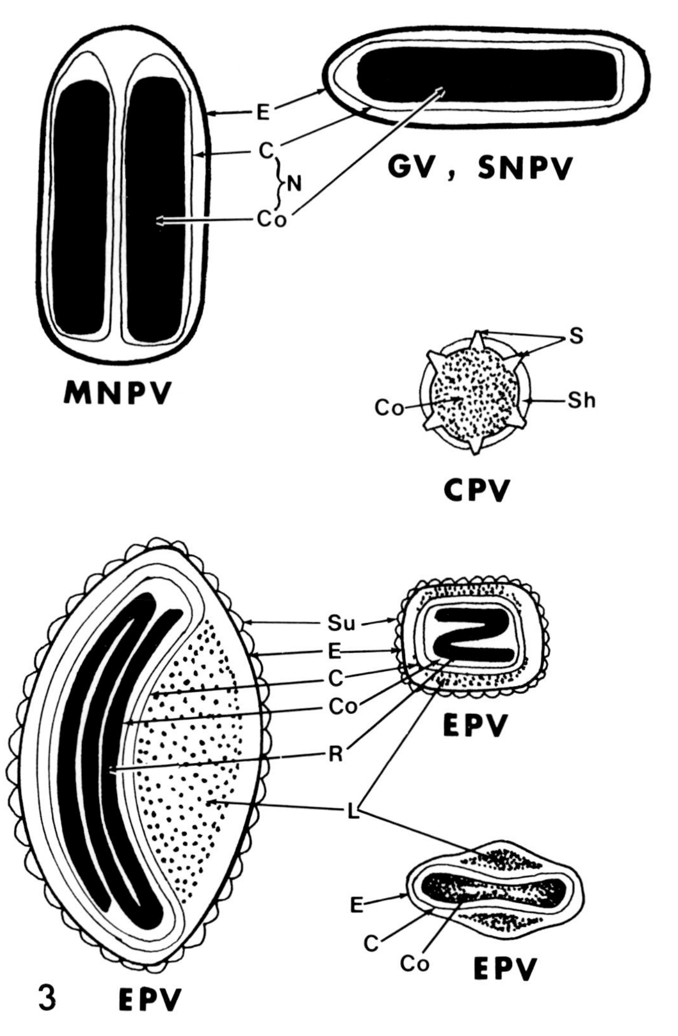

FIGURE 3. Diagrammatic representation of types of virions in occluded insect viruses. C, capsid; Co, core; CPV, cytoplasmic polyhedrosis virus; E, envelope; EPV, entomopoxviruses; GV, granulosis virus; L, lateral body; MNPV, multiple nucleocapsides per envelope nuclear polyhedrosis virus; N, nucleocapsid; R, coiled rodlike structure; S, spikes; Sh, shell, SNPV, single nucleocapsid per envelope nuclear polyhedrosis virus; Su, surface units.

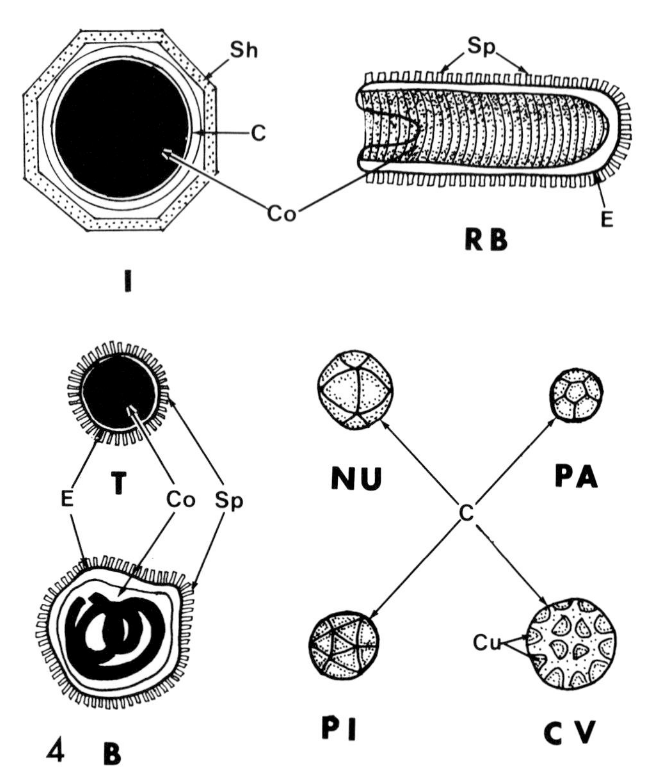

FIGURE 4. Diagrammatic representation of nonoccluded insect viruses. B, Bunyaviridae; C, capsid; Co, core; Cu, cups; CV, Caliciviridae; E, envelope; I, Iridoviridae; NU, Nudaurelia; PA, Parvoviridae; PI, Picornaviridae; RB, Rhabdoviridae; Sh, shell; Sp, surface projections; T, Togaviridae.

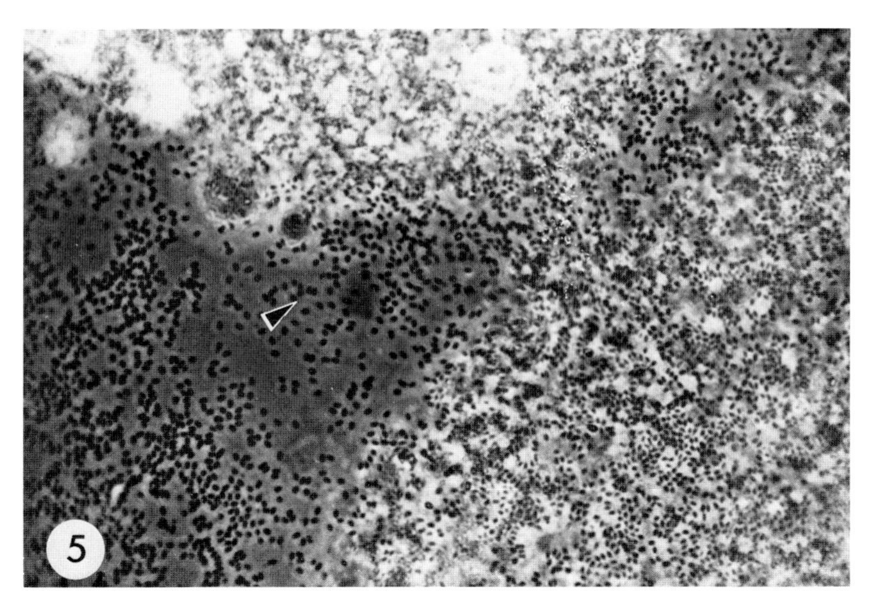

FIGURE 5. Fat body infected with a granulosis virus; note liberated capsules (arrow). ×640.

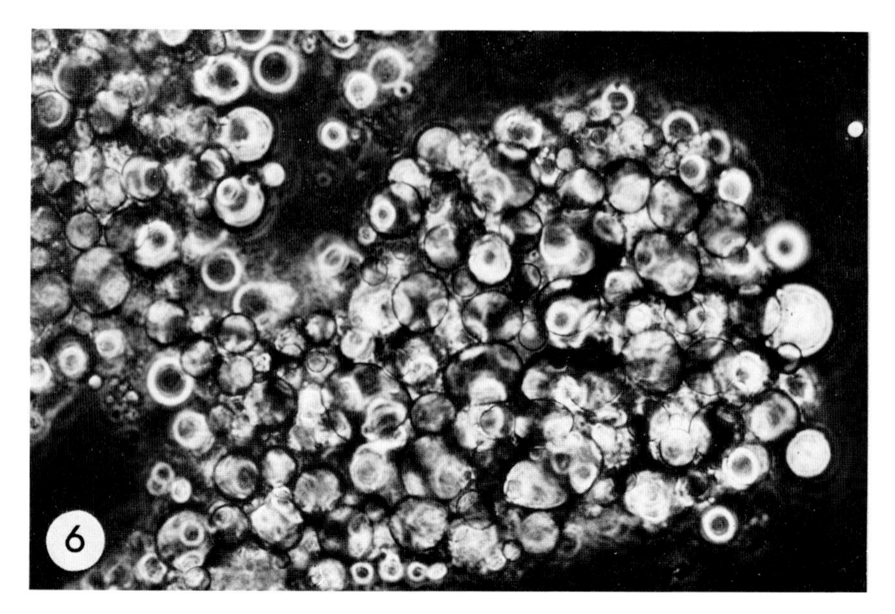

FIGURE 6. Normal fat body. ×640.

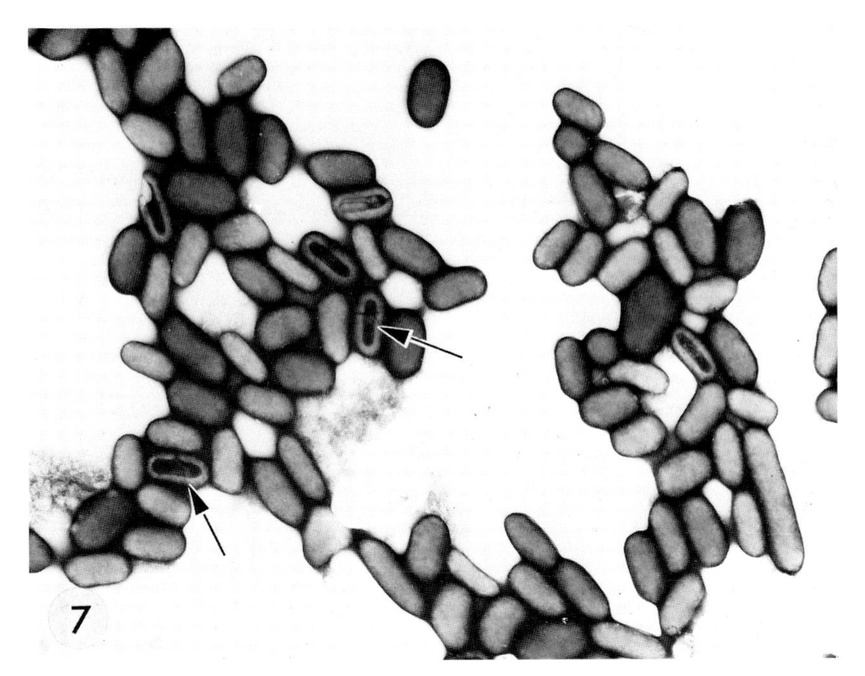

FIGURE 7. Transmission electron micrograph (TEM) of a granulosis virus. A negatively stained preparation showing ovocylindrical capsules. In some cases the stain shows virus rods within the capsules (arrows). (Courtesy of Roberta Hess.) ×27,500.

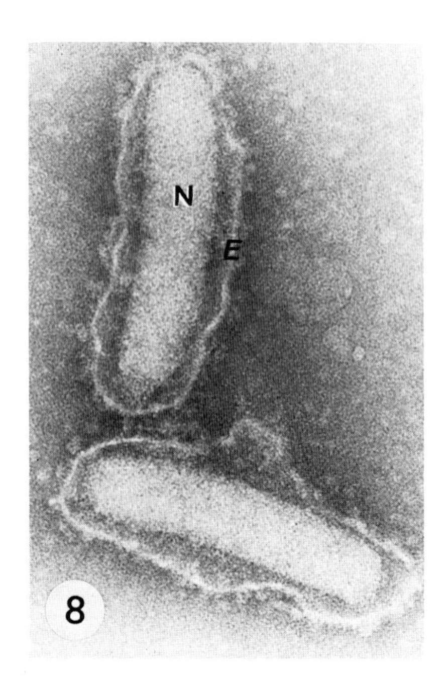

FIGURE 8. TEM of a density-gradient-purified granulosis virus in which the capsule has been removed. Shown are the envelope (E) and the nucleocapsid (N). The preparation is negatively stained. (Courtesy of Roberta Hess.) ×140,000.

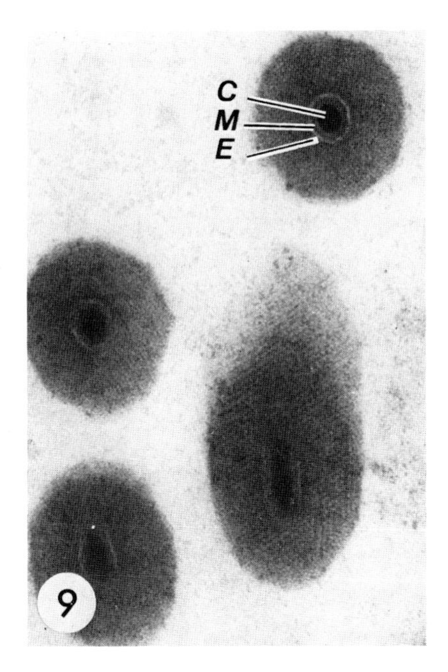

FIGURE 9. TEM section which illustrates the paracrystalline nature of the capsule granulin. A cross section shows the virus core (C), capsid (M), and envelope (E). (Courtesy of Roberta Hess.) ×40,000.

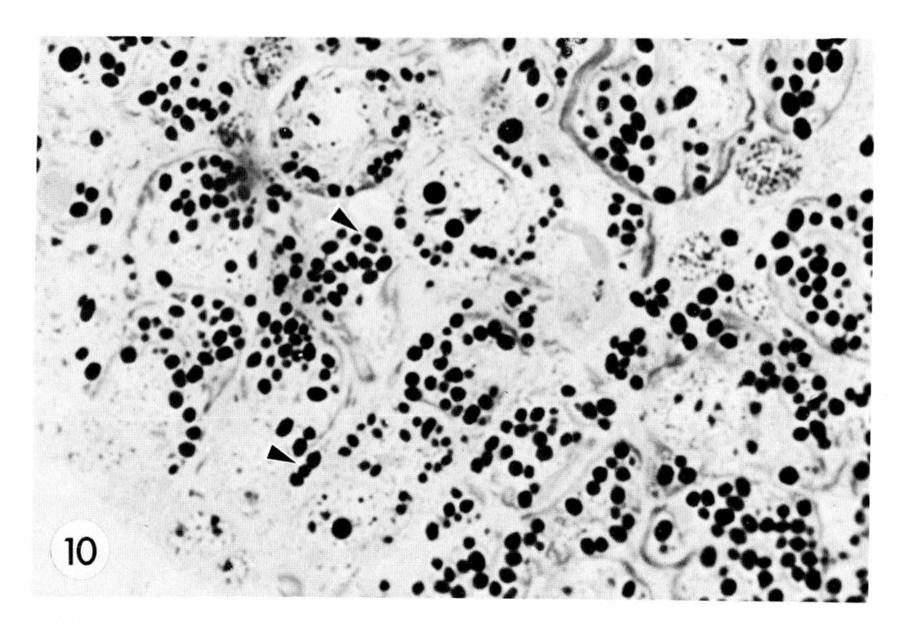

FIGURE 10. Sectioned adipose tissue of *Choristoneura fumiferana* infected with entomopoxvirus. The fat cells lose their granules and numerous spheroids (arrows) appear. (Courtesy of E. Durstak.) $\times$2000.

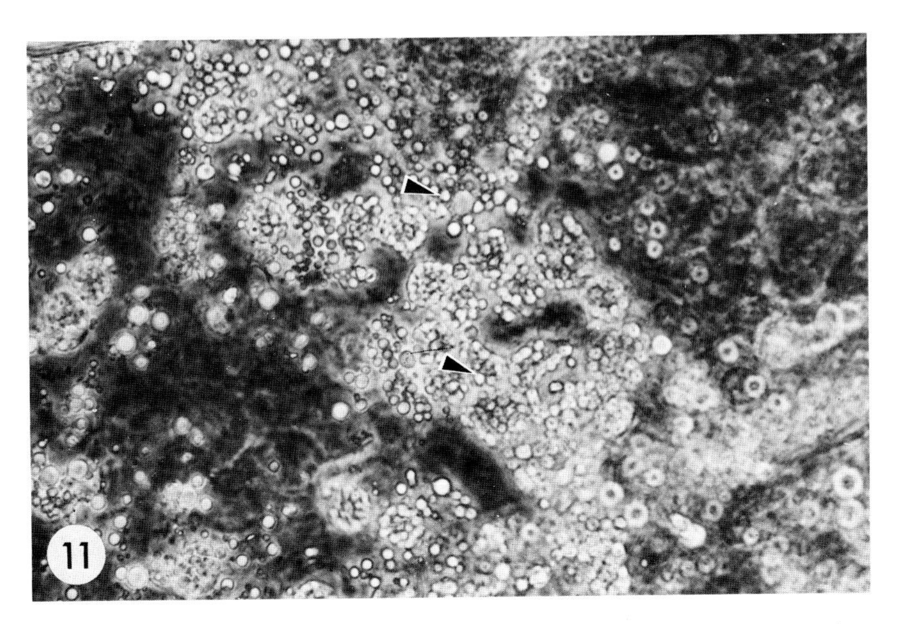

FIGURE 11. A squash mount of the scarabaeid grub *Cyclocephala immaculata* infected with an entomopoxvirus. Note spheroids (arrows). ×2000.

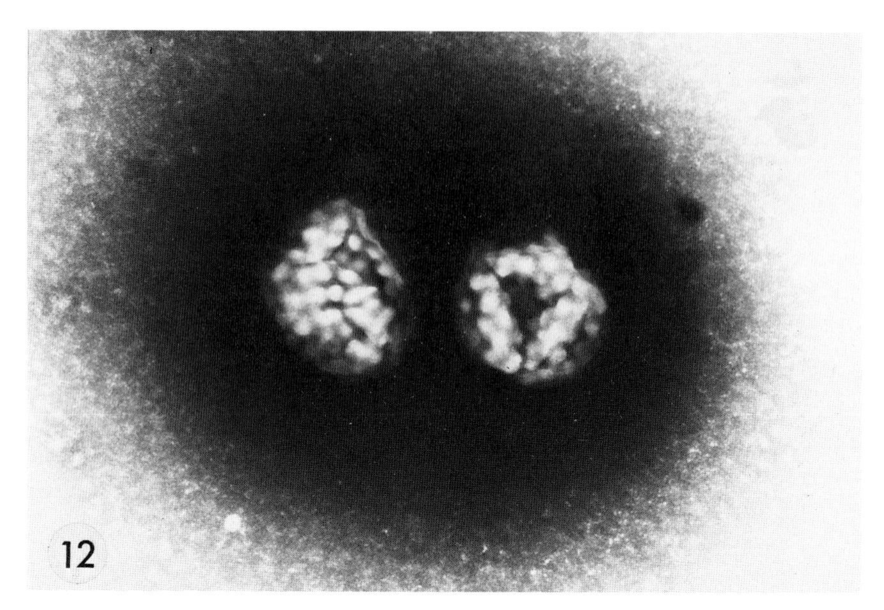

FIGURE 12. *Amsacta moorei* (Lepidoptera) entomopoxvirus particles negatively stained with phosphotungstic acid. Note mulberrylike appearance of the virions. (Courtesy of R. Granados.) ×81,000.

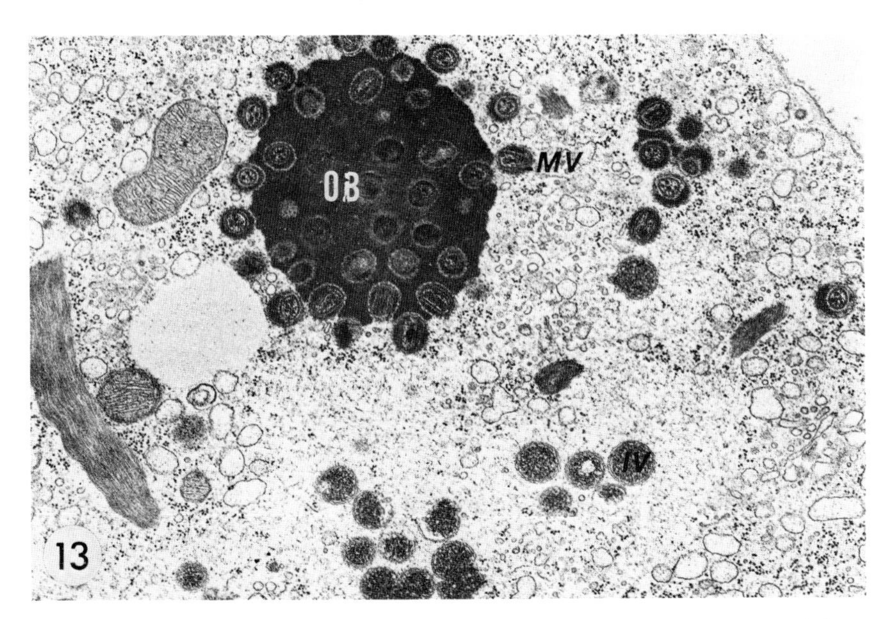

FIGURE 13. Ultrathin section of cultured cell infected with *Amsacta moorei* (Lepidoptera) entomopoxvirus. OB, occlusion body containing numerous mature virus particles; MV, mature virion; IV, immature virion. (Courtesy of R. Granados.) ×15,300.

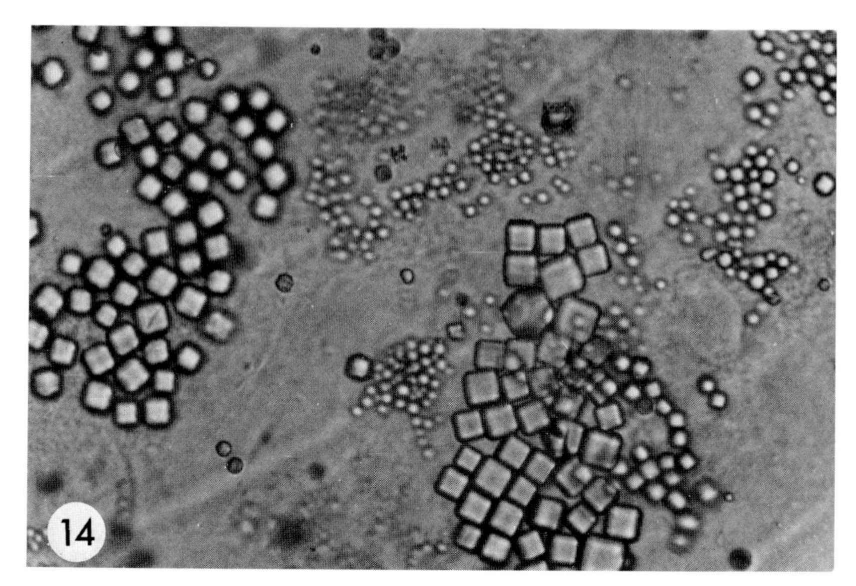

FIGURE 14. Light micrograph preparation of polyhedra of a cytoplasmic polyhedrosis from the H strain of *Bombyx mori* (Lepidoptera). (Courtesy of Tosihiko Hukuhara.) $\times 700$ (approximate).

FIGURE 15. Epithelian midgut cells of *Bombyx mori* (Lepidoptera) infected with cytoplasmic polyhedrosis virus showing polyhedra (P) in the midgut lumen and within the cytoplasm. (Polyhedra stain well with orange G or aniline blue after hydrolysis for 5 min in 1 N HCl at 60°C.) (Courtesy of Yoshimitsu Iwashita.) ×700 (approximate).

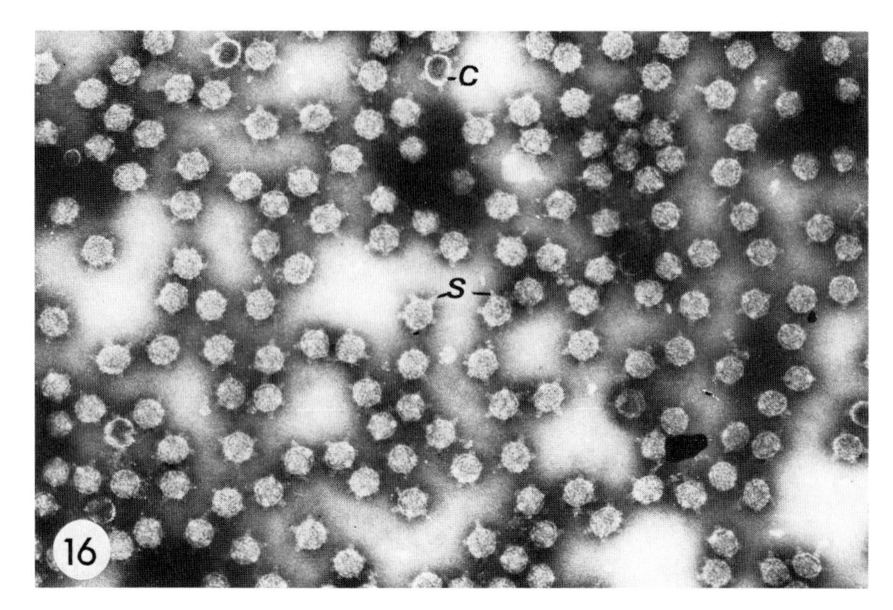

FIGURE 16. A negatively stained preparation of cytoplasmic polyhedrosis virus from *Phalera bucephala* (Lepidoptera) showing spikes (S). Empty particles show capsids (C). (Courtesy of C. Payne.) ×72,000.

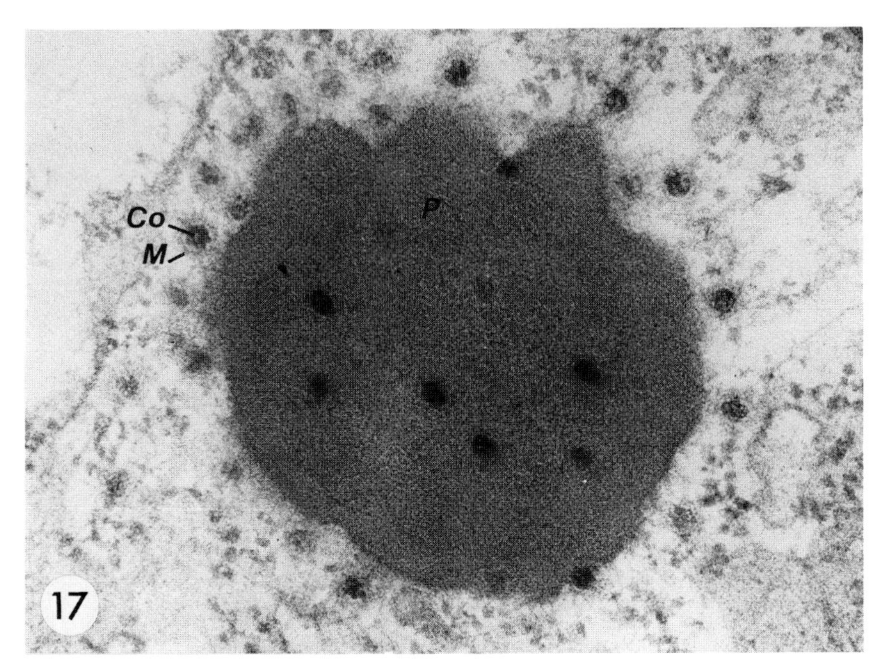

FIGURE 17. Section of cytoplasmic polyhedrosis viruses in the process of occlusion within a polyhedra (P). Easily recognizable are the capsid (M) and electron-dense core (Co.) (Courtesy of Roberta Hess.) $\times$82,000.

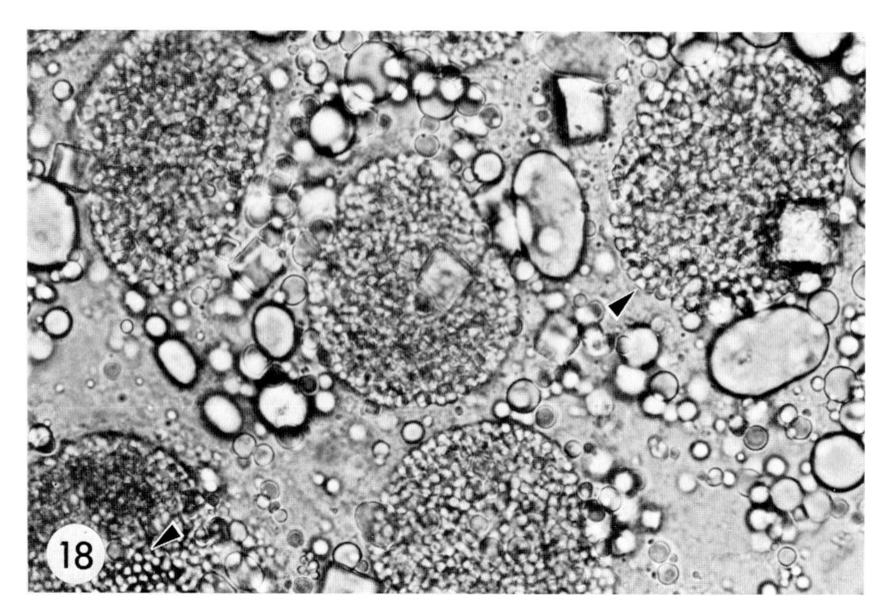

FIGURE 18. Squash of *Autographa californica* (Lepidoptera) nuclear-polyhedrosis-virus-infected tissue of *Trichoplusia ni* illustrating polyhedra (arrows). (Courtesy of B. A. Keddie.) ×680.

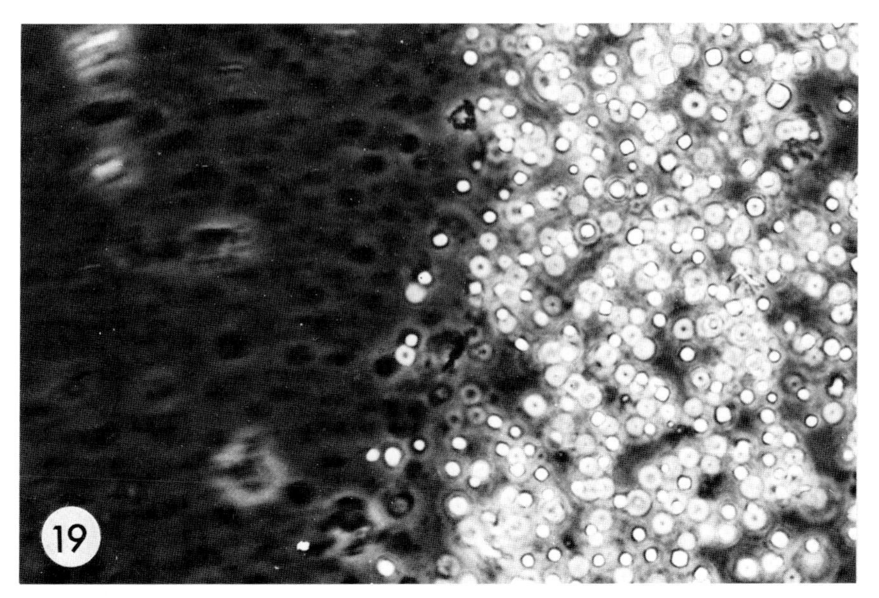

FIGURE 19. Polyhedra of a nuclear polyhedrosis virus being dissolved with NaOH. ×500.

FIGURE 20. Electron micrograph of polyhedra of a nuclear polyhedrosis virus, illustrating polyhedral shape. (Courtesy of Roberta Hess.) ×2000.

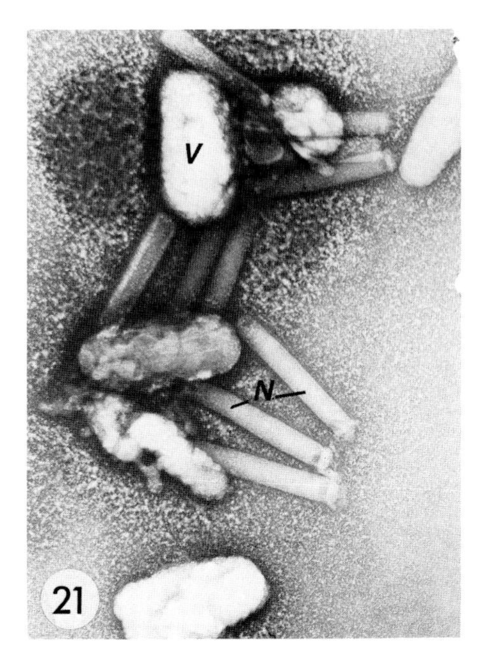

FIGURE 21. Enveloped virions (V) of the MNPV type and nucleocapsids (N) of gradient-purified nuclear polyhedrosis virus. Negatively stained. (Courtesy of Roberta Hess.) ×60,000.

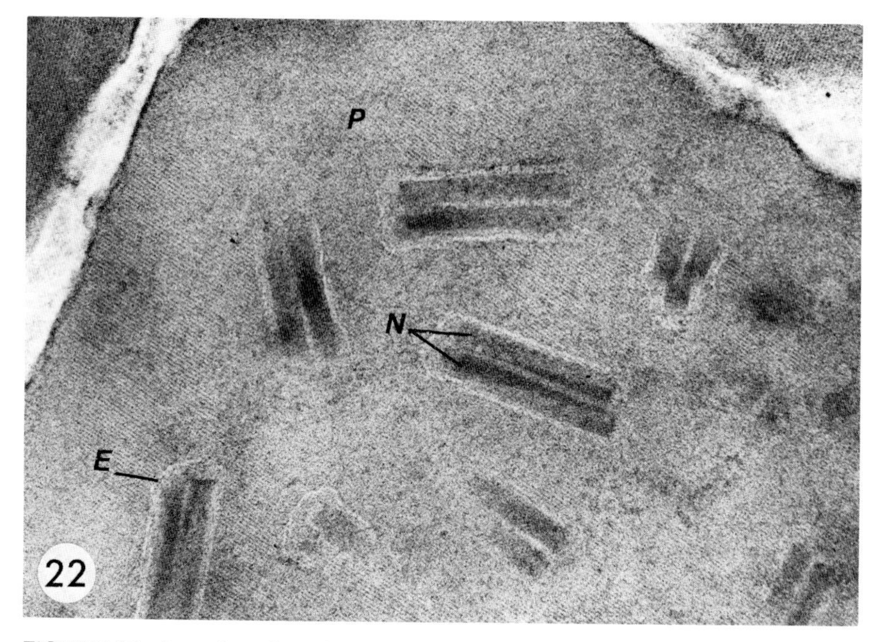

FIGURE 22. A section of a polyhedra (P) showing the paracrystalline nature of the polyhedrin. The nuclear polyhedrosis virus is of the MNPV type and the section shows typically two nucleocapsids (N) per envelope (E). (Courtesy of Roberta Hess.) ×80,000.

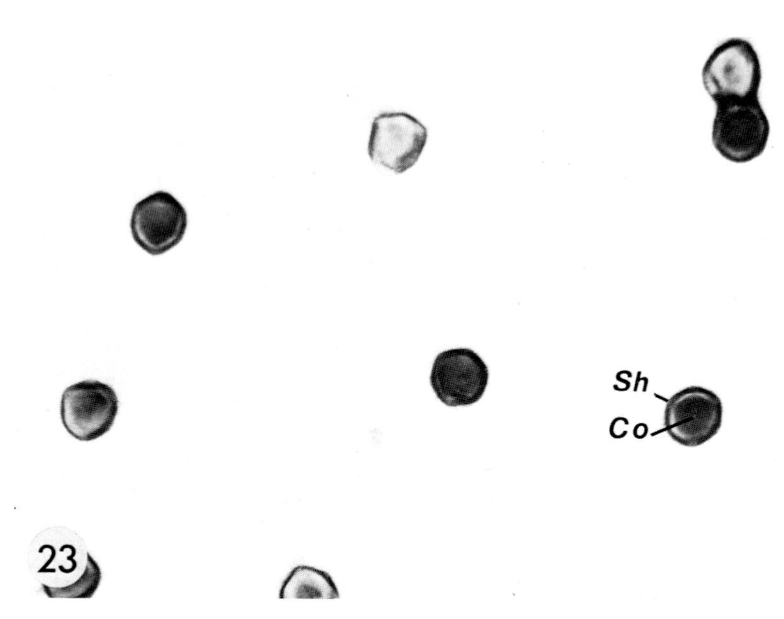

FIGURE 23. Negatively stained preparation of an iridovirus. Co, core; Sh, shell. (Courtesy of Roberta Hess.) ×60,000.

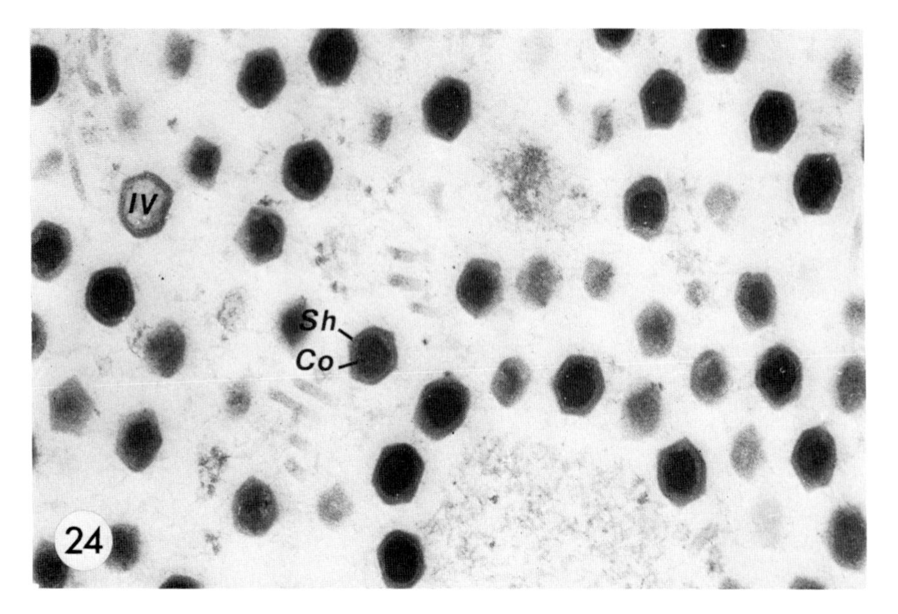

FIGURE 24. Sectioned material of the same iridovirus as in Fig. 23, showing the core (Co) and shell (Sh); IV, immature virus. (Courtesy of Roberta Hess.) ×70,000.

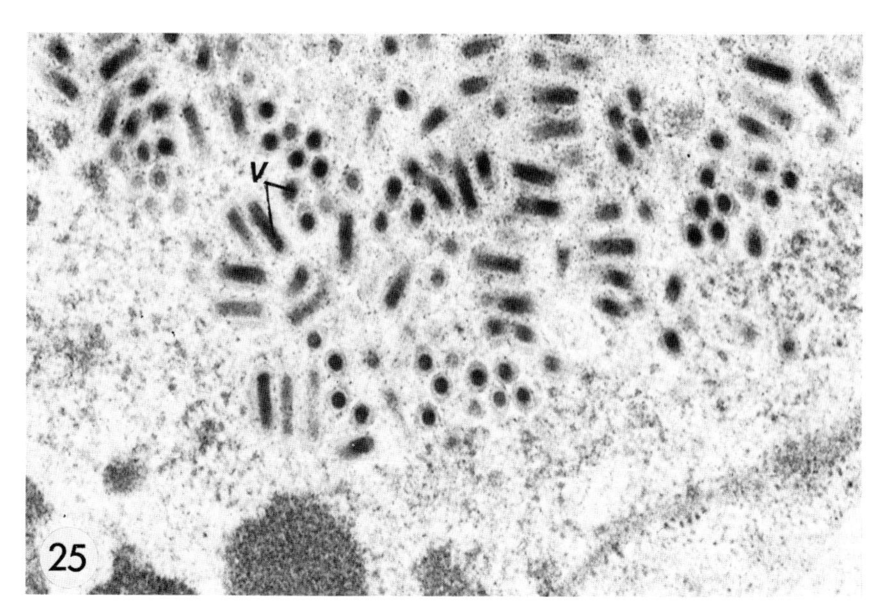

FIGURE 25. A nonoccluded nuclear rod-shaped virus (baculovirus) from *Oryctes rhinoceros,* showing enveloped virus (V) in the nucleus. (Courtesy of Roberta Hess.) ×32,000.

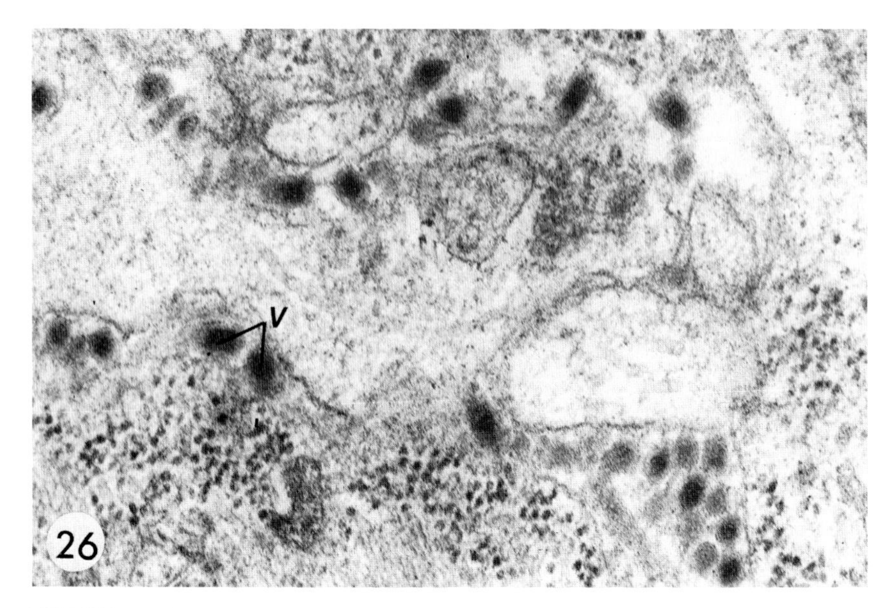

FIGURE 26. A nonoccluded nuclear virus (V) with a polydisperse DNA genome originating from the calyx of the braconid wasp *Phanerotoma flavitestacea* invading tissue of the host, *Amyelois transitella* (Lepidoptera). (Courtesy of Roberta Hess.) ×66,000.

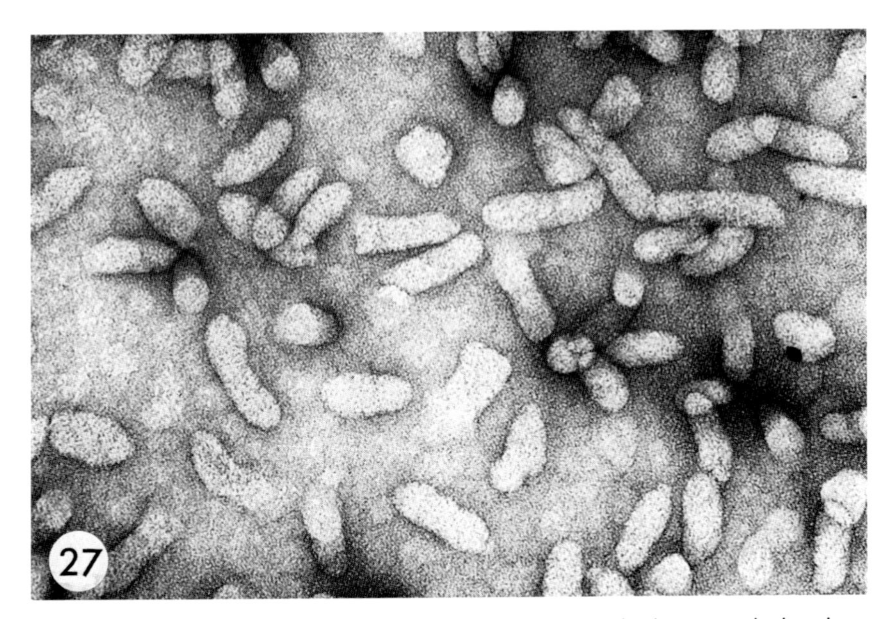

FIGURE 27. Negatively stained preparation of chronic bee paralysis virus, illustrating the anisometric shape (mainly ellipsoidal). (Courtesy of L. Bailey.) ×200,000.

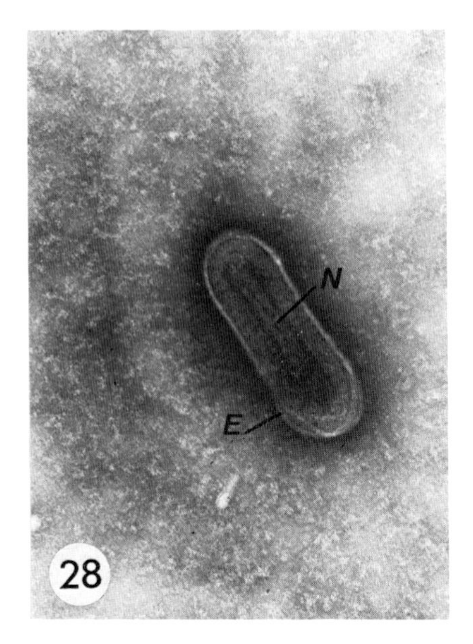

FIGURE 28. Negatively stained preparations of the filamentous virus of the honeybee. The coiled nucleocapsid (N) is seen within the envelope (E). (Courtesy of L. Bailey.) ×60,000.

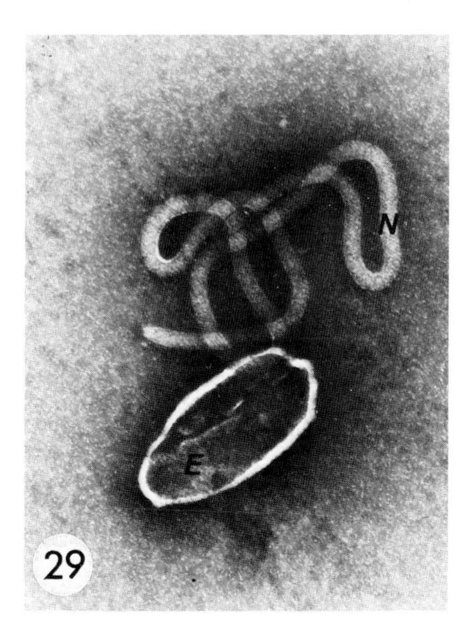

FIGURE 29. Same virus as in Fig. 28 with an empty envelope (E) and released nucleocapsid (N). (Courtesy of L. Bailey.) ×60,000.

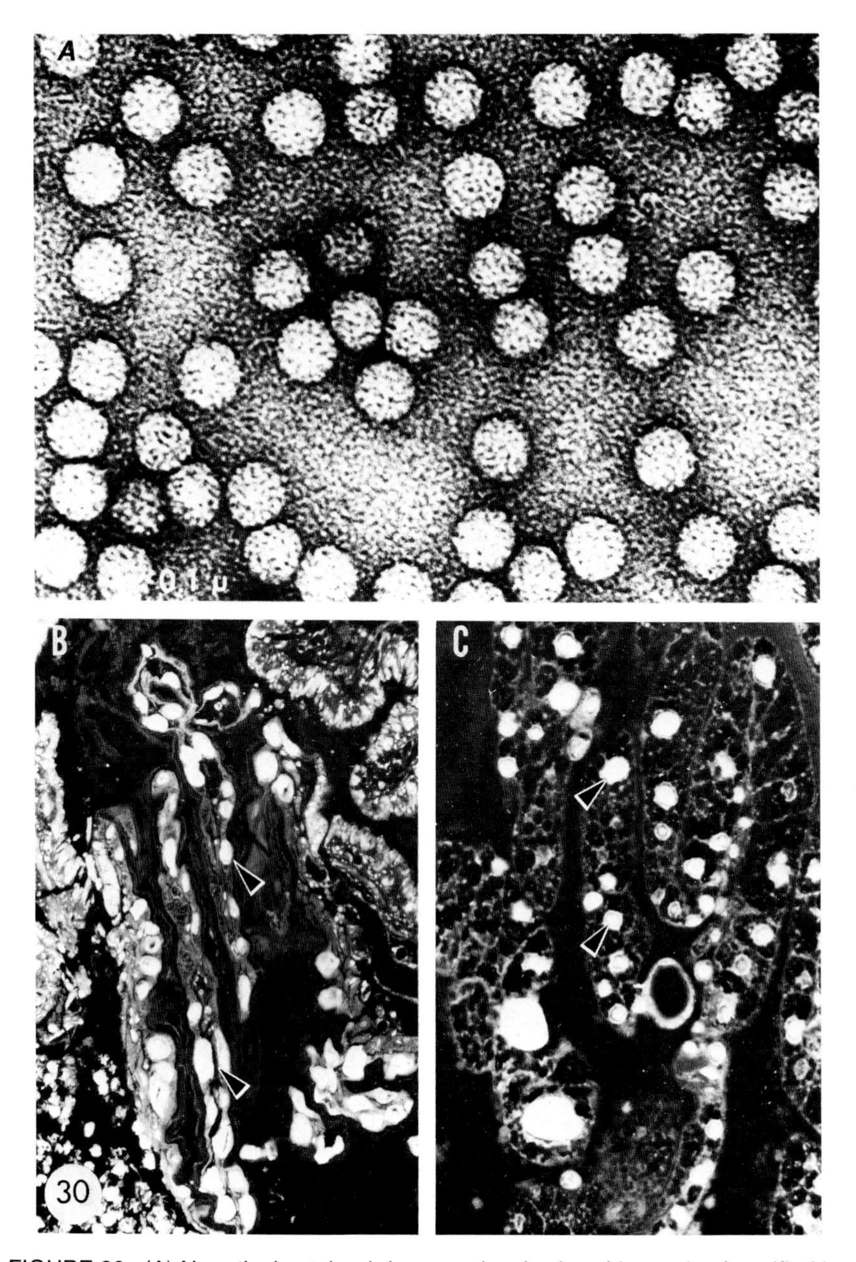

FIGURE 30. (A) Negatively stained densonucleosis virus (densovirus) purified by chromatography on Sephadex G-200, treated with afrine, and stained with 2% phosphotungstic acid. (Courtesy of E. Kurstak.) ×308,000. (B) Densovirus in the digestive system of *Galleria mellonella* (Lepidoptera) stained with acridine orange, pH 3.8. Arrows indicate virus in nuclei. (Courtesy of E. Kurstak.) ×1000. (C) Densovirus in adipose tissue of *Galleria mellonella* (Lepidoptera) stained with acridine orange, pH 3.8. Arrows indicate nuclear site of virus replication. (Courtesy of E. Kurstak.) ×1200.

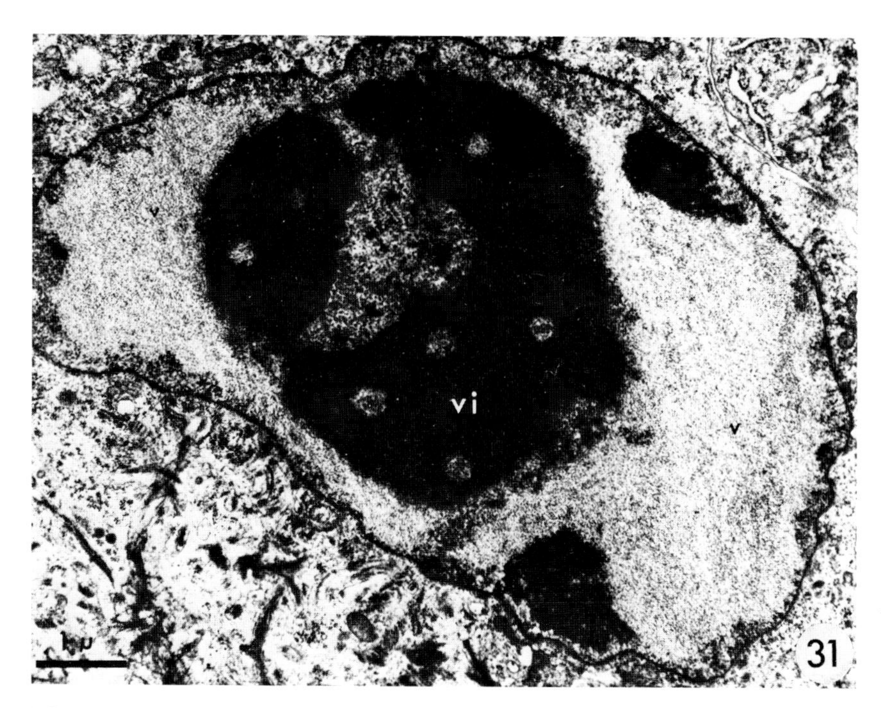

FIGURE 31. Transmission electron micrograph of nucleus of *Galleria mellonella* infected with densonucleosis virus (densovirus), showing the viral inclusions (vi) or virogenic area as an electron-dense mass. Small virions are seen in the remaining nucleoplasm. (Courtesy of E. Kurstak.) ×12,000.

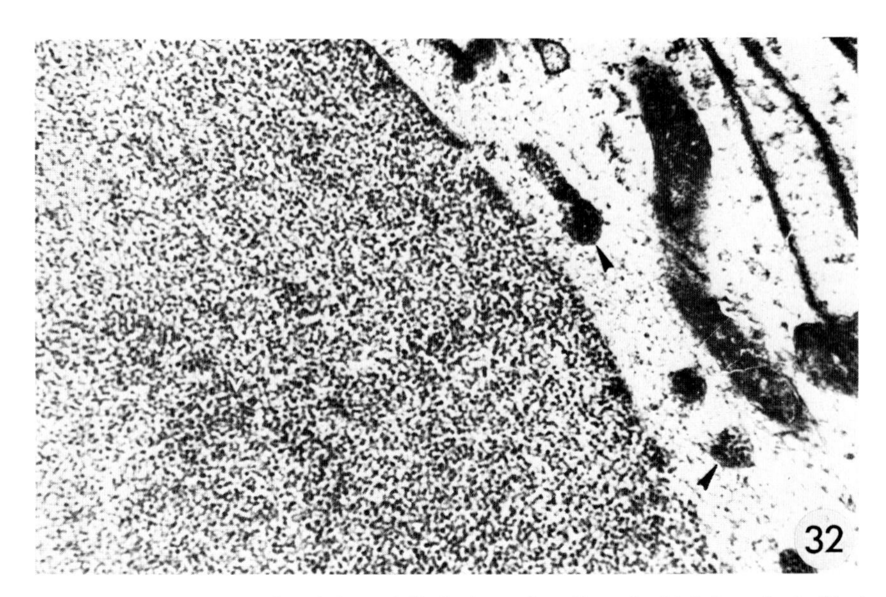

FIGURE 32. Nucleus of an infected *Galleria mellonella* cell which is entirely filled with densovirus. Virions are seen in vesicles in the cytoplasm (arrows). (Courtesy of E. Kurstak.) ×27,000.

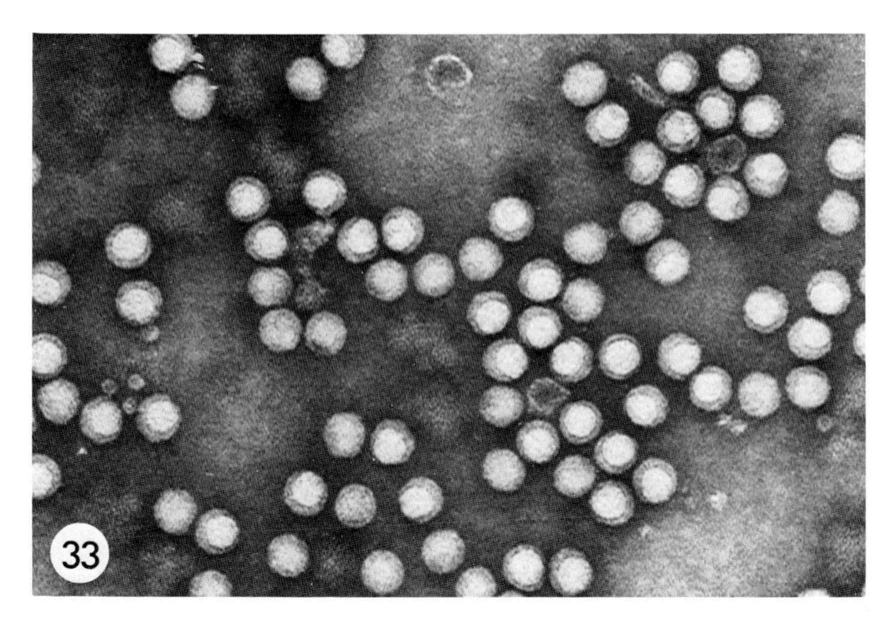

FIGURE 33. Purified preparation of sacbrood virus negatively stained in neutral sodium phosphotungstate. (Courtesy of L. Bailey.) ×200,000.

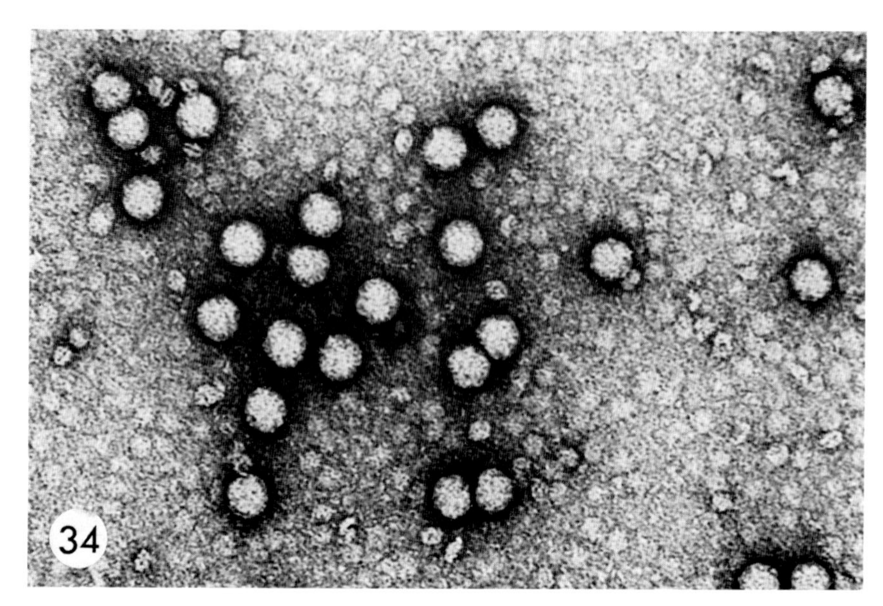

FIGURE 34. Cricket paralysis virus, a member of the Picornaviridae not yet assigned to a genus. (Courtesy of B. Hillman.) ×200,000.

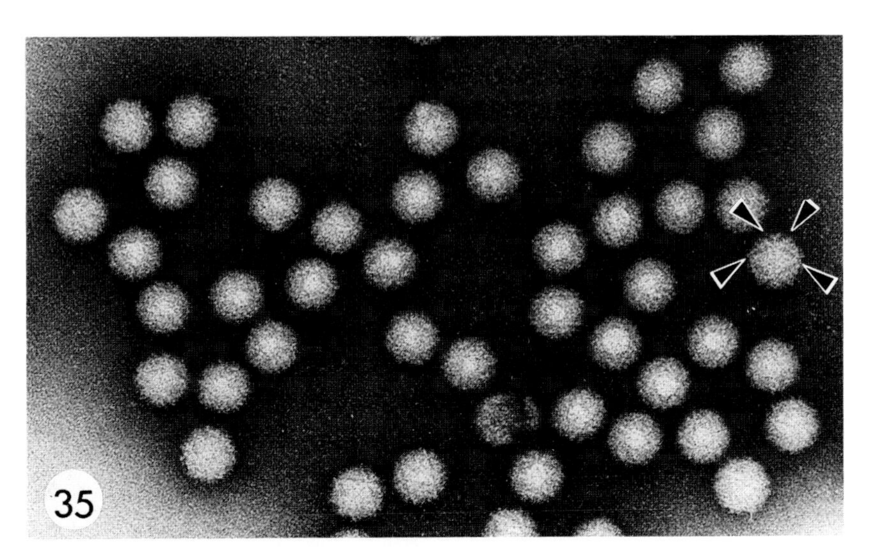

FIGURE 35. Transmission electron micrograph of a virus from the Nudaurelia β virus group. A negatively stained preparation of purified *Trichoplusia* RNA virus showing the virus core and capsid structure with the stain accumulating along fivefold vertices (arrows). (Courtesy of Roberta Hess.) ×200,000.

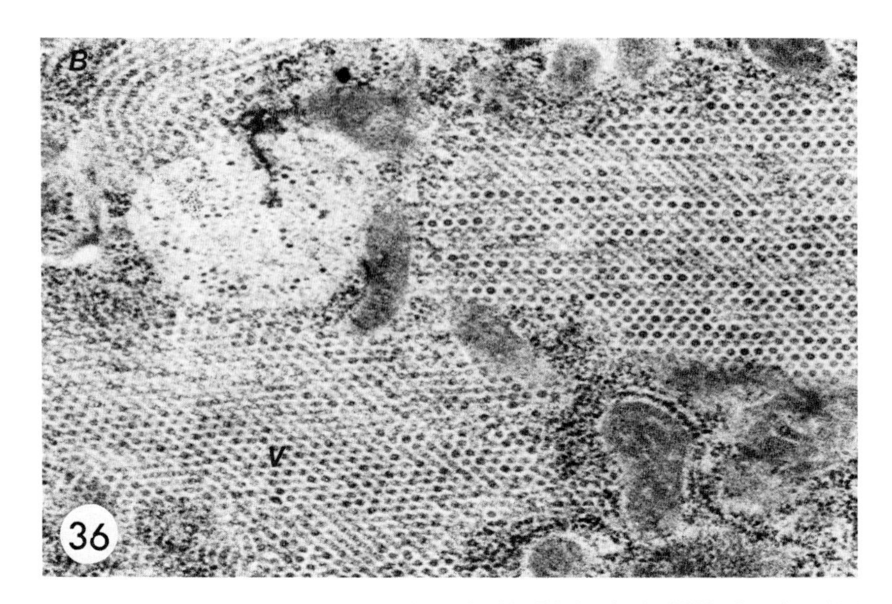

FIGURE 36. Section of an insect infected with *Trichoplusia* RNA virus in which paracrystalline arrays of virus particles (V) fill the cytoplasm. (Courtesy of Roberta Hess.) ×50,000.

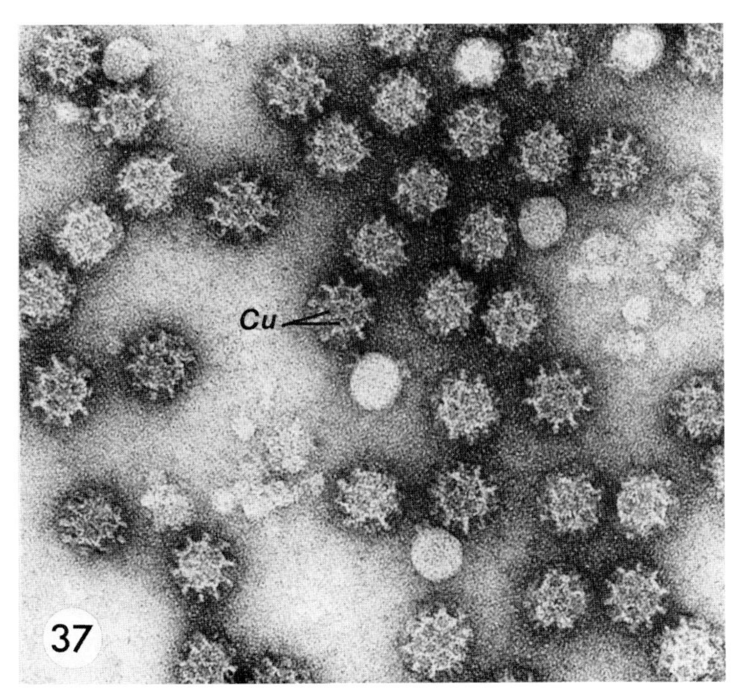

FIGURE 37. Transmission electron micrograph of a calicilike virus isolated from the navel orangeworm. Negatively stained preparation shows the cup-shaped depression (Cu) on the capsid. (Courtesy of B. Hillman.) ×260,000.

RICKETTSIAS

INTRODUCTION

The rickettsias associated with insects may be intracellular, epicellular (living on the surface of cells), or extracellular, and they may have phoretic, mutualistic, or pathogenic relationships with their hosts. For example, *Rochalimaea quintana* (Schmincke), the causal agent of trench fever in man, grows nonpathogenically in the gut lumen of the human louse, *Pediculus humanus*. On the other hand, *Rickettsia prowazekii* da Rocha-Lima, the causal agent of typhus fever in man, is pathogenic to its vector, *P. humanus*. Examples of Rickettsias which are considered essential for development and reproduction of the host are *Symbiotes lectularius* in mycetomes of the common bedbug (*Cimex lectularius*) and *Blattabacterium* sp. in mycetocytes of cockroaches. Rickettsias belonging to the genus *Rickettsiella* are pathogenic to their insect hosts.

Hosts of rickettsias range from trematodes to man. Many of the vertebrate forms transmitted by arthropods are pathogenic to both vector and host. Because of this wide host range, some consider all rickettsias as potentially dangerous to humans, and they should be handled with care. The number of insect hosts and/or vectors of rickettsias is small at present. However, it may be proven that many insects carry these microorganisms somewhere in their bodies.

TAXONOMIC STATUS

The infective stages of rickettsias are minute microorganisms, about the size of granulosis virus inclusion capsules (0.2–0.6 μm). The majority are rod-shaped, coccoid, or pleomorphic and are bound by a typical cell wall containing muramic acid. They do not possess flagella and are gram negative (with one exception). They contain both RNA and DNA and possess metabolic enzyme systems which can be inhibited by chemotherapeutic and antibiotic agents. With the exception of those few forms that multiply epicellularly or extracellularly, the rickettsias multiply within their host cells.

The Eighth Edition of *Bergey's Manual of Determinative Bacteriology* divides the class, Rickettsias, into two orders, the Rickettsiales and the Chlamydiales. Only members of the order Rickettsiales are known to be pathogenic to insects; the Chlamydiales have been recovered from insect tissues but at most appear to restrict the host's growth. The taxonomic scheme for the insect-associated rickettsias in the order Rickettsiales, as modified from *Bergey's Manual,* is outlined in Table 2.

LIFE CYCLE

After ingestion and penetration through the insect's midgut epithelium, the infective stages of insect pathogenic rickettsias invade the host cell and develop into dividing vegetative stages. This division may be by simple binary fission or through the production of secondary cells which then produce smaller cells that develop into the thick-walled infective stages. Several different morphological types of cells are often encountered during rickettsial development. There are small, tubular infective stages; larger, often pleomorphic, rod-shaped vegetative stages; and spherical cells. The latter two often occur in vesicles or vacuoles, which frequently harbor crystal-like inclusion bodies. The most commonly encountered pathogens of insects belong to the genus *Rickettsiella,* which develop intracellularly in various tissues, especially the fat body.

Infection of fleas or lice with *Rickettsia prowazekii* and *R. typhi* is initiated when the insects feed upon an infected vertebrate. In the insect, the *Rickettsia* multiplies in the cytoplasm of the gut epithelial cells. Heavily infected cells are discharged with the feces and are the source of vertebrate infection.

Rochalimaea quintana is normally ingested by its louse vector while feeding on an infected human. The organism grows epicellularly on the gut epithelial cells (Krieg, 1963) and in the gut lumen (Weiss and Moulder, 1974). However, if *R. quintana* is injected intrahemocoelically into

Table 2. Systematic Arrangement for the Order Rickettsiales Associated with Insects

Family	Tribe	Genus	Species	Insect association
Rickettsiaceae	Rickettsiae	*Rickettsia*	*prowazekii* da Rocha-Lima	Vector and pathogen
			typhi (Wolbach and Todd) Philip	Vector and pathogen
		Rochalimaea	*quintana* (Schmincke) Kreig	Vector
	Wolbachieae	*Wolbachia*	*pipientis* Hertig	Mild pathogen
			melophagi (Noller) Philip	Commensal
		Enterella	*stethorae* and other species	Pathogen
		Symbiotes	*lectularius* (Arkwright *et al.*) Philip	Mutualist
		Blattabacterium	*cuenoti* (Mercier) Hollande	Mutualist
		Rickettsiella	*popilliae* (Dutky and Gooden) Philip and other species	Pathogen
Bartonellaceae		*Bartonella*	*bacilliformis* (Strong *et al.*) Strong, Tyzzer, and Sellards	Vector
Anaplasmataceae		*Anaplasma*	*marginale* Theiler	Phoretic vector
			ovis Lestoquard	Phoretic vector
		Haemobartonella	*muris* (Mayer) Tyzzer and Weinman	Vector
		Eperythrozoan	*coccoides* Shilling	Vector
			ovis Neitz	Vector
			parvum Splitter	Vector

the louse, it will first multiply in the hemolymph and later infect the gut epithelial cells (Krieg, 1963). The feces of the louse become heavily contaminated with *Rochalimaea* cells and are the source of human infection when scratched into the skin.

Weiss (1974) comments that species in the genus *Wolbachia* associated with insects are either extracellular or intracellular but do not develop inside mycetomes and are seldom pathogenic for their hosts. *Wolbachia pipientis* may damage cells of the gonads in the mosquito *Culex pipiens* and of the gut epithelium of other insects. The method of transmission from one individual to another is unknown, but one might suspect transovarial transmission in *C. pipiens* and ingestion of contaminated food in other insects.

Wolbachia melophagi occurs epicellularly on gut epithelial cells of the sheep ked, *Melophagus ovinus*. There is no evidence of injury to the host, and its almost universal presence suggests a mutualistic relationship. Although no information on transmission is available, the fact that the microorganisms have been found in smears from host eggs (Weiss, 1974) suggests the possibility of transovarial transmission.

The obligate intracellular symbiotes in the genera *Symbiotes* and *Blattabacterium* are transmitted via the host's ovaries to the embryo. Members of the genus *Symbiotes* characterisically occur in mycetomes of bedbugs (*Cimex* sp.), while the genus *Blattabacterium* occurs in mycetocytes in abdominal fat body, germ tissue, and oocytes of cockroaches (M. A. Brooks, 1974).

The normal infection route of the pathogen *Rickettsiella popilliae* is peroral or possibly through wounds of Japanese beetles. Larvae can be infected by feeding, by holding them in soil infested with the organisms, or by injection. Injection is the most effective method since less than six organisms constitute an LD_{50} (Weiss, 1974). Once the organisms have invaded the hemocoel, infection spreads from the fat body to the blood and other organs.

CHARACTERISTICS OF INFECTED INSECTS

Lice infected with *Rickettsia prowazekii* and *R. typhi* show no initial symptoms, but as the infection progresses the gut wall is irreparably damaged. A few hours before death the lice turn reddish owing to the ingested human blood entering the hemocoel. Fleas infected with *R. typhi* become sluggish and die, but do not turn red.

No particular symptoms occur in insects infected with *Wolbachia pipientis,* since the effects of this organism are extremely mild.

The most striking symptom in insect larvae infected with *Rickettsiella popilliae* is a bluish discoloration within the fat body cells, thus the common name "blue disease." Infected grubs of *Melolontha* spp. show a change in behavior (Niklas, 1957). While reduced temperatures cause healthly larvae to move deeper into the soil, infected grubs come to the surface. Other insects infected with *Rickettsiella* may become sluggish, exhibit a swollen abdomen owing to pathogen multiplication in the fat body, and turn whitish in color. *Enterella* infections are marked by dysentery, coupled with muscular contractions.

Large, distinct crystalline bodies accompany rickettsial infections in some insects (Fig. 38). The significance of these is unknown.

FACTORS AFFECTING NATURAL INFECTIONS

The insect pathogenic rickettsias fall into two categories, those with and those without alternate hosts. Those with alternate hosts include only *Rickettsia prowazekii* and *R. typhi.* The human louse becomes infected with *R. prowazekii* by feeding on blood from an infected human, and the incidence of infection in the louse population is dependent on the intensity and duration of the rickettsemia in the human population. Likewise, spread of the disease in the human population is dependent on the extent of infection in the louse population. Since both hosts are killed by the disease, epizootics tend to be cyclic and density dependent. Rickettsemias of *R. typhi,* on the other hand, tend to be enzootic because latent infections occur in the Rodentia, which have a rapid rate of multiplication. Consequently, this disease tends to be less cyclic.

Insect pathogenic rickettsias without an alternation of hosts include *Wolbachia pipientis, Rickettsiella popilliae,* and *Enterella* spp. Little is known about the epizootiology of *W. pipientis,* probably because the pathogen has so little effect on the host. Natural infections of *R. popilliae* in *Melolontha* larvae and *Enterella* in insect hosts probably occur through ingestion of soil particles contaminated with diseased larvae; consequently, the higher the population, the greater is the chance for infection.

METHODS OF EXAMINATION

Material for microscopic examination may be prepared either as fresh tissue smears or as fixed histological sections. Since rickettsias are minute microorganisms close to the resolution of the light microscope, unstained smears should be examined under dark field or phase contrast. Rickettsias appear as minute refringent flecks under dark field and as gray flecks under phase contrast. They are often in rapid brownian movement very similar to the inclusion bodies of granulosis viruses. Rickettsias sometimes appear in chains, a character distinguishing them from granulosis virus capsules.

Rickettsias can be stained with analine dyes, but the most frequently used stains are the Macchiavello method for tissue smear (blue color indicates positive reaction) and the Giemsa method for fixed specimens (see Techniques). With the latter method, rickettsias are stained red, while granulosis capsules and bacteria are stained blue. Another technique for differentiating rickettsias from granulosis capsules is Giemsa staining after HCl hydrolysis (Krieg, 1963) (see Techniques). Rickettsias react strongly positive while granulosis capsules give negative results. This method works well for detecting the presence of rickettsias in tissue smears.

Multiplication of *Rickettsiella popilliae* results in vacuoles or vesicles filled with rickettsias and crystalline bipyramidal bodies. These vesicles are sometimes referred to as NR bodies because they may be stained with neutral red dye. They are also termed globules of spheroidocytes. The bipyramidal crystals closely resemble albuminoid crystals found in late larval and pupal stages of normal insects. In fixed tissue sections stained with buffered Giemsa, rickettsias, crystals, and globules of spheroidocytes are red, while nuclei, albuminoid crystals, and bacteria are blue.

ISOLATION AND CULTIVATION

Rickettsias rarely can be grown on normal culture media like bacteria. They are best grown in insects inoculated per os, peranal, or intrahemocoelically.

Inocula may be prepared by isolating rickettsias as outlined by Krieg (1963). Infected insects are surface sterilized by treatment with 70% ethanol or ethyl ether, decontaminated in aqueous 0.01% merthiolate, and rinsed in sterile water. The infected organ or insect is homogenized in sterile water. After removal of large particles by low-speed centrifugation (1000*g* for 30 min), the supernatant with the rickettsias can be sedimented by high-speed centrifugation (7500*g* for 1 hr). Small types of rickettsias (*Coxiella* spp. and *Rickettsiella* spp.) may be purified by filtration through membranes of cellulose nitrate with a mean porosity of 0.6 μm.

IDENTIFICATION

To identify insect rickettsias, the type of host association, tissue affinities, gross symptoms, morphology, and staining characteristics should be considered. Because of their size, most rickettsias can be measured accurately only with the electron microscope.

TESTING FOR PATHOGENICITY

Since the normal route of infection for the insect pathogenic rickettsias is through the intestine, pathogenicity tests should be conducted by per os inoculation. This can be accomplished by force-feeding with a microsyringe or by feeding contaminated food. Inocula can be prepared by the method of Krieg (1963) outlined previously. When an infected vertebrate is available, infection of the arthropod can be achieved by feeding it the blood of the infected host.

STORAGE

Rickettsias may be stored in infected hosts under refrigeration, by freezing, or by freeze-drying. They should be stored dry since physiological saline and distilled water cause inactivation within 2–6 hr at room temperature. The hydrogen ion concentration of the medium is very important, and the optimum pH lies near the neutral point (pH 6.4–7.2). Most infected insects can be stored at 4°C for several days. Deep-frozen rickettsias may remain viable for several months: *Rickettsia prowazekii* kept at −20°C remained viable for eight months, and *Rickettsiella popilliae* maintained at −80°C lasted for more than three years (Krieg, 1963). Lyophilization for preservation is practical, but the ampules must be stored frozen, at −10 to −20°C. To protect purified rickettsias from inactivation during freezing, protective colloids such as skimmed milk (pH 7.6), albumin, or peptone can be added.

LITERATURE

A comprehensive account of rickettsias associated with insects is given by Krieg (1963). The most recent review of the rickettsias is found in Part 18 of the Eighth Edition of *Bergey's Manual of Determinative Bacteriology* (Buchanan and Gibbons, 1974). A short summary of the rickettsias associated with insects is given by Vaughn (1974), and their possible use as microbial control agents is discussed by Krieg (1971).

KEY TO RICKETTSIAS ASSOCIATED WITH INSECTS

1. Develop intracellularly; may or may not be pathogenic (extracellular forms may occur owing to degeneration of heavily infected cells)—**9**
1. Develop epicellularly or extracellularly; usually nonpathogenic—**2**

2. Develop epicellularly on cells of gut epithelium—**3**
2. Found free in the gut lumen, not attached to epithelial cells—**4**

3. Found in the gut of the human louse, *Pediculus humanus* (L.); develops on epithelial cells and free in the gut lumen; cells 0.2–0.5 × 1.0–1.6 μm—*Rochalimaea quintana* (Schmincke). See Weiss (1974) for a discussion of this genus.
3. Found in the gut of the sheep ked, *Melophaga ovinus* (L.); develops in closely packed rows in epithelial cells; cells 0.3 × 0.6 μm—*Wolbachia melophagi* (Noller). See Weiss (1974) for a discussion of this genus.

4. Associated with flies—**5**
4. Associated with lice on pigs or rodents—**7**

5. Associated with sand flies (*Phlebotomus* spp.)—*Bartonella bacilliformis* (Strong *et al.*). Known to be established only in South and perhaps Central America. See Weinman (1974) for a discussion of this genus.
5. Associated with horseflies (Tabanidae)—**6**

6. Organisms round (0.3–0.4 μm in diameter); 1–8 occur in inclusion bodies (0.3–1.0 μm in diameter)—*Anaplasma* Theiler. See Ristic and Krier (1974) for a discussion of this genus.
6. Cells ring- or disc-shaped, 0.5–1.0 μm in diameter; inclusion bodies absent—*Eperythrozoon ovis* Neitz. See Ristic and Krier (1974) for a discussion of this species.

7. Associated with the pig louse, *Haematopinus suis* (L.)—*Eperythrozoon parvum* Splitter. See Ristic and Krier (1974) for a discussion of this species.

7. Associated with rodent lice—**8**

8. Associated with the mouse louse, *Polyplax serrata* (Burm.); cells coccoid, 0.4–0.5 μm long—*Eperythrozoon coccoides* Schilling. See Ristic and Krier (1974) for a discussion of this species.

8. Associated with the rat louse, *Polyplax spinulose* (Burm.); cells rod-shaped, 0.1 × 0.3–0.7 μm—*Haemobartonella muris* (Mayer). See Ristic and Krier (1974) for a discussion of this genus.

9. Found in gut epithelium of lice and fleas; may be pathogenic—*Rickettsia* da Rocha-Lima (Figs. 39–41)—**10.** See Moulder (1974) for a discussion of this genus.

9. Not found in lice or fleas—**11**

10. Found in the human louse, *Pediculus humanus humanus*; causes typhus fever in man—*R. prowazekii* da Rocha-Lima. Infected lice become red a few hours before death, at which time rickettsias can be found free in the gut lumen and hemocoel.

10. Found in the human louse, *P. humanus,* human flea, *Pulex irritans* (L.), rat louse, *Polyplax spinulosa* (Burm.), and rat flea, *Xenophylla cheopis* (Roth.). Causes murine (endemic) typhus in man (incidental host) and rats and other rodents (primary reservoir). Cells distinguishable from *R. prowazekii* only by serological techniques—*R. typhi* (Wolbach and Todd) (known also as *R. mooseri*).

11. Nonpathogenic mutualists found in special cells (mycetocytes) or special organs (mycetomes), especially in bedbugs (*Cimex* spp.) or cockroaches—**12**

11. Not found in mycetocytes or mycetomes; slightly to highly pathogenic—**13**

12. Mutualists found in paired mycetomes in most tissues of bedbugs (*Cimex* spp.)—*Symbiotes* Philip. See M. A. Brooks (1974) for a discussion of this genus.

12. Mutualists found in mycetocytes of the fat body, germ tissues, and oocytes of cockroaches—*Blattabacterium cuenoti* (Mercier). See M. A. Brooks (1974) for a discussion of this genus.

13. Occur in gonads of *Culex pipiens* and gut epithelium of *C. fatigans* and other diptera; slightly to mildly pathogenic—*Wolbachia* Hertig (Figs. 42, 43).

13. Occur in fat body and blood cells of Coleoptera and Diptera; often highly pathogenic; infected larvae may turn white or bluish-white—*Rickettsiella* Philips (Figs. 44–48). Weiss (1974) considers all insect-pathogenic species in this genus as belonging to *R. popilliae* (Dutky and Gooden). In the same paper he lists the hosts of this rickettsia. Kellen *et al.* (1972) discuss the life cycle and developmental stages of a *Rickettsiella.* Also in this couplet occur members of the genus *Enterella,* which are found in Lepidoptera, mosquitos (Culicidae), and certain beetles (*Stethorus* spp.; Coccinellidae). Members of this genus generally occur in the host's gut epithelium, but can also be isolated from the fat body and hypodermis. See Entwistle and Robertson (1968) for a discussion of this genus.

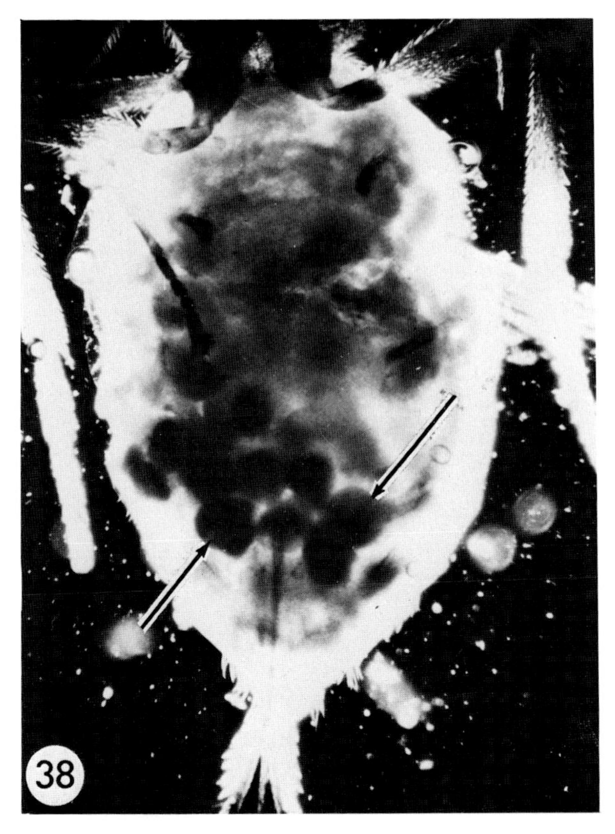

FIGURE 38. Large crystalline bodies (arrows) in the hemocoel of *Bracon hebetor* suffering from a rickettsial infection. (Courtesy of W. R. Kellen.) ×25.

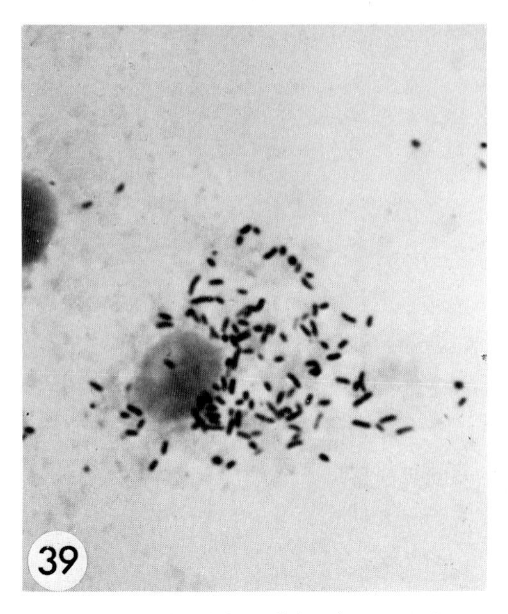

FIGURE 39. Cells of *Rickettsia rickettsii* in the cytoplasm of hemocytes of *Dermacentor andersoni*. (Courtesy of W. Burgdorfer.) ×1400.

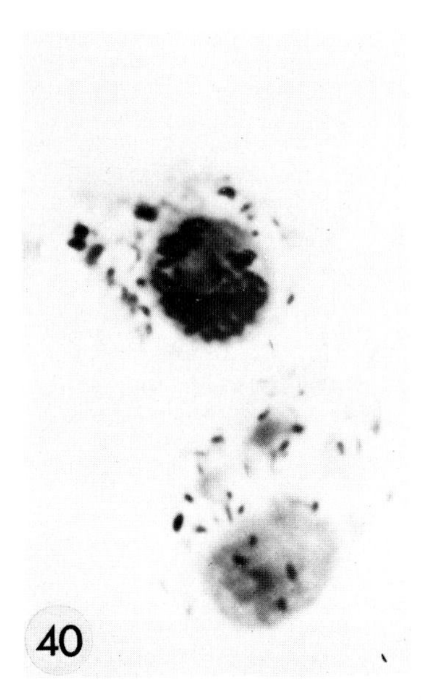

FIGURE 40. Cells of *Rickettsia rickettsii* in the nucleus of hemocytes of *Dermacentor andersoni*. (Courtesy of W. Burgdorfer.) $\times 1400$.

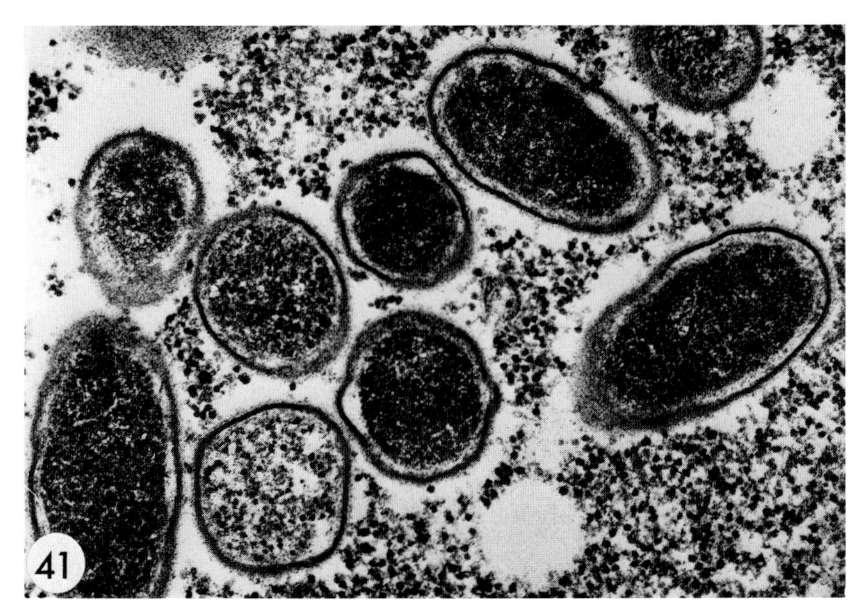

FIGURE 41. Electron micrograph of *Rickettsia rickettsii* in ovarial tick tissue. (Courtesy of L. P. Bronton and W. Burgdorfer.) ×42,000.

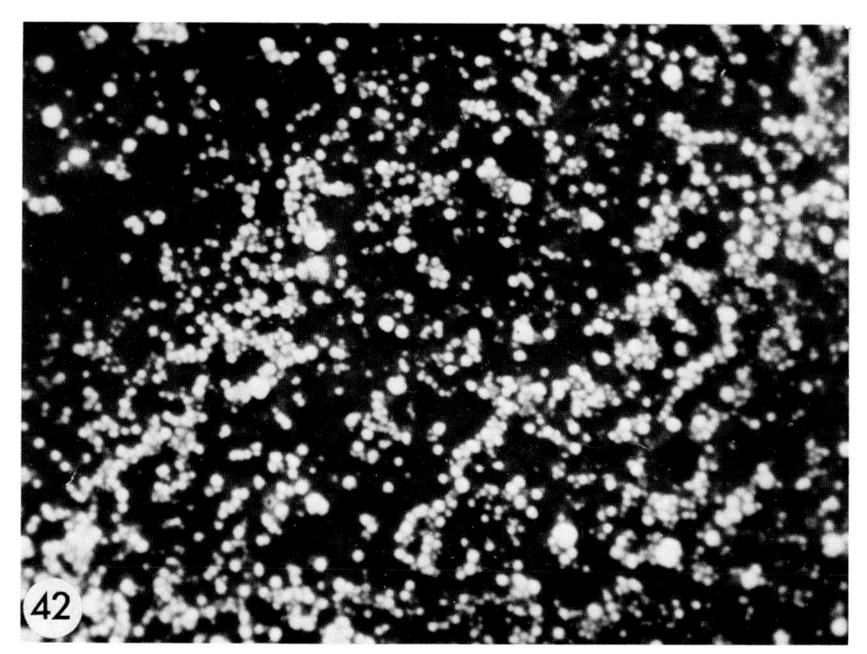

FIGURE 42. *Wolbachia* cells (coccoid forms) in ovarial tissue of *Dermacentor andersoni*. (Courtesy of W. Burgdorfer.) ×1300.

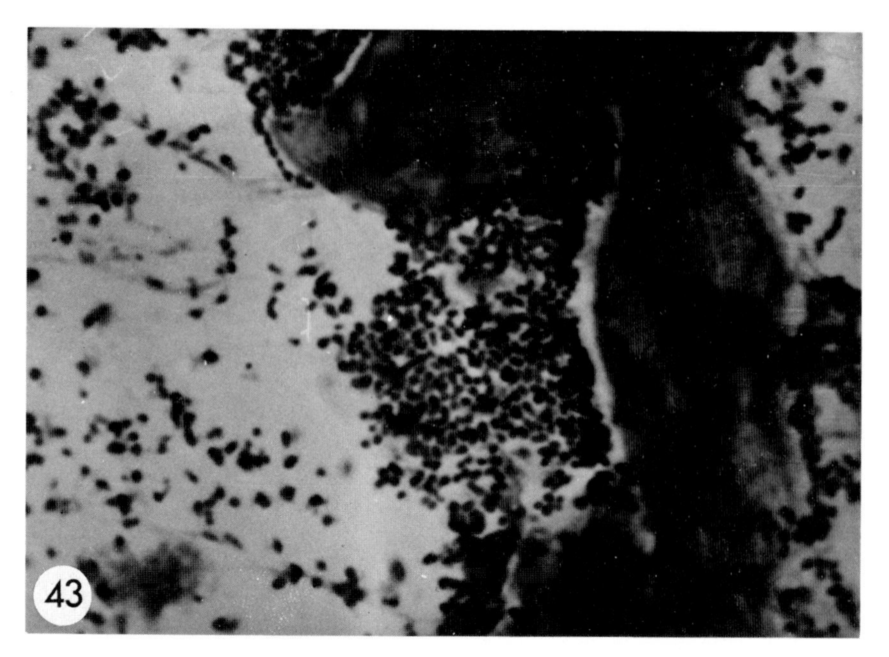

FIGURE 43. *Wolbachia* cells (rod-shaped forms) in ovarial tissue of *Dermacentor andersoni*. ×1300.

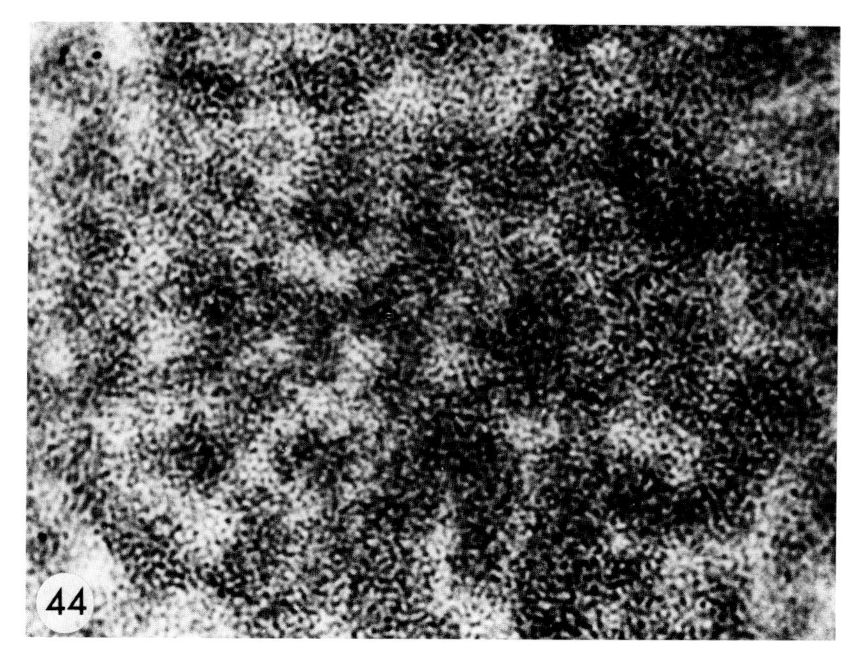

FIGURE 44. Infective stages of a *Rickettsiella* sp. from the larva of a naval orangeworm, *Amyelois transitella.* ×2000.

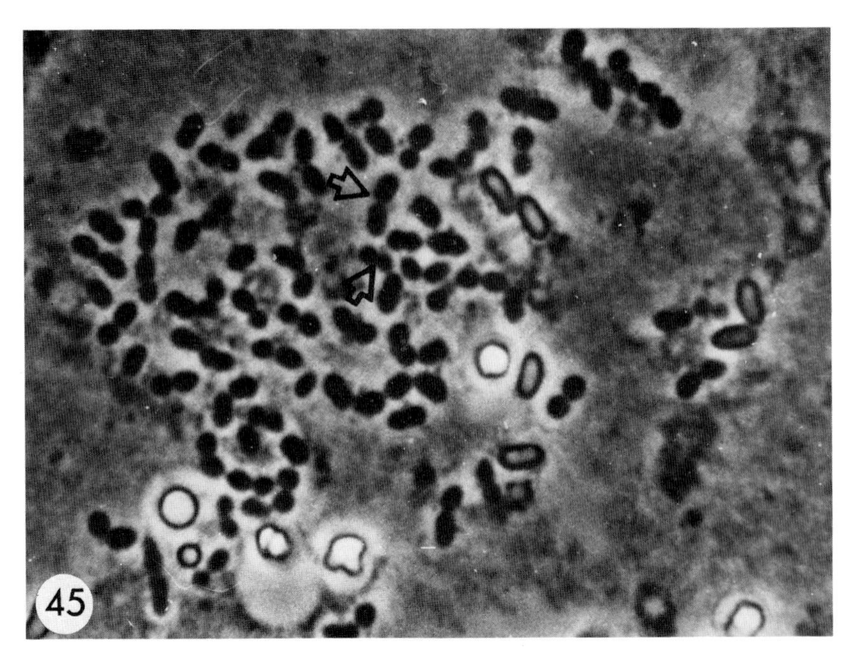

FIGURE 45. Electron micrograph of the developing "daughter cells" (arrows) of a *Rickettsiella* sp. from the navel orangeworm. (Courtesy of W. R. Kellen and D. F. Hoffman.) ×4800.

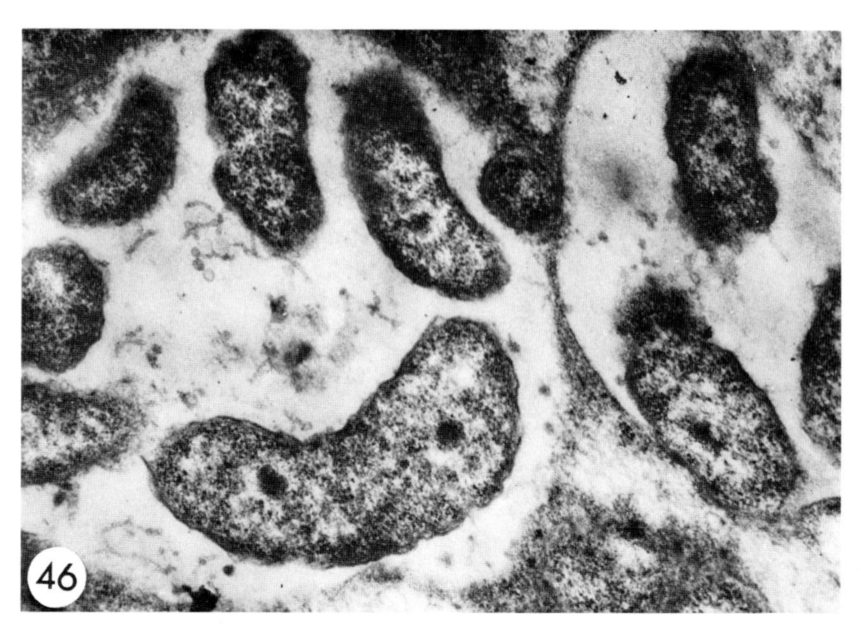

FIGURE 46. Electron micrograph of developing "vesicles" with "secondary cells" in a *Rickettsiella* sp. infection of the navel orangeworm. (Courtesy of W. R. Kellen and D. F. Hoffman.) ×30,250.

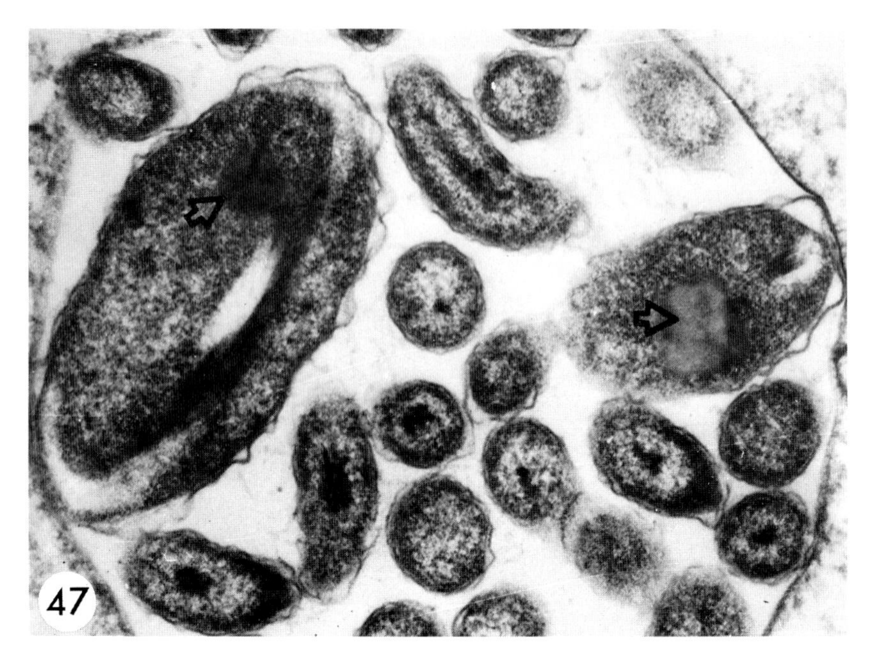

FIGURE 47. Electron micrograph of a large "vesicle" containing "secondary cells" with inclusion bodies (arrows) in a *Rickettsiella* sp. infection of the navel orangeworm. (Courtesy of W. R. Kellen and D. F. Hoffman.) ×38,500.

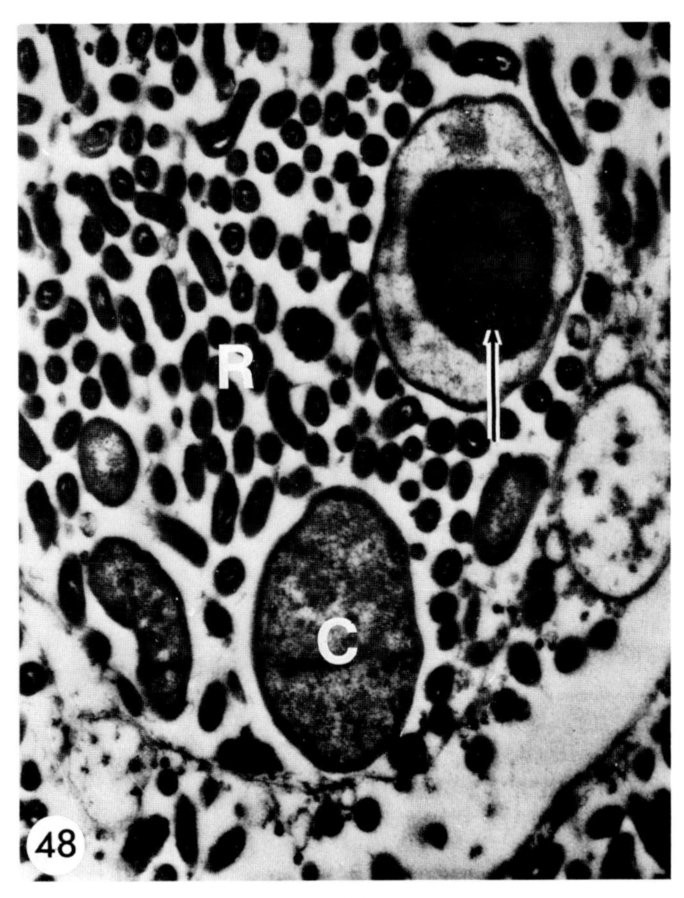

FIGURE 48. Electron micrograph of a *Rickettsiella* sp. infection in *Melolontha melolontha.* Standard-type rickettsiae (R) occur in "vacuoles" along with secondary cells (C) and developing crystal (arrow). (Courtesy of A. M. Huger.) ×25,500.

BACTERIA

INTRODUCTION

The ubiquitous nature of bacteria makes them the most abundant type of microorganism associated with insects. Thus, saprophytic, symbiotic, or pathogenic species of bacteria may be associated externally or internally with insects.

An example of the magnitude of these associations is shown by the relationships just between flies and bacteria (Greenberg, 1971). Then there are special associations, such as the one which results in the mortality of waterfowl that ingest fly maggots containing *Clostridium botulinum* (Hunter, 1970).

In this guide, we are concerned only with two groups of bacteria capable of causing disease. The first consists of the "true" pathogens, which cause infection whenever they are ingested. The second group includes the potential pathogens, which are omnipresent but cause infection when introduced into the hemolymph or when the insect is weakened or under stress.

TAXONOMIC STATUS

Previously placed in the class Schizomycetes, the bacteria have now been made a division in the kingdom Procaryotae, along with the blue-green algae (Buchanan and Gibbons, 1974). The main characteristics of this kingdom are the lack of a membrane separating the nucleoplasm from the cytoplasm, the lack of unit-membrane-bounded cytoplasmic organelles, and the presence of peptidoglycan in the cell walls.

Bacteria pathogenic to insects are all single-cell organisms that reproduce primarily by fission. They may or may not be motile or form spores and most can be grown on artificial media.

LIFE CYCLE

In most infections, bacteria invade the host's hemocoel through the alimentary tract. This is certainly true for all entomogeneous species of the genus *Bacillus,* in which spores serve as the infectious agents. For the potential pathogens, any break in the body wall or alimentary tract will serve as a port of entry.

Bacillus thuringiensis contains parasporal crystals which paralyze the gut of many lepidopterous larvae, thus allowing easy penetration by the vegetative cells. The known insect pathogens in the genus *Clostridium* cause disease by multiplying only in the insect gut and never enter the host's hemocoel.

Bacterial cells also may be introduced into the insect hemocoel by parasites or predators. The case of *Xenorhabdus nematophilus* is an excellent example. This bacterium occurs in the body of insects attacked by certain neoaplectanid nematodes and in the gut of the infective-stage nematodes. The bacterial cells have no invasive power of their own and are introduced into the insect's hemocoel by infective-stage neoaplectanid nematodes (Poinar and Thomas, 1967). Other genera and species of nematodes are also associated with related species of bacteria and have a similar relationship with insects.

CHARACTERISTICS OF INFECTED INSECTS

Since the gut is usually the initial organ affected in bacterial infections, the first signs of disease are related to feeding and assimilation. Loss of appetite, cessation of feeding, diarrhea, gut paralysis, and regurgitation are characteristic initial stages in many bacterial infections. Later the insect may appear sluggish (rarely irritable), have convulsions, and become uncoordinated; a general paralysis may set in, accompanied by septicemia and death. In a few instances, infected insects may show behavioral changes by, for example, moving to an elevated position or seeking refuge under leaves.

Certain bacteria impart a characteristic color to the cadaver. For example, a red color suggests the presence of *Serratia marcescens*. Bee larvae infected with *Bacillus alvei* become yellow or gray, while those containing *B. larvae* become dark brown. The posterior portion of Japanese beetle grubs infected with *B. popillae* turns white. Most other bacterial infections turn the host brown-black, which is a color associated with bacterial decomposition.

FACTORS AFFECTING BACTERIAL INFECTIONS

With the obligate bacterial pathogens, certain conditions are necessary for successful infections. The so-called "milky disease" *Bacillus* species are infective only to members of the family Scarabaeidae. However, *B. thuringiensis* is known to infect insects in at least four orders, although lepidopterous larvae with a gut pH ranging from 9.0 to 10.5 are most susceptible. *Bacillus thuringiensis* var. *israelensis* is particularly adapted to infecting Culicidae and Simuliidae.

Abiotic factors may also regulate infections. It has been known for some time that ultraviolet light is deleterious to spores and crystals of *B. thuringiensis*. Temperature, humidity, and formulation (when used as a biological insecticide) also affect the stability of *B. thuringiensis*. Environmental factors are especially important in governing infection by potential pathogens. A drop in temperature, starvation, stress resulting from crowding—all of these can make an insect more susceptible to pathogens that need the right "opportunity" to initiate infection.

METHODS OF EXAMINATION

A simple method of determining whether an insect has a bacterial infection is to examine a drop of hemolymph under the microscope. Although bacterial rods and spores can be seen with bright field, most stages are more distinct with phase contrast.

Infections with bacterial pathogens are best diagnosed in the early stage of infection (before the insect has started to decompose), since then the majority of bacteria present in the blood or tissues will be those of the pathogen. The presence of bacteria of various sizes and shapes indicates the presence of saprophytic forms, and such cases are usually the most difficult to diagnose. It could be an infection by one of the potential pathogens, or the terminal stage of a virus or fungal infection. If a bacteriosis is suspected, then all bacteria present should be isolated and cultured. After pure cultures are obtained, identifications and pathogenicity tests can be conducted.

ISOLATION AND CULTIVATION

The first step in isolating bacteria from a diseased insect is to externally disinfect or "sterilize" the specimen to remove contaminating saprophytic forms. The anterior and posterior orifices of the specimen can be ligatured to keep the sterilizing solutions from entering the intestine and penetrating into the hemocoel. The following method has been used with satisfactory results. The specimen first should be dipped into 70% or 95% ethanol (wetting agent) for 2 sec. It can then be transferred to a solution of sodium hypochlorite [household bleach (5.25%) is suitable] for 3.5 min, then placed for an equal period in 10% sodium thiosulfate to remove the free chlorine.

After the external surface has been sterilized, the specimen is rinsed in three changes of sterile distilled water and placed on a sterile dissecting dish. The specimen is then opened with sterile scissors or a scalpel by cutting the integument along a longitudinal dorsal or lateral line, with care being taken not to cut into the gut epithelium (all instruments should be sterilized, either by autoclaving or by periodically dipping them into 70% ETOH and flaming off the alcohol). Blood and body fluids may be sampled with a sterile capillary tube, diluted in 2 ml sterile HOH or Ringer's solution, and then placed on NA, BHIA, or other suitable bacterial media by the streak plate method (see Techniques). Tissue may be examined by removing a small sample, placing it in 1 ml of sterile HOH or Ringer's, and then triturating it with a sterile glass rod. The suspension is then streaked on a plate of nutrient or brain–heart infusion agar. The investigator may also want to inoculate AC medium if the presence of anaerobic pathogens is suspected. The agar plates may be inverted and incubated at room temperature or 30°C overnight, and then examined for well-isolated colonies. These colonies should represent pure bacterial strains and can be described on the basis of the following characteristics (Fig. 49):

1. *Form* (of colony)
   a. Punctiform—under 1 mm in diameter, but visible to the naked eye.
   b. Circular—over 1 mm in diameter, round with smooth edges.
   c. Filamentous—growth consisting of interwoven or irregularly placed threads.
   d. Rhizoid—growth spreading out in an irregularly branched or rootlike manner.
   e. Spindle—usually subsurface colonies, larger in the middle than at the ends.
   f. Irregular—periphery is variable and nonuniform.
2. *Elevation*
   a. Flat—colonies are thin and raised little above agar surface.
   b. Raised—colonies are thick and noticeably raised above the agar surface.
   c. Convex—colonies are curved.
   d. Pulvinate—cushion-shaped colonies.
   e. Umbonate—colonies have a raised center; knoblike.
3. *Surface*
   a. Smooth—surface is even.
   b. Contoured—surface is irregular and smoothly undulating, similar to a relief map.
   c. Radiately ridged—ridges extend out from the center of the colony.
   d. Concentrically ringed—surface is marked with rings, one inside the other.
   e. Rugose—surface is wrinkled in appearance.
4. *Margin*
   a. Entire—a smooth margin.
   b. Undulate—wavy.
   c. Lobate—with rounded projections.
   d. Erose—irregularly notched.
   e. Filamentous—with long interwoven threads.
   f. Curled—composed of parallel chains of threads in wavy strands.
5. *Density*
   a. Opaque—light does not pass through.
   b. Translucent—light passes through, but is diffused so that objects beyond the colony cannot be discerned.

6. *Chromogenesis*—Refers to the production of pigment, such as the greenish pigment of some strains of *Pseudomonas aeruginosa* and the red and orange-red pigments of *Serratia marcescens.*

IDENTIFICATION

Practical methods of bacterial identification include both descriptive morphology and biochemical tests for the detection of, for example, metabolic products and enzymes. Highly specialized techniques such as genetic analysis, serology, bacteriophage typing, and esterase patterns are beyond the scope of most insect pathology laboratories and are not covered here.

Some groups of bacteria can be tentatively identified with a fair probability of success by their colony characteristics, such as the small gray punctiform colonies of *Streptococcus* or the typical rough colonies of *Bacillus*. However, caution must be exercised since bacteria exhibit colonial variations, such as the rough–smooth variations of some *Bacillus* spp. and the colonial dimorphism exhibited by *Xenorhabdus* spp. An example of the latter is *X. luminescens,* in which the primary form produces colonies more mucoid, smaller, and more highly pigmented than the secondary forms. In addition the secondary colonies are flatter with spreading, irregular margins (Fig. 50).

Because of time and material limitations, certain shortcuts for identification can be followed. Thus, a Gram-positive, motile, spore-forming rod isolated from the hemocoel of a diseased insect is almost certainly a *Bacillus*. Likewise, a polar-flagellate, oxidative, Gram-negative rod which produces a green fluorescent pigment is probably a *Pseudomonas* species and requires little further testing for genus.

Morphological examination is very important. Phase contrast microscopy provides a rapid method for detecting cell shape as well as motility. The Gram reaction is probably the most important staining procedure in bacteriology (see Techniques) and should be used routinely. Tests for flagellation and acid-fastness are also very useful.

Special diagnostic media may be very useful. On Tergitol-7 medium with triphenyltetrazolium chloride, *Escherichia coli* forms yellow colonies against a yellow background, while *Pseudomonas aeruginosa* and other Enterobacteriacae form red colonies against a blue background. A diagnostic medium is available for *Streptococcus faecalis* (Meade, 1963), while *P. aeruginosa* can be determined on regular agar plates with the cytochrome oxidase test (Schaefer, 1961).

There are several species of *Bacillus* and *Clostridium* which do not grow well or at all on artificial media. Thus, careful comparison should be made with the bacteria found in the insect and the isolates growing on culture media to be certain the latter are not contaminants.

TESTING FOR PATHOGENICITY

Experimental infections are often necessary to establish the pathogenicity of an unknown isolate. These can best be done by introducing the bacteria in question into the gut of test insects. A glass-tipped syringe can be used for force-feeding or the inoculum can be mixed with food. Koch's rules of pathogenicity can be followed if the results are successful.

STORAGE

From the practical standpoint, the most important entomogenous bacteria belong to the genus *Bacillus*. The spores of most of these species are resistant and survive well if stored in a cool, dry location. Non-spore-forming bacteria can be maintained on simple bacteriological media or lyophlyized.

LITERATURE

The Eighth Edition of *Bergey's Manual* (Buchanan and Gibbons, 1974) contains the latest views on bacterial classification and includes keyes to the genera of bacteria.

For entomogenous bacteria, Steinhaus (1949) provides a good general introduction to the topic, and more specific coverage is provided by Bucher (1963), Dutky (1963), Heimpel and Angus (1963), and Lysenko (1963). The use of bacteria for insect control was discussed by Falcon (1971), while other subjects on the toxins and host spectrum of *B. thuringiensis* were included in Burges and Hussey's *Microbial Control of Insects and Mites* (1971) and updated in Burges (1981). Faust (1974) provides a thorough review of the bacterial diseases of insects, and Afrikian (1973) covered entomogenous bacteria. Davidson (1981) contains six chapters dealing with entomopathogenic bacteria.

KEY TO COMMON GENERA AND SPECIES OF BACTERIA

1. Vegetative cells coccoid in shape (Fig. 51), young cultures Gram positive, spores absent—**2**

1. Vegetative cells rod-shaped (may be very short); Gram negative, positive, or variable; spores present or absent—**3**

2. Cells often forming chains in nutrient broth; catalase negative—*Streptococcus* Rosenbach (Fig. 51). *S. faecalis* Andrews and Horder is a commonly encountered coccus causing insect disease. See Doane and Redys (1970) for a discussion of this species. *S. pluton* (White) is associated with European foulbrood of honeybees. See Bucher (1963) for a discussion of this genus as related to insect disease.

2. Cells occurring singly or in pairs, tetrads, irregular clusters, or cubical packets; catalase positive—*Staphylococcus* Rosenbach and *Micrococcus* Cohn. Members of these genera may occur in diseased insects, but have never been implicated as causal agents.

3. Spores present; young cultures Gram positive (*Bacillus sphaericus* is gram variable)—**4**

3. Spores absent; young cultures Gram negative—**10**

4. Crystalline parasporal body present (may be absent in some strains) (Fig. 53)—**5**

4. Crystalline parasporal body absent (Fig. 54)—**6**

5. Good growth on artificial media under aerobic conditions; catalase-positive; found mainly in larvae of Lepidoptera—*Bacillus thuringiensis* Berliner (Figs. 52, 53). See de Barjac and Bonnefoi (1973) and Heimpel (1967) for a discussion of this group.

5. Poor growth on artificial media under aerobic conditions; catalase-negative; found in Scarabaeidae—*Bacillus popillae* Dutky (Figs. 55, 56). This species contains several varieties, one of which lacks the parasporal body (Wyss, 1974). See Dutky (1963) for a discussion of this species and Wille (1956) for a discussion of *B. fribourgensis*.

6. Little or no growth on artificial media under aerobic conditions (obligate or facultative anaerobes); bacteria restricted to gut of diseased insect; thus far natural infections found in *Malacosoma* sp. (Lepidoptera)—*Clostridium* Prozmoski (Fig. 57). See Bucher (1961) for a discussion of two species in this genus that cause insect diseases.

6. Good growth on artificial media under aerobic conditions; bacteria occur in hemocoel of diseased insects; found in a variety of insects—**7**

7. Sporangia definitely swollen (Figs. 47, 48); spores oval or spherical—**8**

7. Sporangia not swollen; spores ellipsoidal or cylindrical—**9**

8. Spores nearly spherical; pathogenic for mosquito larvae—*Bacillus sphaericus* Neide (Fig. 58). See Kellen *et al.* (1965) for a discussion of this species.

8. Spores oval; cause of American foulbrood in honeybees—*Bacillus larvae* White (Figs. 59, 60). See Heimpel and Angus (1963), Cantwell (1974), and Glinski (1968) for a discussion of this species.

9. Acetylmethylcarbinol not produced [Voges Proskauer (VP) test is negative]; thought to cause disease in silkworm and the scarabeid, *Melolontha melolontha*—*Bacillus megaterium* de Bary. See Heimpel and Angus (1963) for a discussion of this species.

9. Acetylmethylcarbinol produced (VP test is positive)—*Bacillus cereus* Frankland and Frankland (Fig. 54). Saprophytic species of *Bacillus* will also key out here, but only *B. cereus* has been associated with insect disease. See Heimpel and Angus (1963) for a discussion of this species.

10. Cells polar flagellate; colonies on agar give a rapid positive reaction to the cytochrome oxidase test (see Techniques)—*Pseudomonas aeruginosa* (Schroeter) (Fig. 61). See Bucher (1963) for an account of the pseudomonads associated with insects.

10. Cells peritrichously flagellate or without flagella; cytochrome oxidase test negative—**11**

11. Bacteria associated with the presence of rhabditoid nematodes in the hemocoel; large rods (1–10 μm long)—*Xenorhabdus nematophilus* Poinar and Thomas (Fig. 62). This bacterium is associated with neoaplectanid nematodes and its nature with insects is discussed by Poinar and Thomas (1967). A closely related species (*X. luminescens*) (Fig. 50) is associated with nematodes of the genus *Heterorhabditis* and turns infected insects a reddish color (color plate, A)

11. Bacteria not associated specifically with rhabditoid nematodes; small rods (0.5–2.0 μm long)—**12**

12. Often producing various shades of orange to red pigments in the insect or on agar plates; flagella close coiled when stained by Leifson's method (see Techniques); does not ferment arabinose—*Serratia marcescens* Bizio (Fig. 63). See Steinhaus (1959) and Bucher (1963) for associations of this species with insects.

12. Nonchromogenic; flagella not close coiled; may or may not ferment arabinose—**13**

13. Phenylalanine deaminase negative; VP usually positive—*Enterobacter* Hormaeche and Edwards. Species in this genus have been reported as potential pathogens of grasshoppers; however, their invasiveness under natural conditions is questionable. See Bucher (1959) and Faust (1974) for a discussion of these forms.

13. Phenylalanine deaminase positive; VP usually negative—*Proteus* Hauser. Species in this genus have been reported as potential pathogens of grasshoppers; however, their invasiveness under natural conditions is questionable. See Faust (1974) for a discussion of these forms.

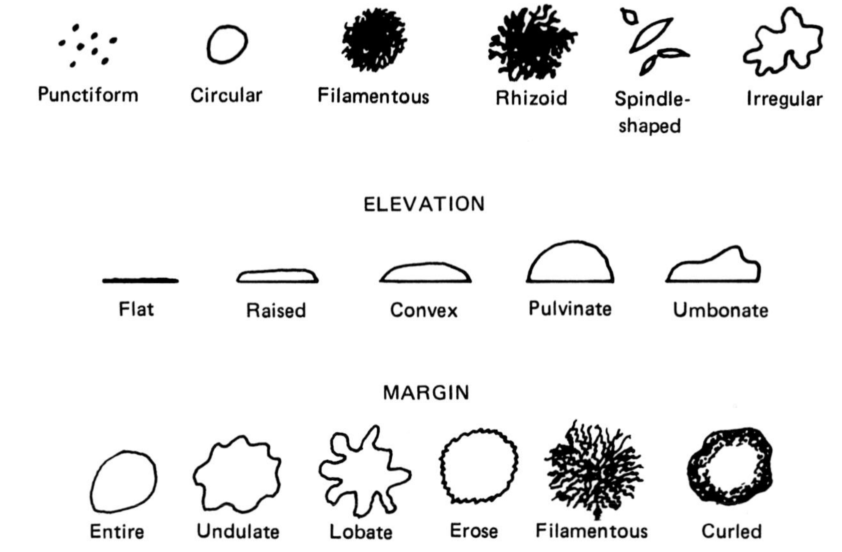

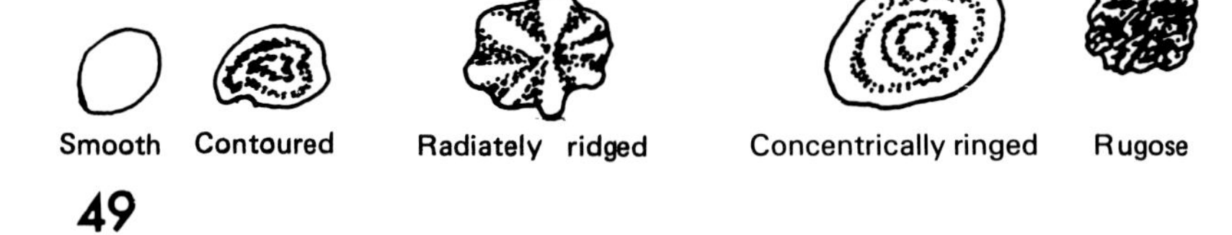

FIGURE 49. Characteristics of bacterial colonies.

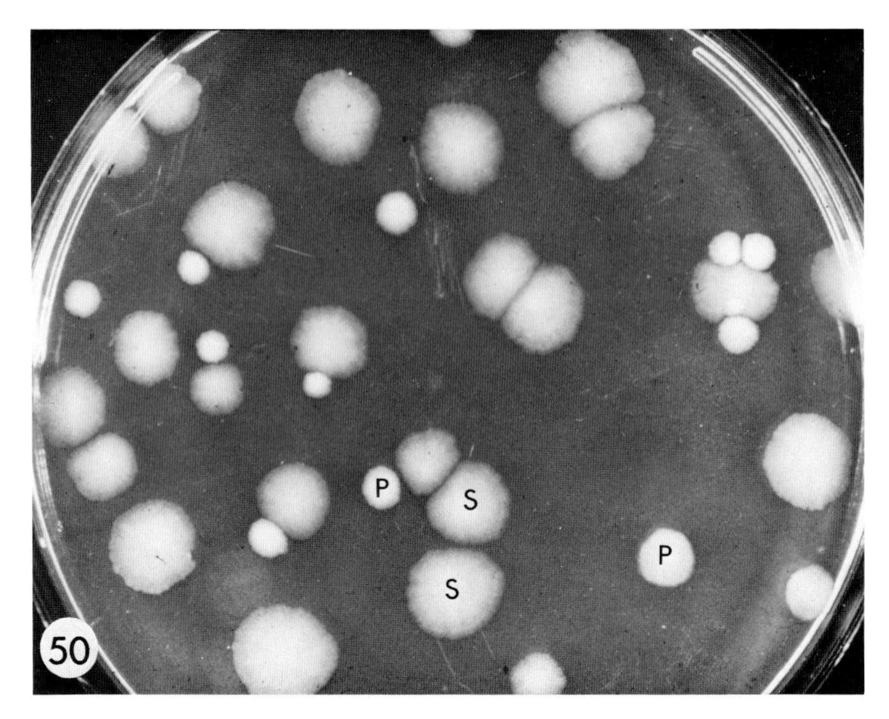

FIGURE 50. Colonies of *Xenorhabdus luminescens* showing the primary (P) and secondary (S) forms.

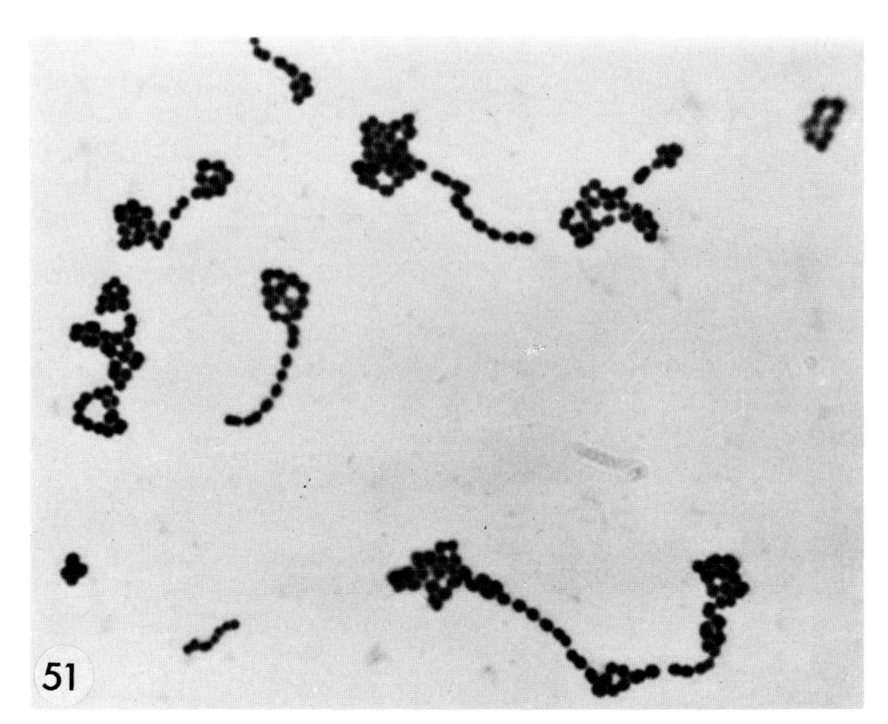

FIGURE 51. Coccoid vegetative cells of *Streptococcus* sp. (methyl blue stain). ×1800.

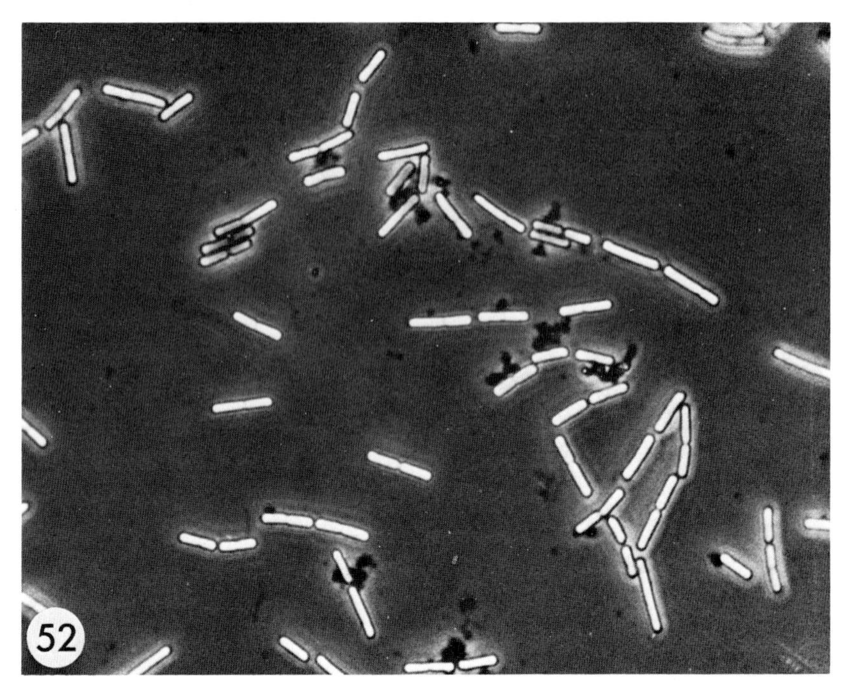

FIGURE 52. Vegetative cells of *Bacillus thuringiensis*. ×1600.

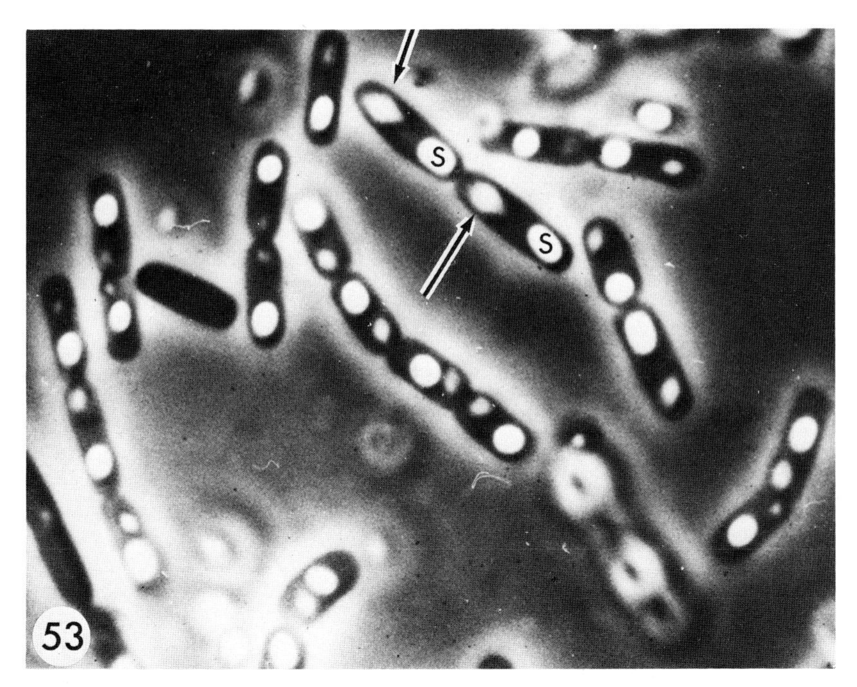

FIGURE 53. Sporangia of *Bacillus thuringiensis* showing spores (S) and crystalline parasporal bodies (arrows). ×2560.

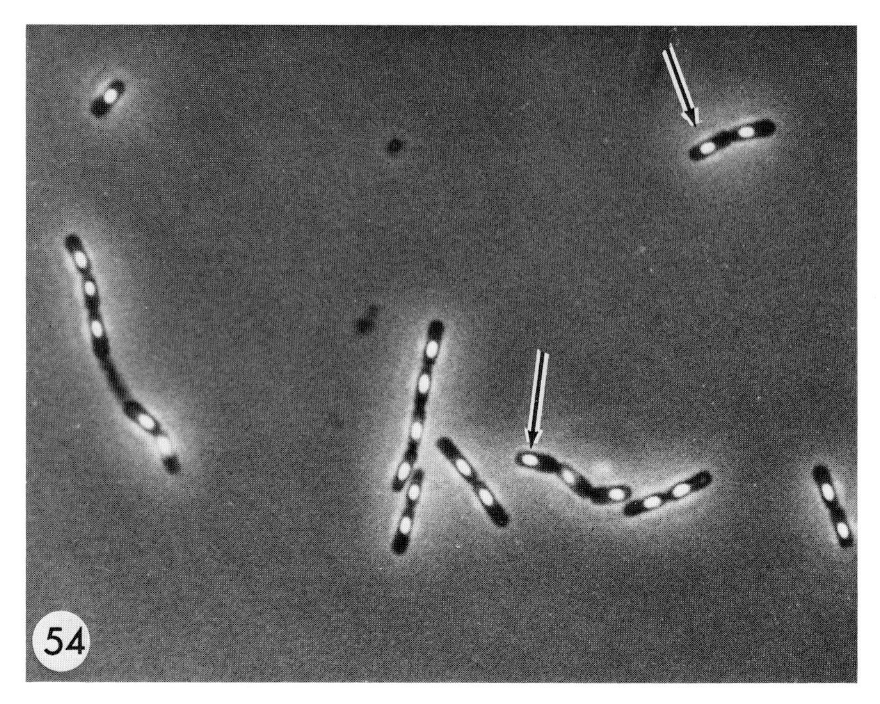

FIGURE 54. Nonswollen sporangia of *Bacillus cereus*. Note spores (arrows). ×1600.

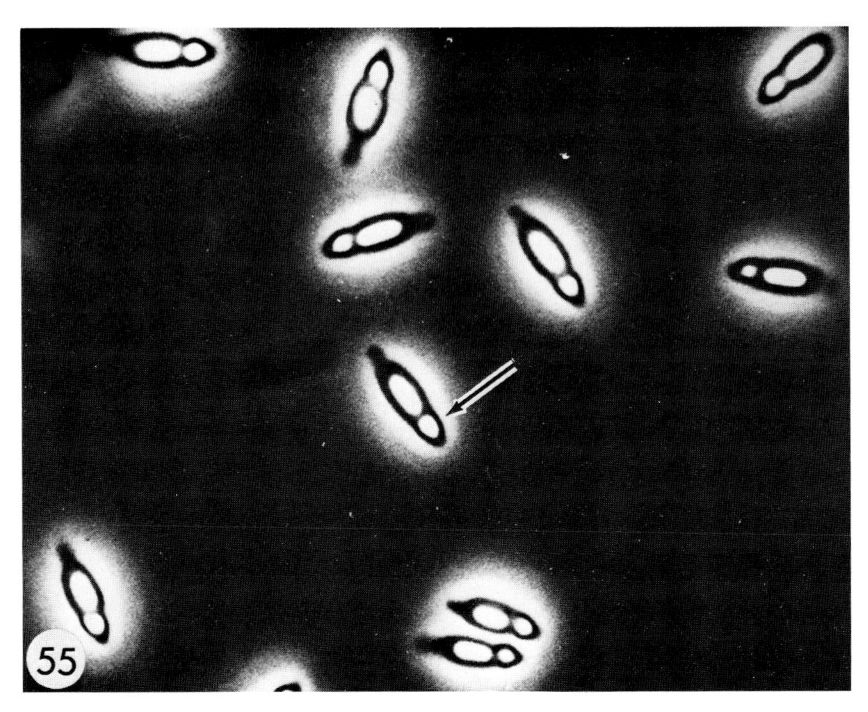

FIGURE 55. Sporangia of *Bacillus popillae* with parasporal body (arrow). ×3400.

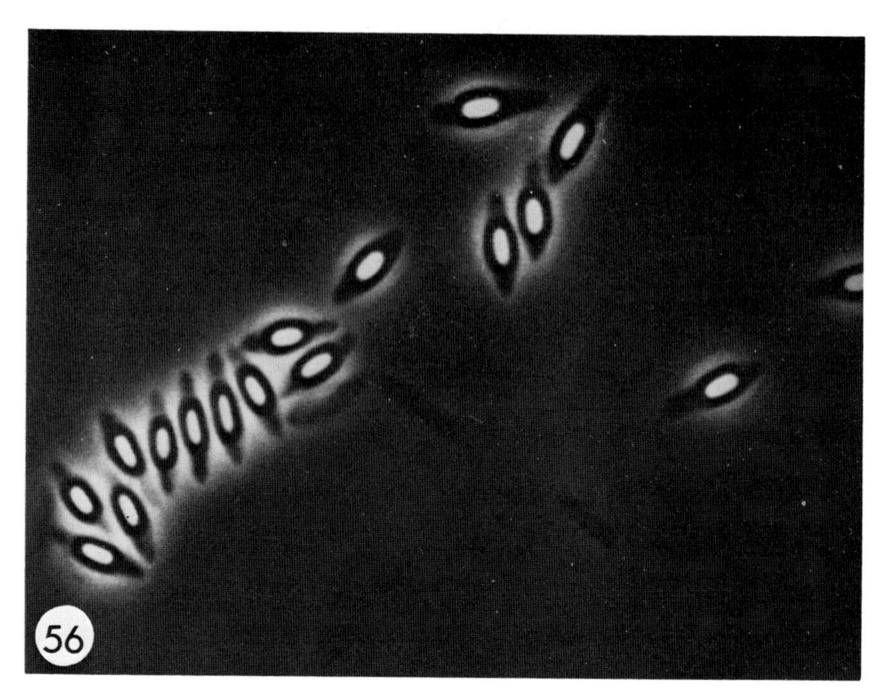

FIGURE 56. Sporangia of *Bacillus popillae* without parasporal body. ×3400.

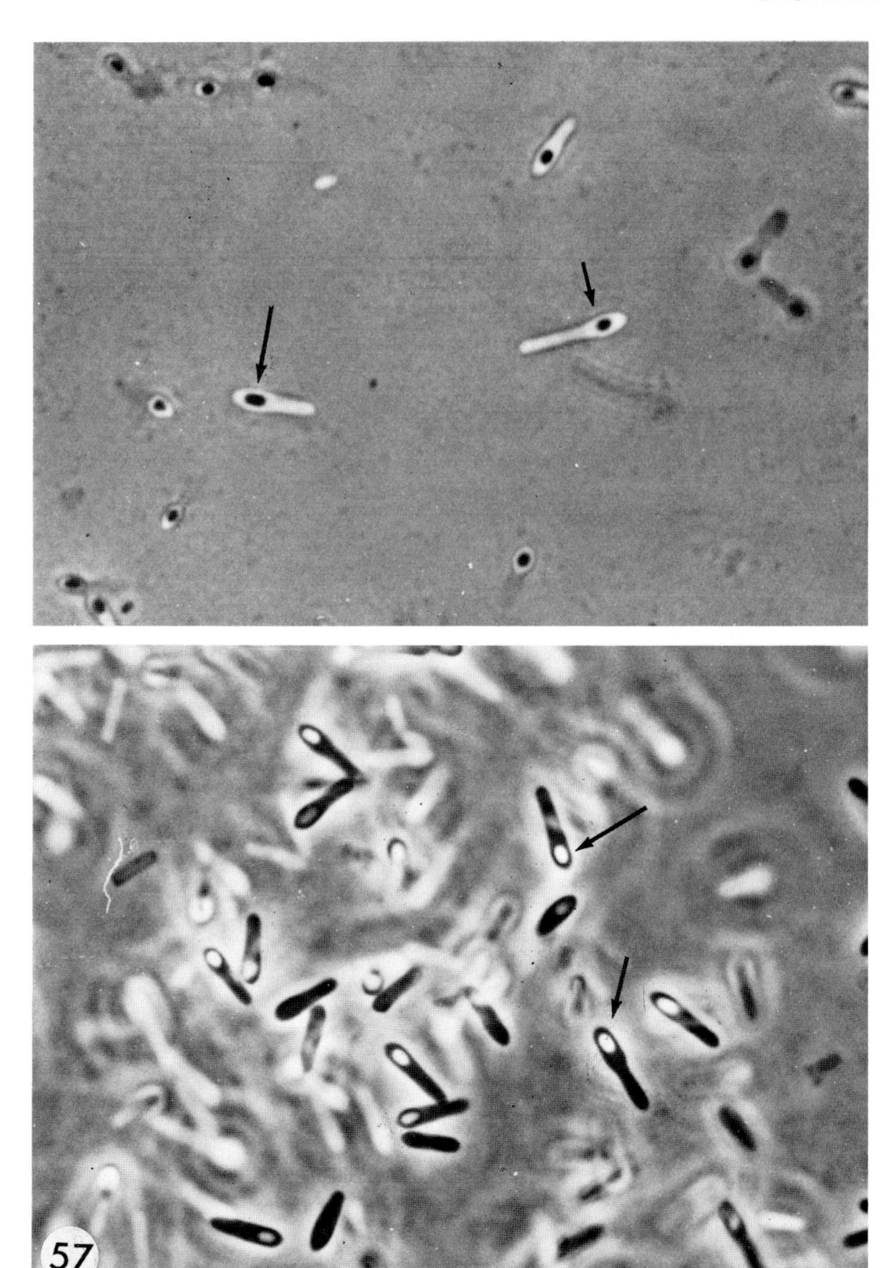

FIGURE 57. Sporangia of *Clostridium sporogenes* with terminal spores (arrows). ×2100.

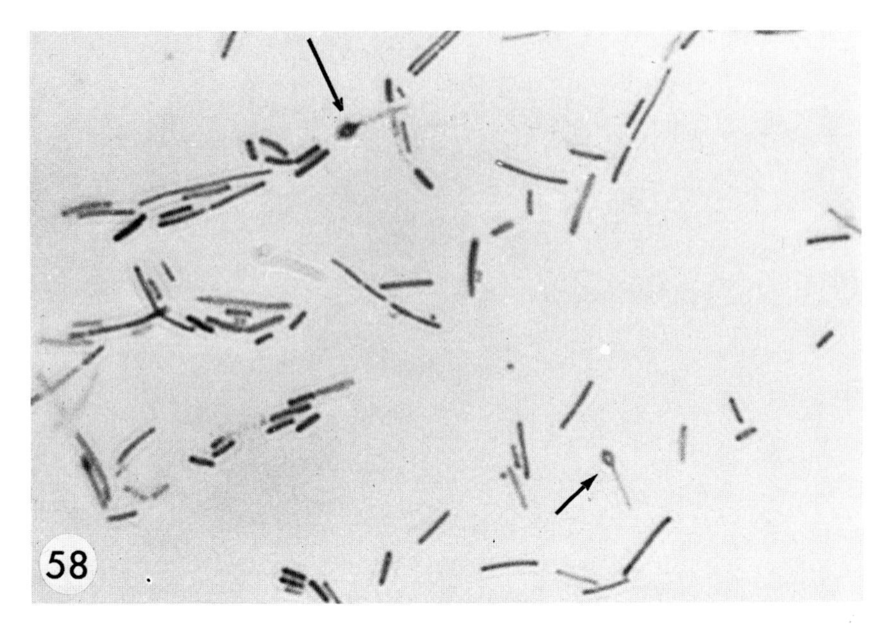

FIGURE 58. Vegetative cells and swollen sporangia of *Bacillus sphaericus*. Note nearly spherical spores (arrows). ×1000.

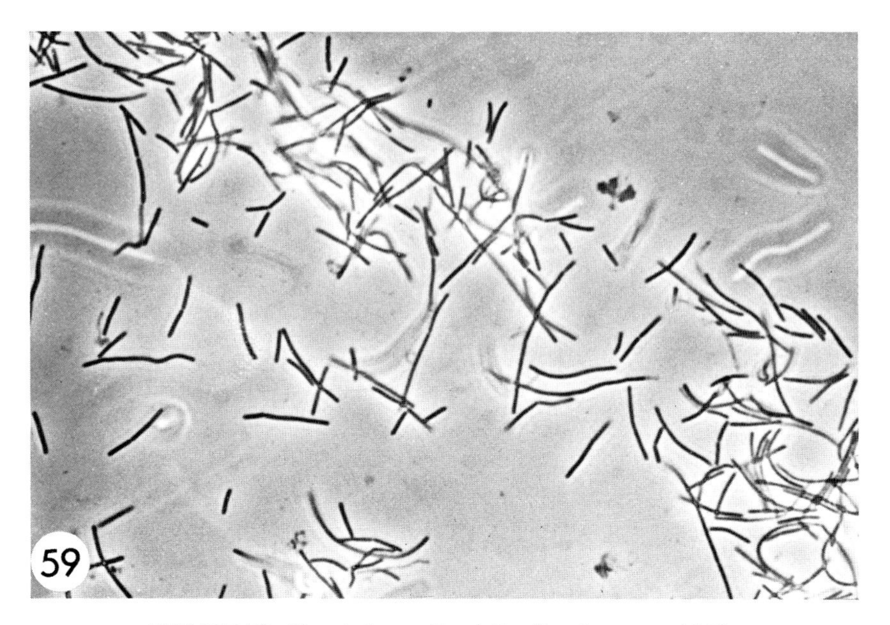

FIGURE 59. Vegetative cells of *Bacillus larvae.* ×1540.

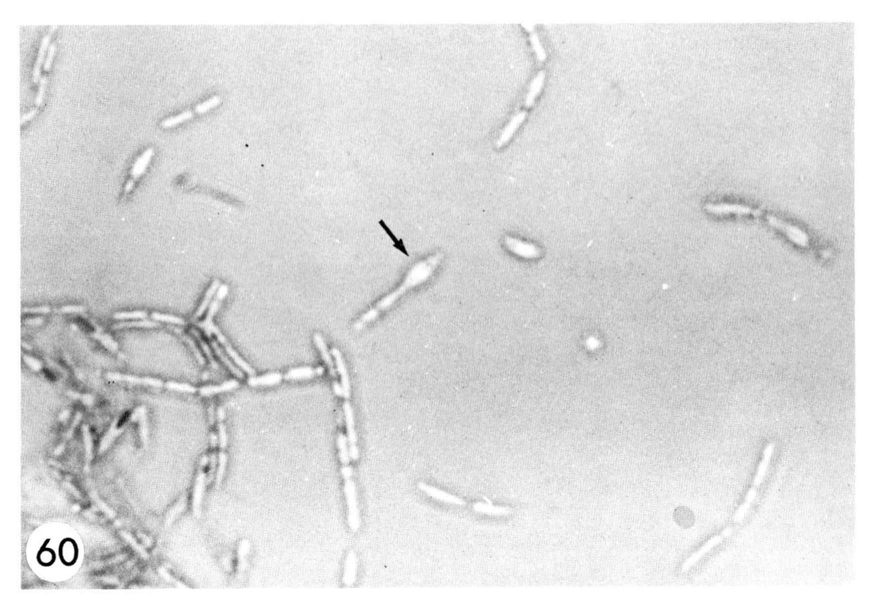

FIGURE 60. *Bacillus larvae* sporophores, showing swollen sporangia and oval spores (arrow). ×1540.

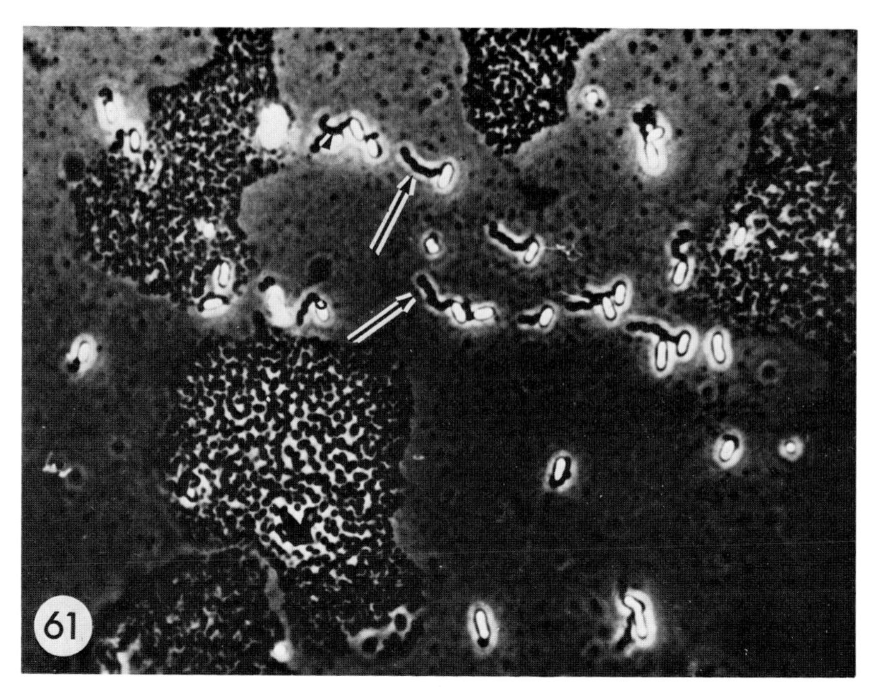

FIGURE 61. Vegetative cells of *Pseudomonas aeruginosa* with polar flagella (arrows). ×1800.

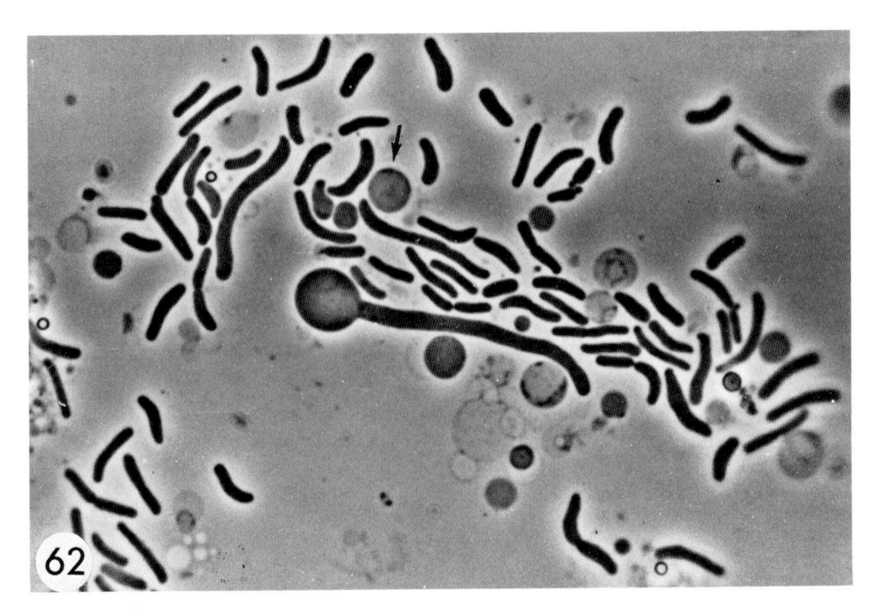

FIGURE 62. Variable-sized vegetative cells and associated protoplasts (arrow) of *Xenorhabdus nematophilus*. ×1500.

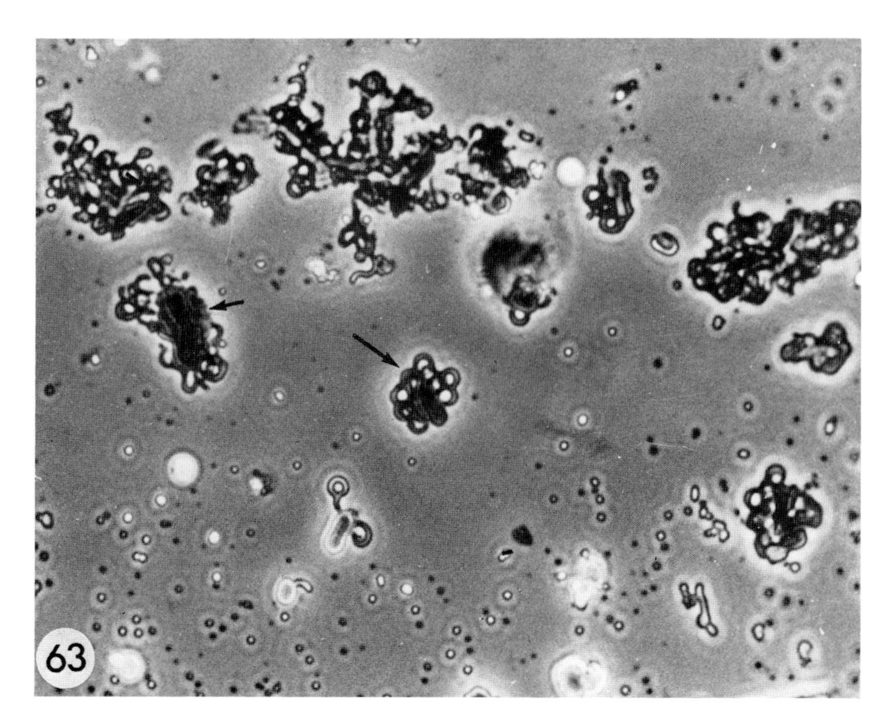

FIGURE 63. Vegetative cells (arrows) of *Serratia marcescens* showing close coiled peritrichous flagella. ×1800.

FUNGI

INTRODUCTION

There are many types of associations between insects and fungi, ranging from inquilinism to obligate parasitism. Some insects, especially representatives of the lower Diptera such as the fungus gnats (Sciaridae, Mycetophilidae), feed on mushrooms and plant tissue invaded by fungi. Many species of bark beetles (Scolytidae) and other wood-boring insects (like members of the Siricidae) carry fungal bodies or spores in specialized pouches (mycetangia, etc.). These fungi are usually symbionts and provide food or a suitable habitat for growth and development of their insect carrier. Some fungi, such as yeasts or highly specialized forms such as the Trichomycetes, occupy certain niches in the alimentary tract of insects. Still others, such as the unique Laboulbeniales, are adapted for survival on the insect's cuticle. Although they are true parasites, they rarely cause disease. True pathogenic fungi, which can eventually destroy the insect host, attack a variety of terrestrial and aquatic insects. Although fungi were responsible for some of the earliest known insect diseases, basic studies on the majority of the entomogenous forms are still incomplete.

Some entomogenous fungi are being commercially produced today for use as biological control agents. Strains of the broad-spectrum species *Metarrhizium anisopliae* are being used to control Homoptera in sugar cane fields in South America. In England a product composed of *Verticullium lecanii* is used in glasshouses against scales and whiteflies. *Beauveria bassiana* and *Nomuraea rileyi* are both undergoing large-scale field tests and may soon be commercially available.

TAXONOMIC STATUS

There is some question concerning the taxonomic position of fungi and whether they warrant a separate kingdom; however, there is general agreement regarding certain basic characteristics. As heterotrophic (lacking photosynthesis) organisms possessing chitinized cells, they are typically nonmotile, although motile stages (zoospores) are sometimes present. Most of the entomogenous fungi contain hyphae (fungal strands developing from a germinating spore) which together constitute a mycelium. Reproduction is mainly by spores, which may arise sexually or asexually. Sexual reproduction involves the union of nuclei belonging to two sex cells or gametes. Asexual reproduction generally results in spores born in sporangia (sporangiospores) or on hyphae (conidia). The most commonly encountered fungi belong to three subdivisions of the Eumycotina. The majority are found in the subdivision Deuteromycotina (Fungi Imperfecti), others in the Mastigomycotina and Ascomycotina. The few members of the Basidiomycotina which form parasitic associations with insects are only rarely encountered.

LIFE CYCLE

Most entomogenous fungi initiate infection by a germinating spore (usually a conidium) which penetrates the cuticle of an insect. The invasive hypha enters the host's tissues (often the fat body is first attacked) and ramifies through the hemocoel. In forms such as *Metarrhizium anisopliae, Beauveria bassiana, Conidiobolus coronata,* and *C. apiculatus,* ''hyphal bodies'' or segments of hyphae break off and circulate in the host's hemocoel during the early stages of infection. This distributes the fungus throughout the host's hemocoel, although there is some question as to whether the large hyphal bodies of *Entomophthora* spp. can actually circulate (Prasertphon and Tanada, 1968). After filling the dying or dead insect with mycelium, emergence hyphae grow out through the insect's integument and produce spores (usually conidia) on the external surface of the host. These spores are dispersed by wind or rain and even by the parasitized insect during feeding or mating. In some species of *Coelomomyces,* which infect mosquito larvae, the life cycle of the fungus requires infection of a copepod as an alternate host (Whistler *et al.*, 1974; Federici, 1975; Weiser, 1976). In some instances the fungi are localized in special organs of the host; e.g., the fungi *Massospora cicadina* and *Strongwellsea castrans* occur only in the abdomen of adult hosts.

Some fungi are host or stage specific, while others (e.g., *Beauveria bassiana* and *Metarrhizium anisopliae*) exhibit a wide host range.

Fungal infections and especially epizootics are largely dependent on a large host population and ideal climatic conditions. Adequate moisture and temperature are usually required for successful sporulation and spore germination. Bell (1974) discusses the environmental factors affecting susceptibility of insects to fungal infections.

CHARACTERISTICS OF INFECTED INSECTS

The most striking characteristic of a fungal infection is the presence of mycelium in or on the affected insect. In the early stages of infection the insect may show general ill effects such as a cessation of feeding, weakness, and disorientation. The host often changes color, and the cuticle may show dark spots indicating areas of fungal penetration. In some *Entomophthora* infections the insect turns yellow when filled with conidia and black when resting spores are present.

Several general characteristics of the infection can give clues to the identity of the fungus. An insect covered with powdery white spores would be suspected of having a *Beauveria bassiana* or *Hirsutella* infection; one covered with green spores would suggest an infection with *Metarrhizium anisopliae* or *Nomuraea rileyi*; with greenish-yellow spores, *Aspergillus flavus*; and with yellowish spores, *Paecilomyces farinosis*. Certain strains of the above-mentioned fungi, however, may not produce this characteristic color and the pigment is not always produced in agar cultures.

Aside from color, the position of the host at death may also be characteristic (e.g., whether the host is fixed to a substrate, as with flies, grasshoppers, and caterpillars attacked by *Entomophthora,* or relaxed on the ground). The consistency of the body may also be a diagnostic character (e.g., whether it is hollow, cheeselike, or hardened).

METHODS OF EXAMINATION

Since it is sometimes difficult to know whether a fungus growing externally on an insect is pathogenic or simply a saprophyte, it is usually best to attempt to identify all fungi present. Some of the mycelium or spores on the surface or inside the host can be transferred to a drop of water or cotton blue stain on a microscope slide. If several fungi are involved or sporulation has not occurred, then the fungi should be cultivated first (see following section). After sporulation, a portion of the agar containing mycelium and spores can be transferred to a microscope slide and examined in bright field or phase contrast. For bright field the specimens can be mounted in cotton blue, analine blue, or Guegen's solution (see Techniques). If a portion of the agar is taken with the fungus material, the agar may be evenly distributed over the slide by gently heating before adding the cover slip.

Characteristics to be noted are the size and shape of the spores, their attachment to the hyphae, and the presence or absence of hyphal septa and clamp connections. When examining insect tissue, remember that fungal spores can sometimes be confused with pollen grains and protozoan spores.

ISOLATION AND CULTIVATION OF ENTOMOGENOUS FUNGI

Before attempting to isolate a fungal pathogen, it is important to keep the specimens fairly dry to avoid further deterioration by growth of bacteria and saprophytic fungi. The following steps are suggested for isolating fungi from diseased insects. Remember that success depends largely on starting out with a fairly fresh specimen and maintaining sterile conditions throughout.

1. Surface sterilize the insect by immersing it in a 5% solution of NaClO or another suitable germicide for several minutes, then rinse it in three changes of sterile water.
2. In a sterile dish open the specimen and transfer a small portion of infected tissue to a sterile culture plate. We use Sabouraud dextrose agar with yeast extract (see Techniques) since it produces quick growth for many entomogenous fungi and the acid reaction (pH 5.6) retards bacterial growth. Other media containing various bactericidal and fungicidal agents for retarding bacteria and saprophytic fungi have been devised (Veen and Ferron, 1966).

Mycelium and spores can also be removed from a fresh specimen and placed directly on the medium; however, this isolate should be compared with that obtained from the infected tissues, since the chance of encountering a saprophytic fungus from the surface of the specimen is much greater than from the internal tissues.

3. The cultures can be placed in a moist incubator at 25°C and examined daily. After growth and sporulation (from 1–2 weeks), preparation can be made for microscopic examination (see previous sections).

IDENTIFICATION

One of the most difficult aspects of identification is the preparation of slides clearly showing the diagnostic characters of the fungus. Characteristics to be noted are the size and shape of the spores, their attachment to the hyphae, and the presence or absence of hyphal septa and clamp connections. Is the mycelium modified into a stroma or synnemata? Are the spores motile, septate, catenulate, or borne in slime drops?

Lastly, many fungi are specific to certain hosts or host groups, and every attempt should be made to identify the diseased insect.

TESTING FOR PATHOGENICITY

In testing the pathogenicity of an unknown fungus, it is best to use insects of the same species originally attacked. If not available, then laboratory test insects such as larvae of *Galleria mellonella* can be used (see Techniques for methods of rearing). Since most fungal pathogens penetrate the host's cuticle, it is usually sufficient to place spores directly on the insects' integument with a sterile instrument. The insects can also be made to walk over a sporulating fungus culture. The treated insects should then be kept in a moist, warm environment, which induces spore germination.

Spores of fungi which invade through the intestine can be collected after washing the culture surface with sterile water and then mixed with the host's food. They can also be introduced into the insect's buccal cavity with a glass-tipped hypodermic syringe.

STORAGE

Many fungi can be stored for years under lyophilization, in liquid nitrogen, or on silica gel [see Bell and Hamalle (1974) for a discussion of the latter process]. More sensitive forms can be maintained on agar slants capped with wax under refrigeration or immersed in mineral oil at room temperature. Of course, periodic transfers must be made.

LITERATURE

For general taxonomic studies on fungi, the treatises of Ainsworth *et al.* (1973) and Barnett and Hunter (1972) are useful. Two books on entomogenous fungi have been written by Evlakhova (1974) and Koval (1974). For host–pathogen relationships, Steinhaus's *Insect Pathology: An Advanced Treatise* (1963) and Madelin's review article (1966) can be consulted. Samson (1981) presents a recent treatment of entomopathogenic Deuteromycetes, while King and Humber (1981) provide a similar treatment for the Entomophthorales, and Bland *et al.* (1981) give an account of *Coelomomyces*. Roberts and Yendol (1971) discuss the use of fungi for insect control, and Bell (1974) presents a general review of insect mycoses. *Ainsworth and Bisby's Dictionary of the Fungi* (Ainsworth, 1961) gives an excellent account of mycological terms and definitions of genera and higher fungal taxons.

KEY TO COMMON GENERA OF FUNGI

1. Mycelium nonseptate (coenocytic), variable in development; motile cells may be present (Mastigomycotina)—**2**

1. Mycelium septate, usually well developed; motile cells absent—**8**

2. Mycelium usually well developed; sexual reproduction results in the formation of nonmotile round zygospores (Fig. 64); asexual reproduction results in the formation of sporangia or conidia, which are produced inside or on the exterior surface of the host (Zygomycetes)—**3**

2. Mycelium usually sparse; reproduction results in the formation of motile zoospores; thick- or thin-walled resting spores or sporangia often present inside the host—**5**

3. Conidia forcibly discharged, produced outside the host—**4**

3. Conidia not forcibly discharged, produced inside the host (in abdominal cavities)—*Massospora* Peck (Fig. 65). For a discussion of this genus, see MacLeod (1963) and Speare (1921).

4. Conidiophores single-celled, arise from hyphae—*Entomophthora* Fres. (Figs. 66, 67) (color plate, F) [for a discussion of this genus see MacLeod (1963), Waterhouse (1973), and King and Humber (1981)] and *Conidiobolus* Bref. [see King and Humber (1981)]. *Strongwellsea* Batko and Weiser (Fig. 68) and *Entomophaga* Bakto are other genera included in this group.

4. Conidiophores linear, four-celled, arise from germinated zygospore on larval Tabanidae—*Tabanomyces* (Dudka and Koval) Couch *et al.* (Fig. 69). See Couch *et al.* (1979). Now in the genus *Meristacrum* Dyech.

5. Hyphae unwalled, become converted into resting spores or sporangia (usually sculptured and pigmented in *Coelomomyces*) (Fig. 70); kidney-shaped, or round zoospores posteriorly uniflagellate (Chytridiomycetes)—**6**

5. Hyphae walled; round zoospores biflagellate, formed in a vesicle at the tip of a discharge tube (Oomycetes)—**7**

6. Attack aquatic insects, especially mosquito larvae; host turns yellow, orange, or brown owing to the color of the mature resting spores—*Coelomomyces* (Fig. 70) pathogenic to mosquitos. The related genus *Coelomycidium* Debaiseux (Figs. 71, 72) will also key out here. This pathogen has been recovered from blackflies (Simuliidae) and often turns the host a pinkish hue.

6. Attack terrestrial insects; transform host tissues into an orange-colored mass or globular resting spores or sporangia, which become powdery when dry—*Myiophagus* Thaxt. (=*Myrophagus*). For an account of this genus, see Karling (1948).

7. Hyphae become broken up into segments, each of which gives rise to distinct sporangia or gametangia—*Lagenidium* Schenk (Figs. 73, 74). For an account of this genus, see Umphlett and Huang (1972).

7. Hyphae mostly vegetative, only certain portions giving rise to sporangia or gametangia—*Pythium* Pringsh. (Fig. 75). Clark *et al.* (1966) have shown that members of this genus can be potential pathogens, similar to *Saprolegnia,* another member of the Oomycetes that has been studied by Rioux and Achard (1956).

8. Sexual reproduction results in the formation of transversely septate basidia which bear four basidiospores. Parasites of scale insects; stroma flat, appressed to bark of tree (Septobasidiales)—**9**

8. Sexual reproduction absent or resulting in the formation of an ascus containing eight ascospores. Asexual reproduction with conidia borne on conidiophores—**10**

9. Epibasidium arises from a hypobasidium (probasidium); parasitizes entire colonies of scale insects—*Septobasidium* Pat. (Figs. 76, 77). For an account of this genus, see Couch (1938).

9. Epibasidium arises from an elongated uredospore; parasitizes single scale insects—*Uredinella* Couch. See Couch (1937) for an account of this genus.

10. Sexual stage resulting in an ascus containing eight ascospores (Ascomycotina)—**11**

10. Sexual stage absent, hyphae and conidia present; mycelium generally found inside and on the surface of the host (Deuteromycotina)—**15**

11. Mycelium scanty or lacking; unicellular, reproducing asexually by budding, fission, or both. When produced, ascospores are born in naked ascus—Yeasts of the class Hemiascomycetes. Some genera in this class are insect pathogens and many others are associated with healthy insects. *Candida* Berkout (Fig. 78) can be pathogenic to insects (see Martignoni *et al.*, 1969). Other yeasts in the genera *Mycoderma, Saccharomyces,* and *Blastodendrion* have been isolated from insects. However, their pathogenicity has not been elucidated; see Steinhaus (1949) and M. W. Miller and van Uden (1970) for coverage of these forms.

11. Mycelium generally present; ascospores borne in ascocarps (fruiting bodies); asexual reproduction, when present, by conidia; budding rare or absent—**12**

12. Mycelium sparse, small bush- or hairlike growths on the surface of the insect's cuticle; thallus usually consisting of a few cells; usually only four ascospores produced (rarely eight); ascospores once septate—Laboulbeniomycetes (Fig. 79). These fungi are generally not pathogenic and are more curiosities than anything else. However, *Hesperiomyces virescens* was reported to be pathogenic to coccinellid beetles (Kamburov *et al.*, 1967).

12. Mycelium extensive; found inside as well as on the external surface of insects; asci with eight (rarely two to four) ascospores (or many ascospores occur in a spore ball); ascospores nonseptate or with two or more septa—**13**

13. Ascospores borne in spore balls in dark-colored cysts which appear as tiny black specks on mummified bee larvae; ascospores nonseptate—*Ascosphaera* Olive and Spiltoir (Figs. 80, 81). Species in this genus cause chalk-brood disease in honey and leafcutter bees. See Skou (1972) for a review of this genus and related species.

13. Ascospores not borne in spore balls in cysts; ascospores with two or more septa—**14**

14. Parasites of scale insects; mycelium forms a black cushion-shaped mat covering one or more scales; asci with thick dark septate ascospores embedded in the fungus stroma—*Myriangium* Mont. and Berk. (Figs. 82, 83). See J. H. Miller (1940) for a discussion of this genus.

14. Parasites of various terrestial insects (rarely scales); the host is filled with septate mycelium, which forms an aerial stroma extending out of the insect's body for some distance; perithecia containing cylindrical asci occur on the fertile portion of the stroma; ascospores filiform and multiseptate—*Cordyceps* (fr.) (Figs. 84–86). See McEwen (1963) for a discussion of this genus.

15. Hyaline, fusoid conidia born in pycnidia formed in a cavity of the fungal stroma; pycnidia usually brightly colored (color plate, H); parasites of whiteflies and scales—*Aschersonia* Mont. (Figs. 87, 88). A monograph on this genus was prepared by Petch (1921), and Mains (1959) discussed the North American species parasitic on whiteflies.

15. Conidia variable, not produced within a pycnidium embedded in a fungal stroma—**16**

16. Conidiophores united into synnemata or elongated hornlike structures arising from the dead insect, bearing a superficial resemblance to *Cordyceps*—**17**

16. Conidiophores variable, but not united into snynnemata—**23**

17. Phialides (bottle-shaped structures which give rise to spores) borne singly on the conidiophores (Fig. 89)—**18**

17. Phialides borne in groups or clusters on the conidiophores (Fig. 90)—**21**

18. Phialides usually short and thick (sometimes nearly spherical); conidia dry; synnemata thick, whitish, usually not branched—*Isaria*

Pers. (Figs. 89, 91, 92). May be an imperfect stage of *Cordyceps* species (McEwen, 1963). See DeHoog (1972) for a recent revision of this genus.

18. Phialides elongated, slender; conidia covered with mucus; synnemata slender, buff colored, usually branched—**19**

19. Phialides enlarged at base; conidia not in heads—*Hirsutella* Pat. (Figs. 89, 93–95) (color plate, G). May be imperfect stage of *Cordyceps* species (McEwen, 1963). See Mains (1951) for a review of the genus.

19. Phialides not enlarged at base; conidia in clusters or heads—**20**

20. Synnemata simple or branched; phialides awl-like, tapering to a long slender neck; conidia covered with mucus with several spores held together in clusters—*Synnematium* Spear (Figs. 96, 97). See Mains (1951) for a review of this genus.

20. Synnemata simple, unbranched; conidiophores with enlarged apex bearing prophialides and phialides forming a globose head on which hyaline conidia are borne—*Pseudogibellula* Samson and Evans (Fig. 98). See Samson and Evans (1973).

21. Phialides obtuse at apex (Fig. 89); conidia borne singly (Fig. 89), not catenulate; synnemata cylindrical—*Hymenostilbe* Petch (Fig. 99). May be an imperfect stage of *Cordyceps* species (McEwen, 1963).

21. Phialides pointed at apex; conidia borne in chains (Fig. 90)—**22**

22. Tips of synnemata swollen (Fig. 90)—*Insectiocola* Mains (Figs. 100, 101). See Mains (1950) for a discussion of this genus.

22. Tips of synnemata not swollen (Fig. 90), but cylindrical or pointed—*Akanthomyces* Leb. May be an imperfect stage of *Cordyceps* species (McEwen, 1963). See Mains (1950) for a discussion of this genus.

23. Conidiophores covering a cushion-shaped stroma; conidia dark; parasitic on scale insects—*Aegerita* Pers. For the biology of this genus, see Morrill and Black (1912).

23. Conidiophores borne over the surface of the host, not arising from a stroma covering the host; conidia usually hyaline; may or may not parasitize scale insects—**24**

24. Two types of conidia present: slender, usually septate, canoe-shaped macroconidia and smaller microconidia (may not occur together); conidia not borne in chains—*Fusarium* Link (Figs. 102, 103). For notes on pathogenesis, see Madelin (1963) and Hassan and Vago (1972).

24. Typically only one type of conidia present—**25**

25. Conidia borne in slime or mucus balls at the tips of phialides (especially in older, mature mycelium)—**26**

25. Conidia not collecting in slime drops at the apex of the phialides—**28**

26. Conidophores slender; phialides awl-shaped, arranged in whorls (verticillate); conidia hyaline, ovoid to elipsoid—*Verticillium* Nees (Fig. 104). See Hall (1981) for a discussion of *Verticillium lecanii.*

26. Phialides flask-shaped (swollen base)—**27**

27. Conidia club-shaped, sticky, borne on clusters of flask-shaped phialides, especially on mosquito larvae and related Diptera.—*Culicinomyces* Couch *et al.* (Fig. 105). See Couch *et al.* (1974), Sweeney (1975), and Sweeney *et al.* (1982).

27. Conidia small, round, hyaline, borne on verticillate, flask-shaped phialides with narrowly tapering, bent neck—*Tolypocladium* Gams (Fig. 106). See Gams (1971) for a discussion of this genus.

28. Conidia borne singly, not catenulate; fertile portion of conidiophore zig-zag in shape and drawn out at tip—*Beauveria* Vuill. (Figs. 107–109). See DeHoog (1972) for a revision of this genus.

28. Conidia borne in chains (catenulate)—**29**

29. Phialides borne in clusters or groups on the often enlarged apex of single conidiophores—*Aspergillus* Mich. (Figs. 110–114). For notes on the pathogenicity of this genus, see Madelin (1963) and Sussman (1951).

29. Not as above—**30**

30. Conidiophores in compact columns or groups; conidia elongate or rod-shaped—*Metarrhizium* Sorok. (Figs. 115–117). The common species *M. anisopliae* is a characteristic green color (color plate, D). For pathogenicity studies, see Zacharuk and Tinline (1968).

30. Conidiophores not in compact groups; conidia variable in shape, usually globose—**31**

31. Phialides in nondiverging clusters, usually green—*Penicillium* Link (Figs. 90, 118). For pathogenicity studies see Sen *et al.* (1970).

31. Phialides divergent in loose groupings—**32**

32. Usually white, yellow, rose, or red colonies—*Paecilomyces* Bain (Fig. 119). See Brown and Smith (1957) for a taxonomic discussion of this genus.

32. Green colonies—*Nomuraea* Maulblanc (=*Spicaria* Harting) (Figs. 120, 121). May be an imperfect stage of *Cordyceps* species (McEwen, 1963).

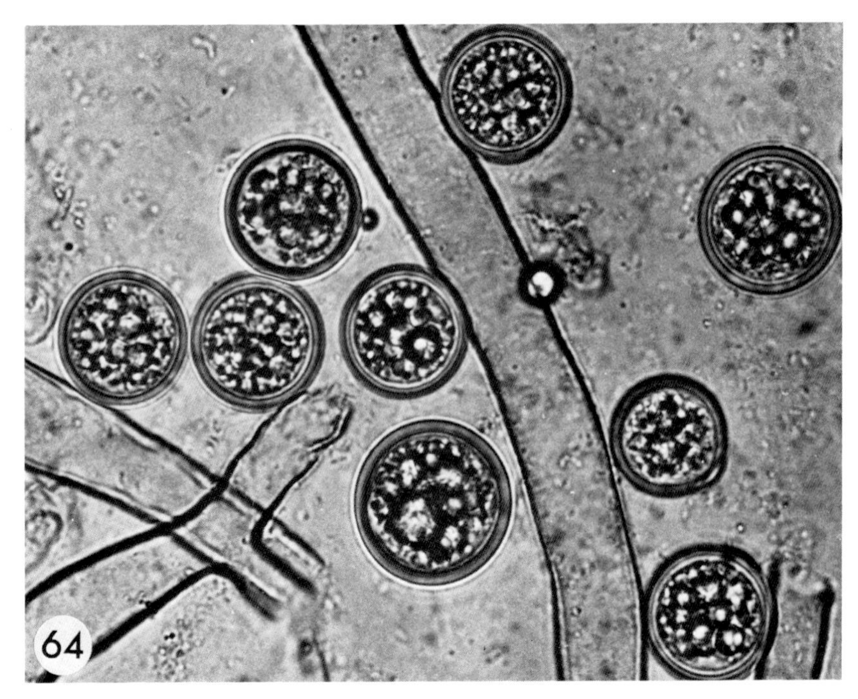

FIGURE 64. Zygospores of an *Entomphthora* sp. ×750.

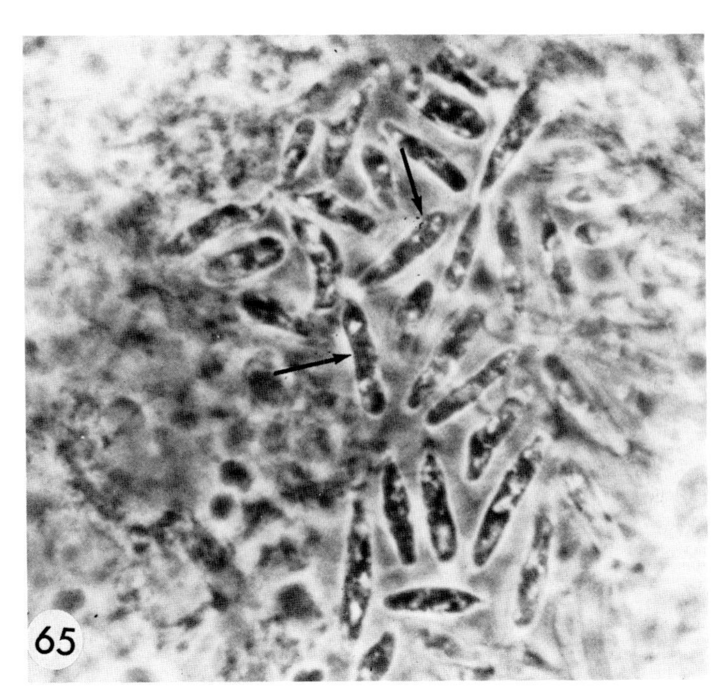

FIGURE 65. Conidia (arrows) of *Massospora* sp. in the abdominal cavity of a cicada. ×2000.

FIGURE 66. The trypetid fly *Anastrepha suspensa* infected with *Erynia echinospora* (top) ($\times 5$), and the aphid *Acyrthosiphum pisum* infected with *E. aphidis* (bottom) ($\times 40$).

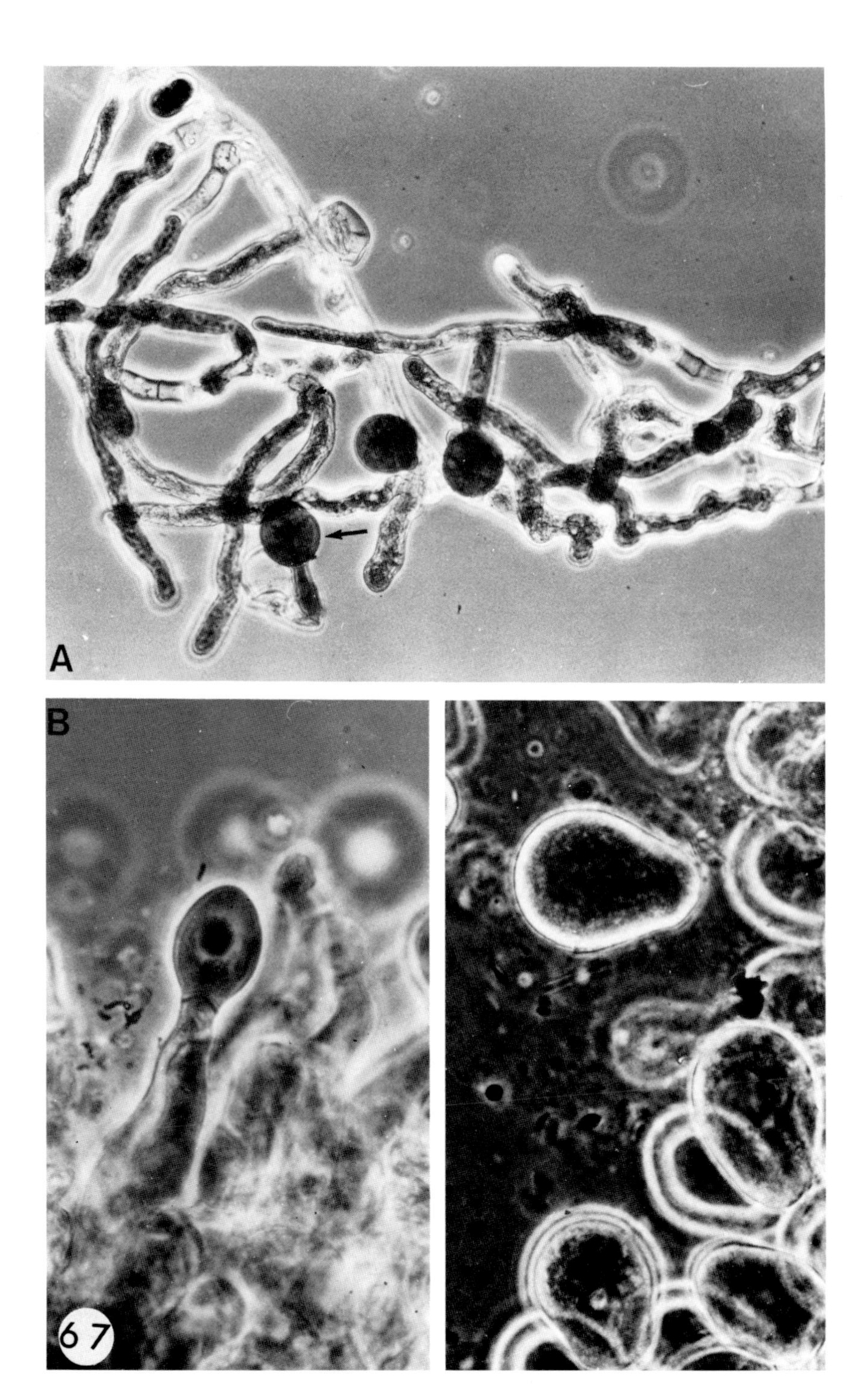

FIGURE 67. (A) Mycelium and spores (arrow) of *Conidiobolus thromboides*. ×213. (B) Conidia of *Entomophthora* sp. (left) (×820), and conidia of *Entomophaga grylli* (right) (×720).

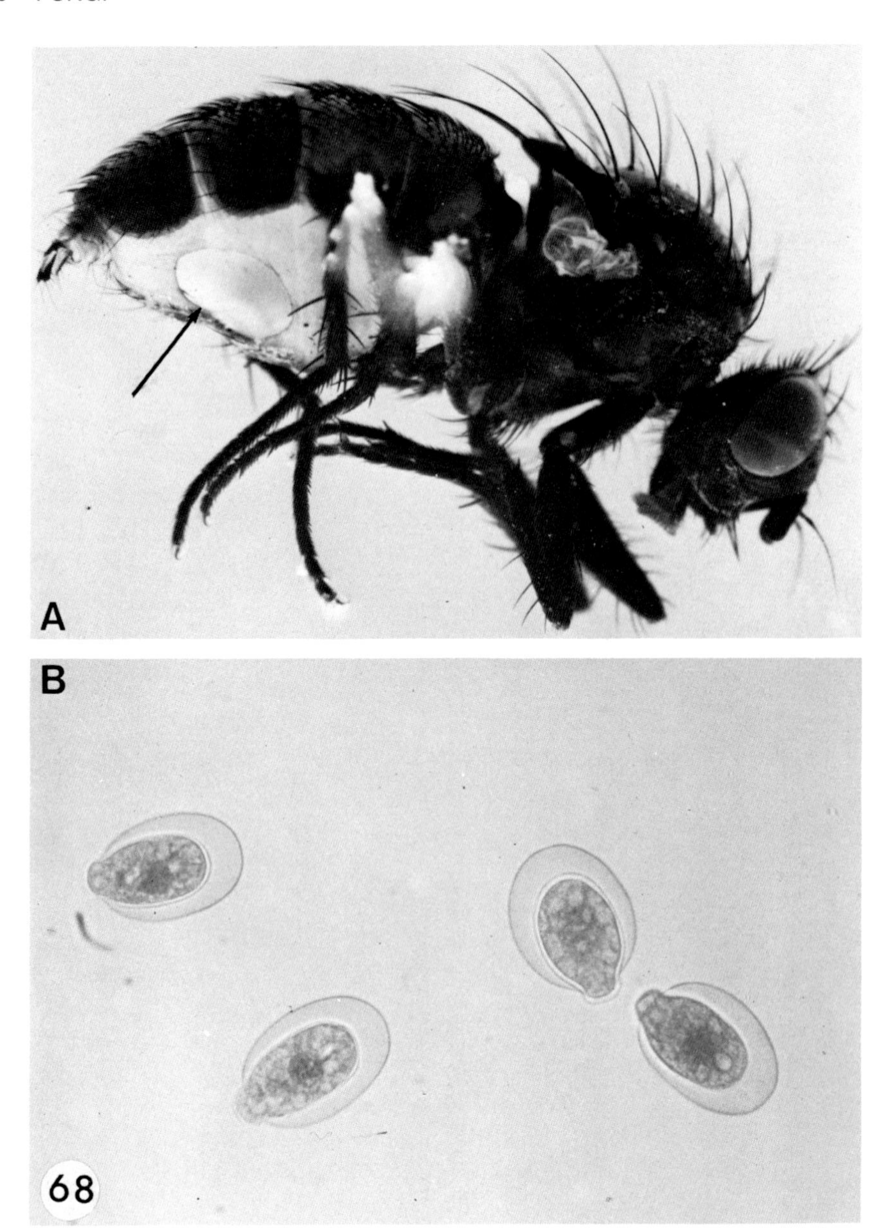

FIGURE 68. (A) A female *Hylemya brassicae* adult showing the abdominal hole (arrow) caused by *Strongwellsea castrans* infection. ×10. (b) Conidia of *S. castrans* shed by a live infected fly. Note the two covering membranes. (Courtesy of K. S. S. Nair.) ×720.

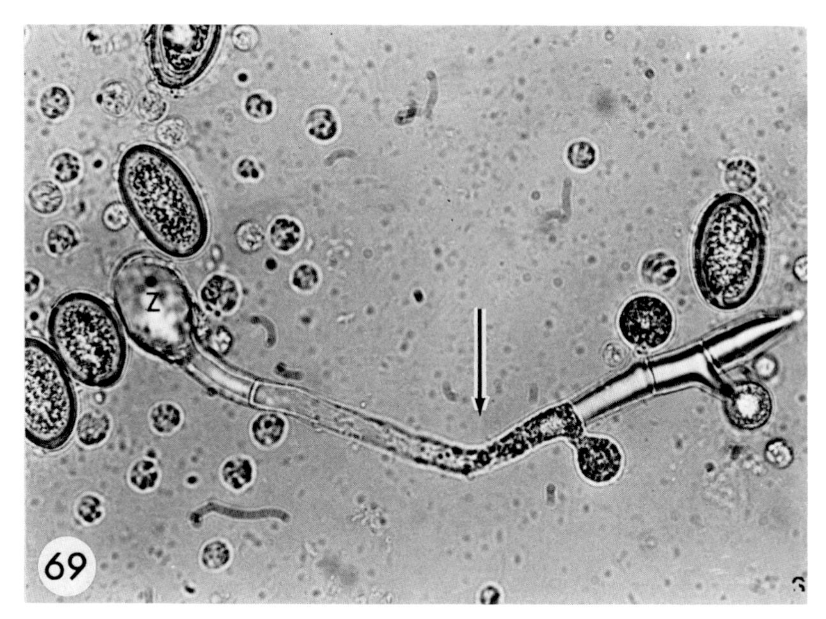

FIGURE 69. Linear, four-celled conidiophore of *Meristacrum milkoi* (arrow), arising from germinated zygospore (Z). (Courtesy of J. N. Couch.) ×418.

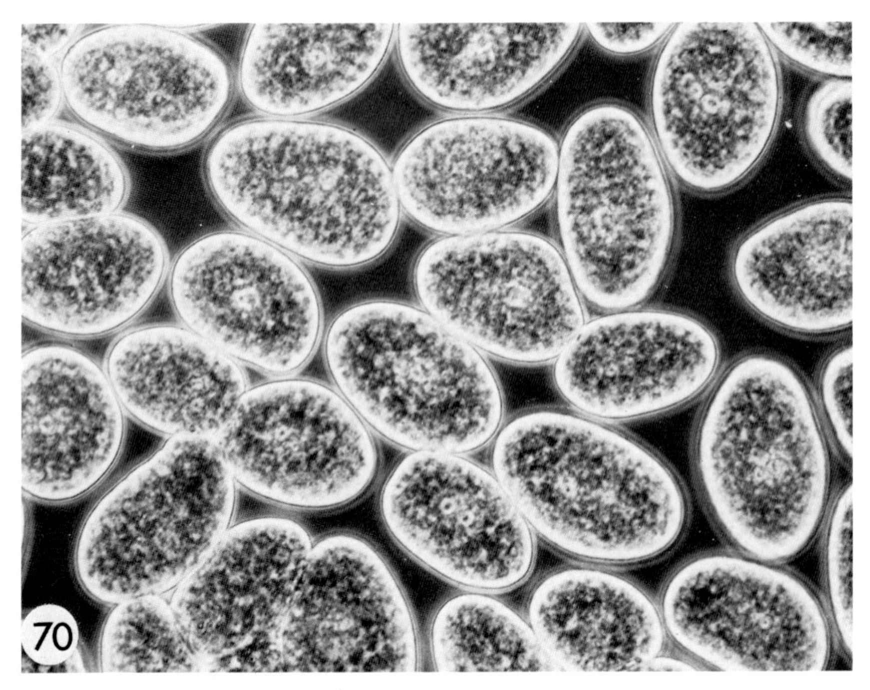

FIGURE 70. Resting spores of *Coelomomyces* sp. from a chironomid larva. $\times 600$.

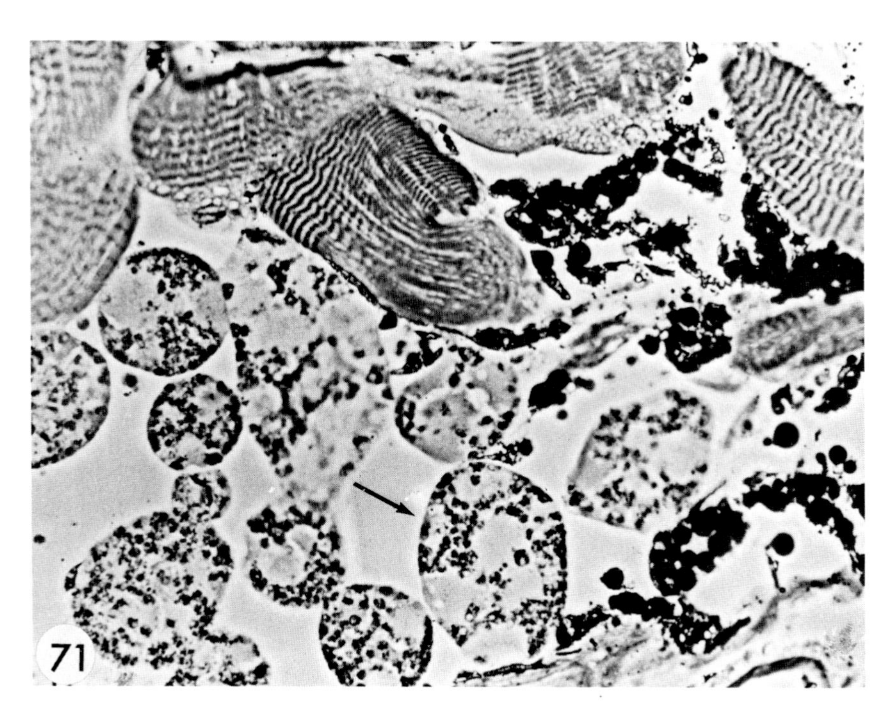

FIGURE 71. Developing sporangia of *Coelomycidium* sp. (arrow) from *Simulium vittatum*. (Courtesy of Brian Federici.) ×700.

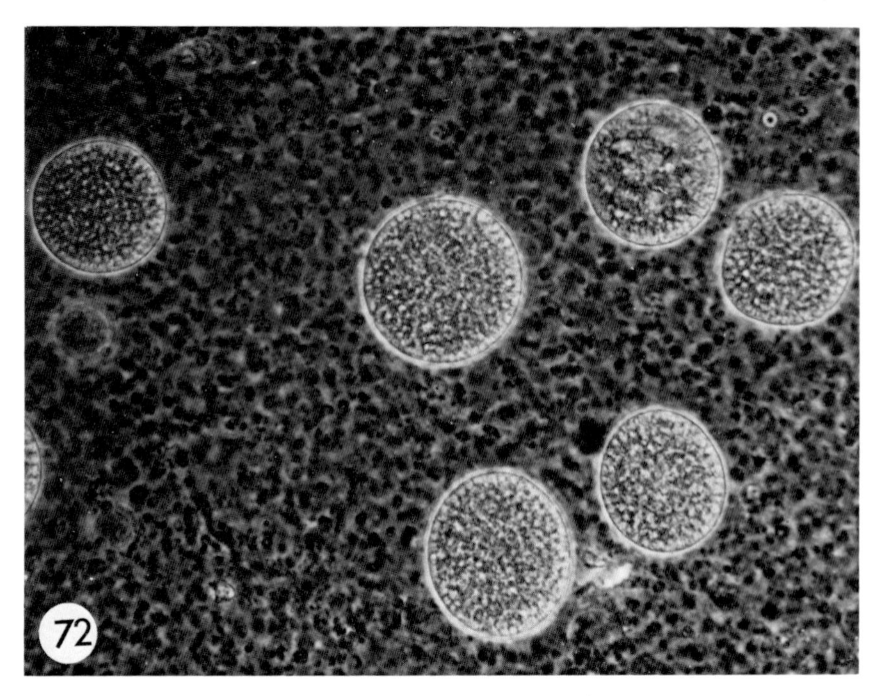

FIGURE 72. Sporangia of *Coelomycidium* sp. from blackflies. ×256.

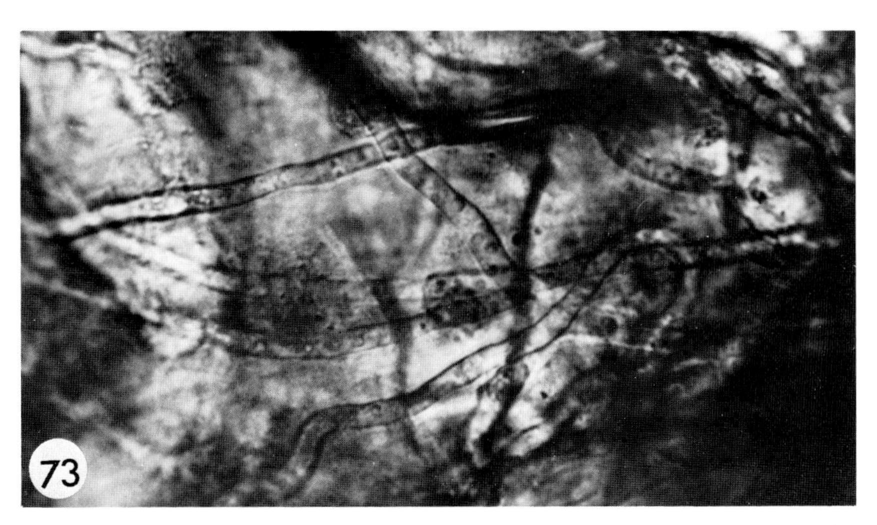

FIGURE 73. Hyphae of *Lagenidium giganteum* in larva of *Culex quinquefasciatus.* (Courtesy of Elmo M. McCray, Jr.) ×500.

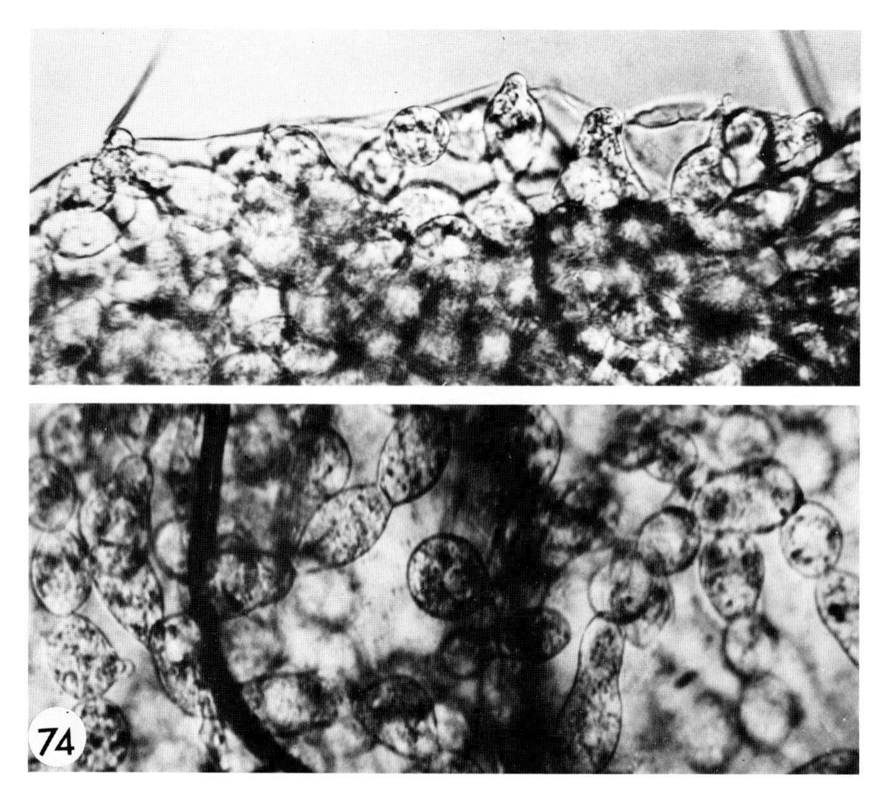

FIGURE 74. Developing sporangia of *Lagenidium giganteum* in larva of *Culex quinquefasciatus.* (Courtesy of Elmo M. McCray, Jr.) ×500.

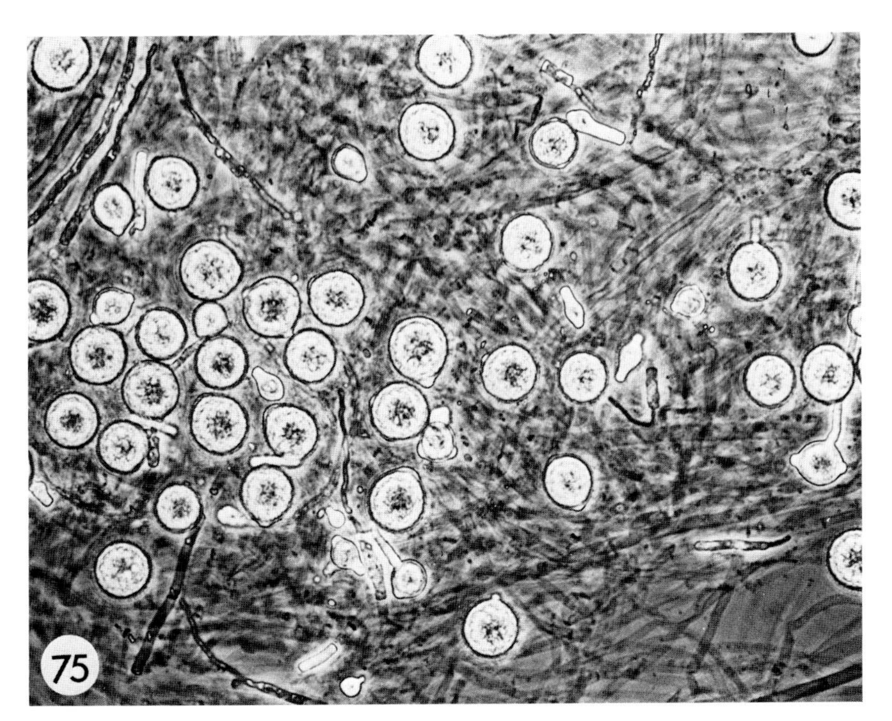

FIGURE 75. Sporangia of *Pythium* sp. (Courtesy of J. Hurlimann.) ×225.

FIGURE 76. *Septobasidium* sp. fungal stroma appressed to tree. (After Couch, 1938.) ×1.35.

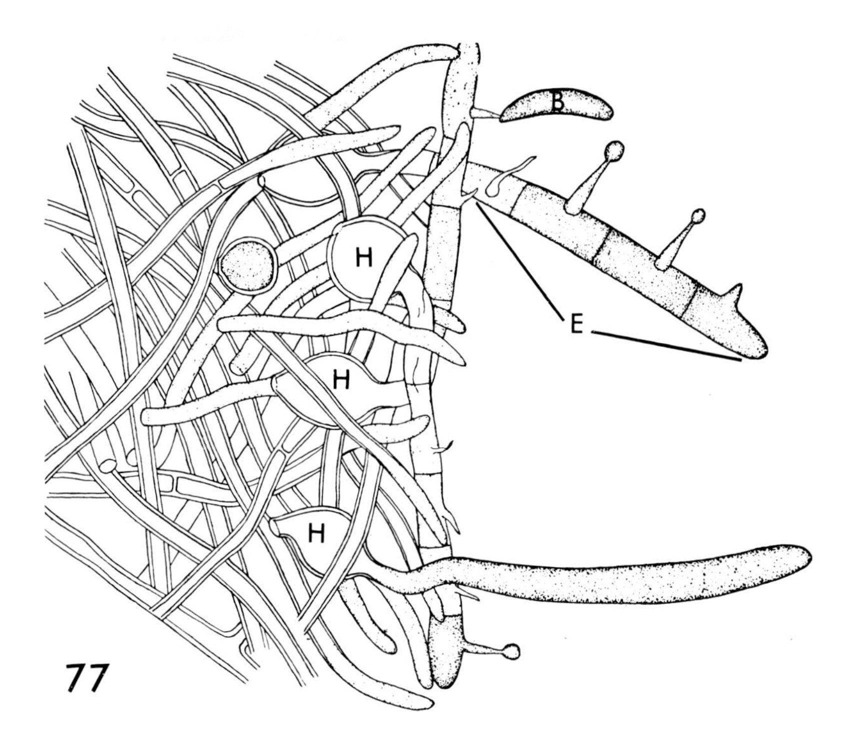

FIGURE 77. *Septobasidium* sp. mycelium and basidium. B, basidiospore; E, epibasidium; H, hypobasidium. ×940.

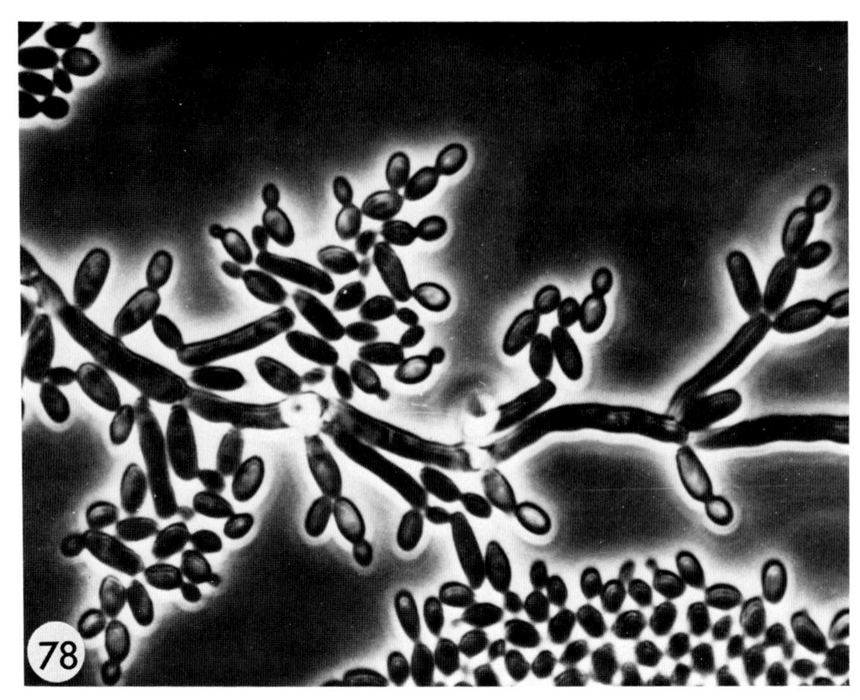

FIGURE 78. *Candida* sp. from a larva of the Douglas fir tussock moth. (Courtesy of M. Martignoni.) ×1000.

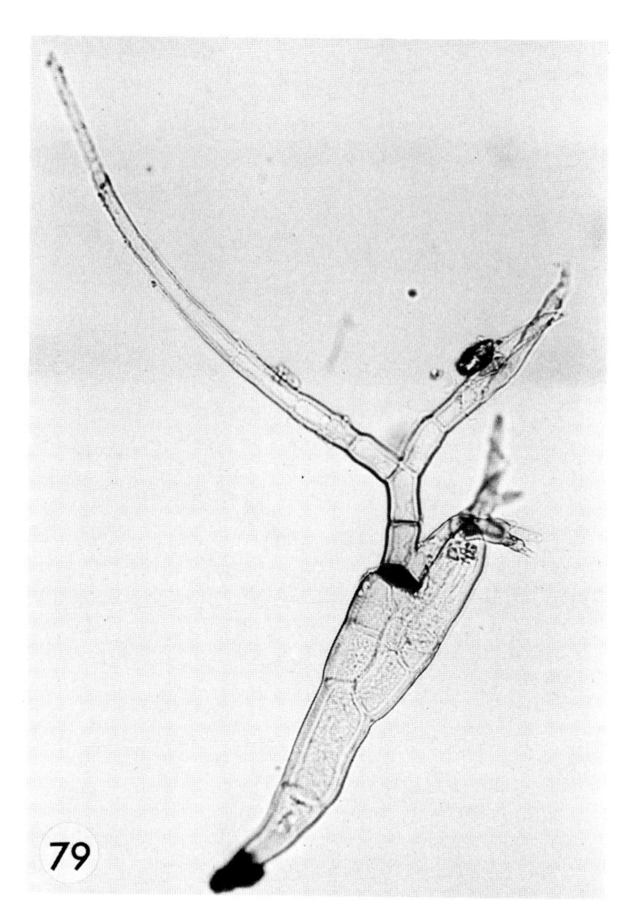

FIGURE 79. A member of the *Laboulbeniales.* (Courtesy of A. Kaplan.) $\times 250$.

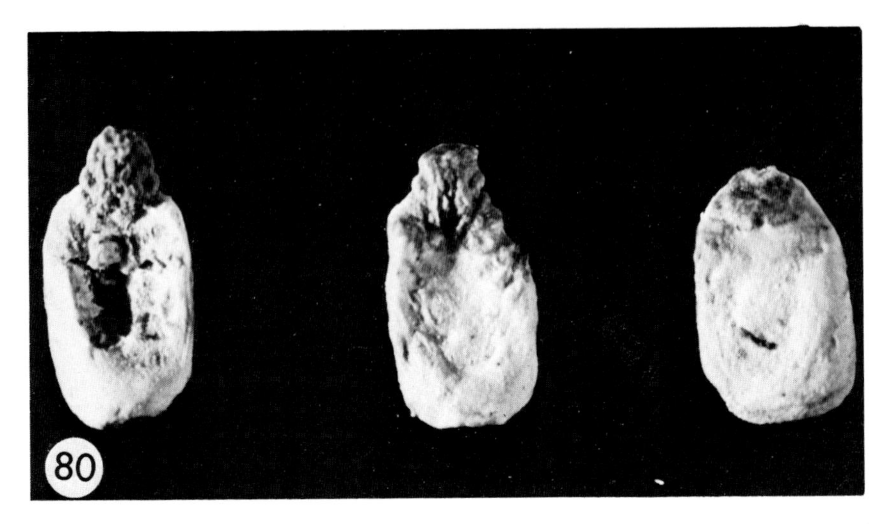

FIGURE 80. Mummified larvae of *Apis mellifera* covered with mycelium of *Ascosphaera apis.* ×6.

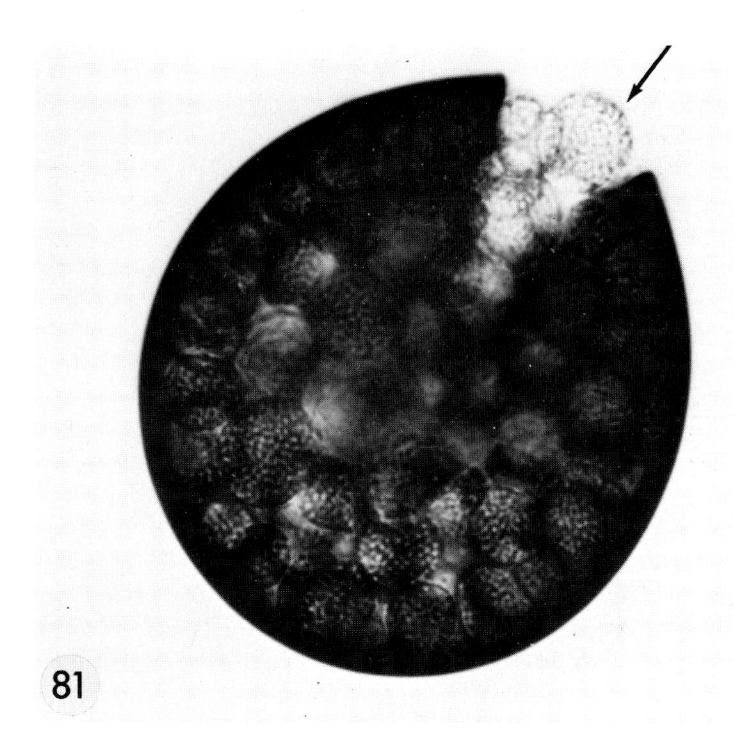

FIGURE 81. Cyst of *Ascosphaera apis* containing spore balls (arrow) filled with ascospores. ×800.

FIGURE 82. Mycelial mat of *Myriangium* sp. ×8.

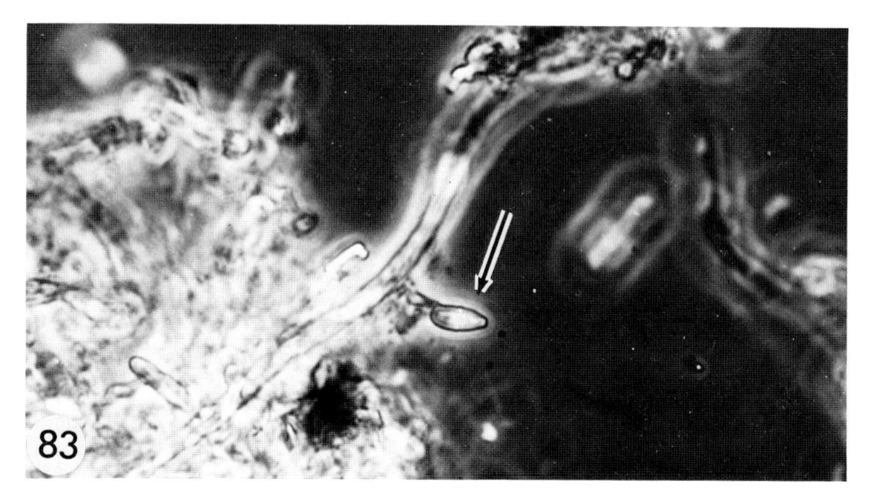

FIGURE 83. Ascospore of *Myriangium* sp. (arrow). ×700.

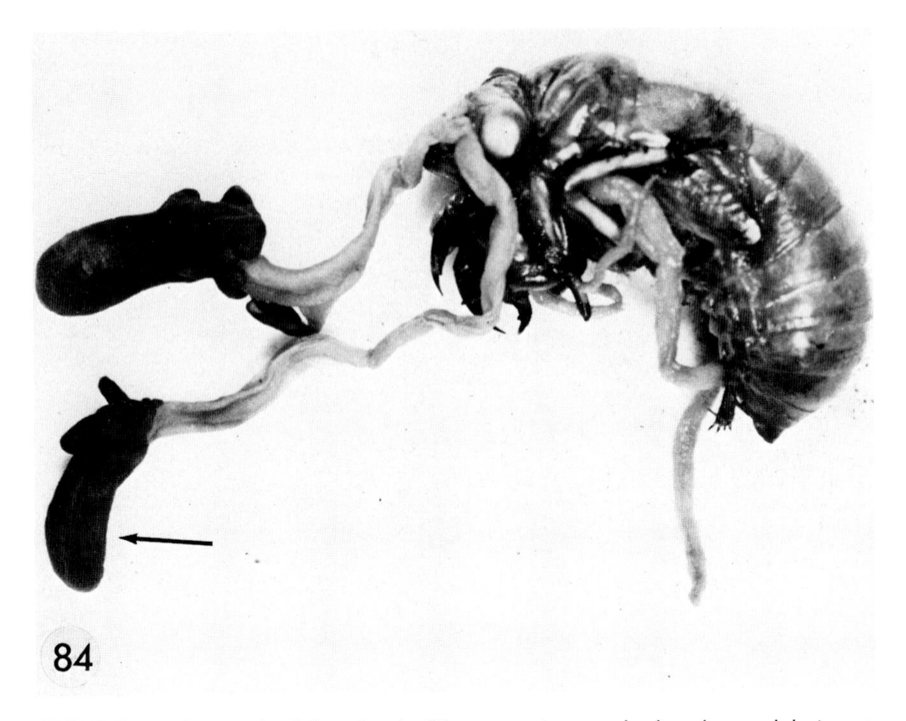

FIGURE 84. A nymph of the cicada *Diceroprocta apache* bearing aerial stroma terminated with perithecia (arrow) of *Cordyceps* sp. (probably *C. sobolifera*). ×3.5.

FIGURE 85. *Cordyceps ravenellii* on a grub of *Polyphaga* sp. ×4.

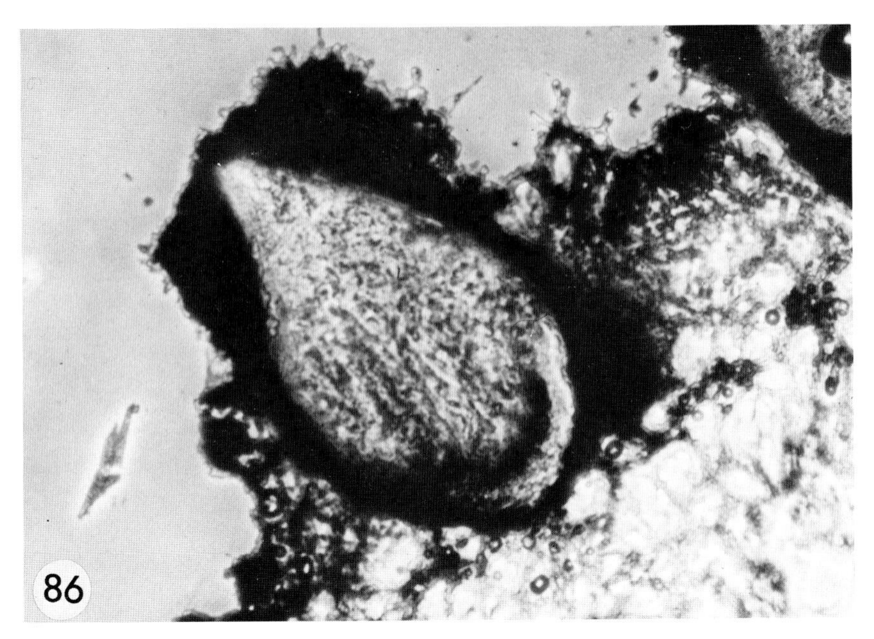

FIGURE 86. Partially embedded perithecium containing ascospores (not completely developed in this section) of *Cordyceps ravenellii*. ×280.

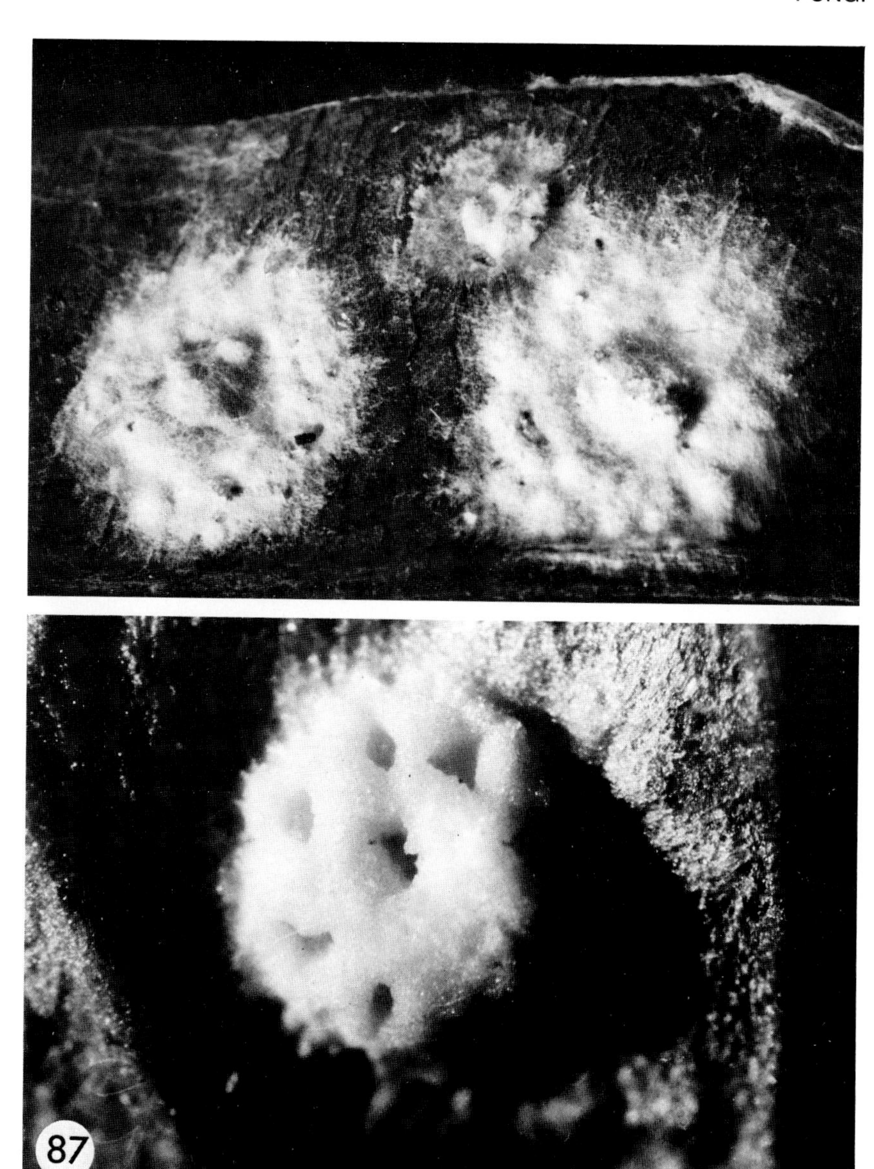

FIGURE 87. Fungal stroma of *Aschersonia* spp. growing on a diaspid scale (upper) (×8), and on a whitefly (lower) (×40).

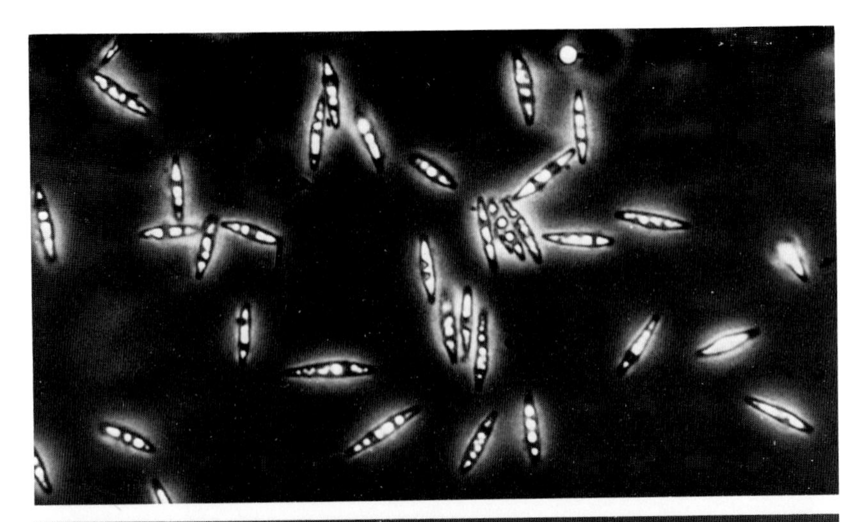

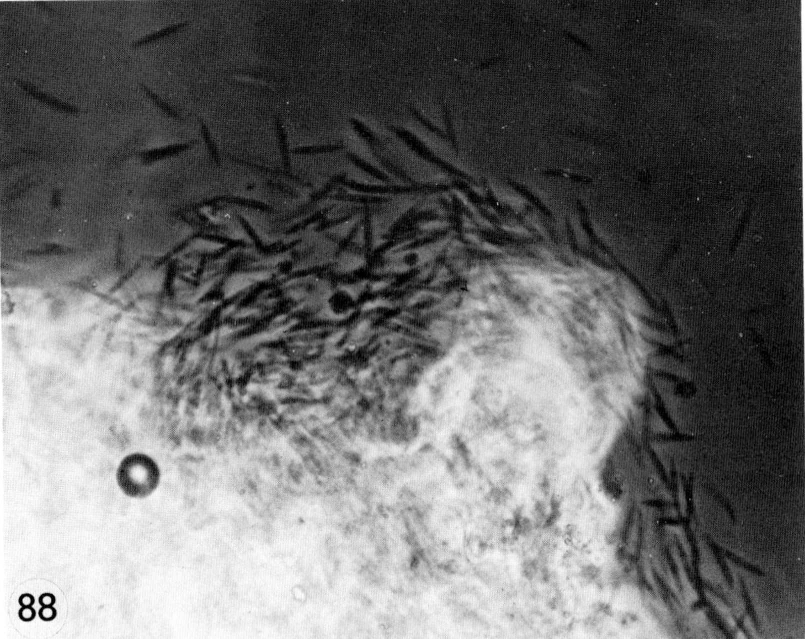

FIGURE 88. *Aschersonia* spp. spores in locule of stroma (lower) (×640), and free in suspension (upper) (×770).

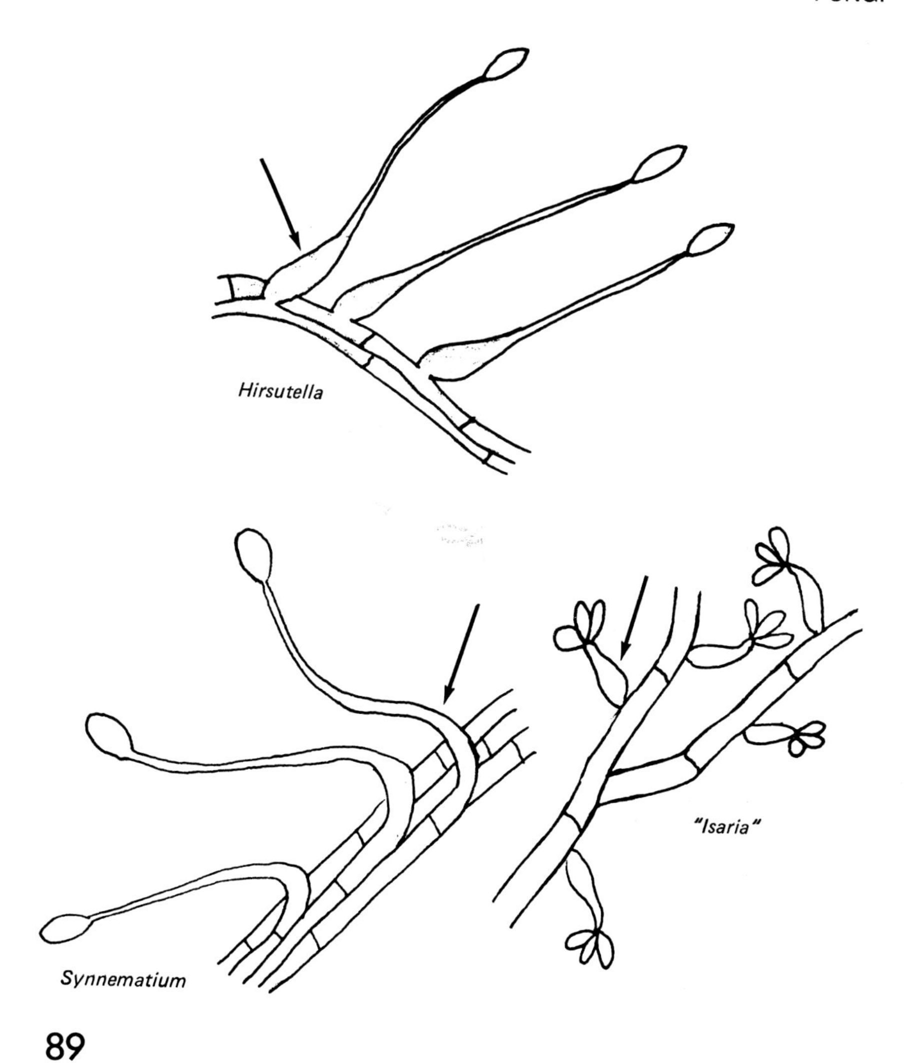

FIGURE 89. Examples of phialides (arrows) born singly on conidiophores.

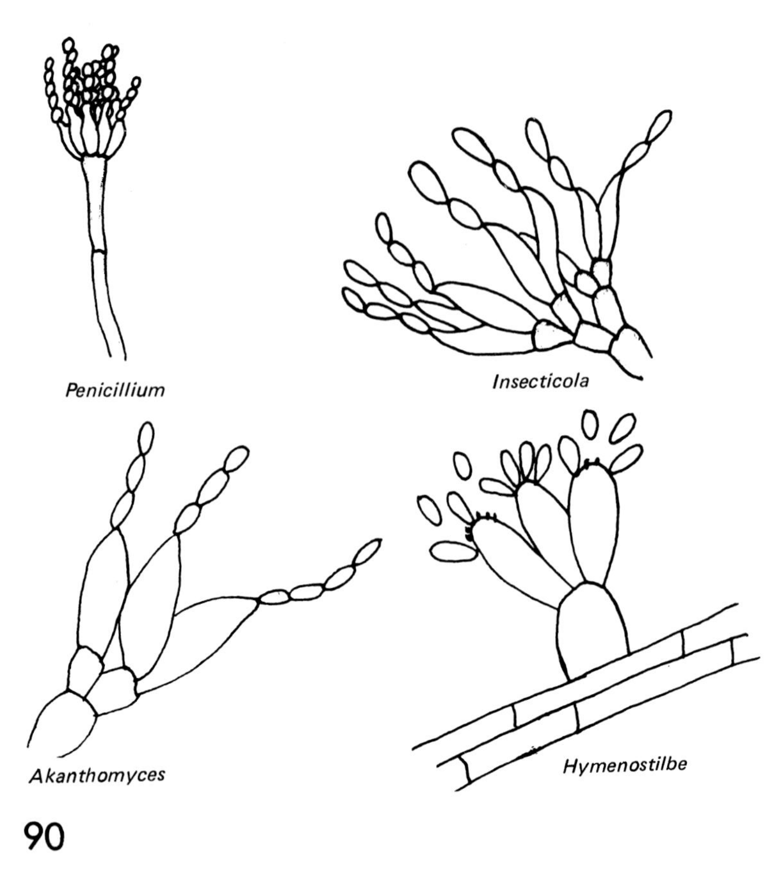

FIGURE 90. Examples of phialides born in groups or clusters on conidiophores.

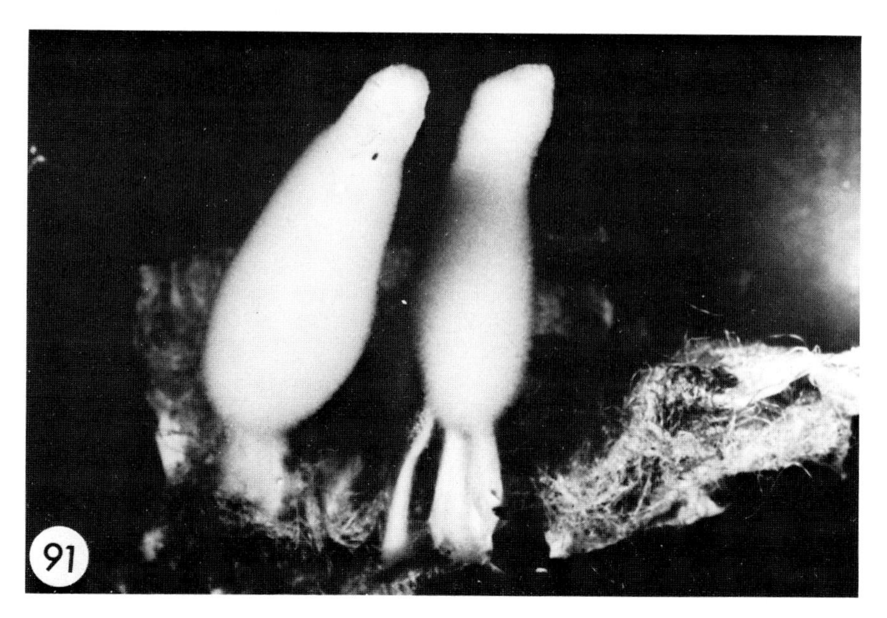

FIGURE 91. Synnemata of "*Isaria*" sp. emerging from a pupa of the codling moth. ×20.

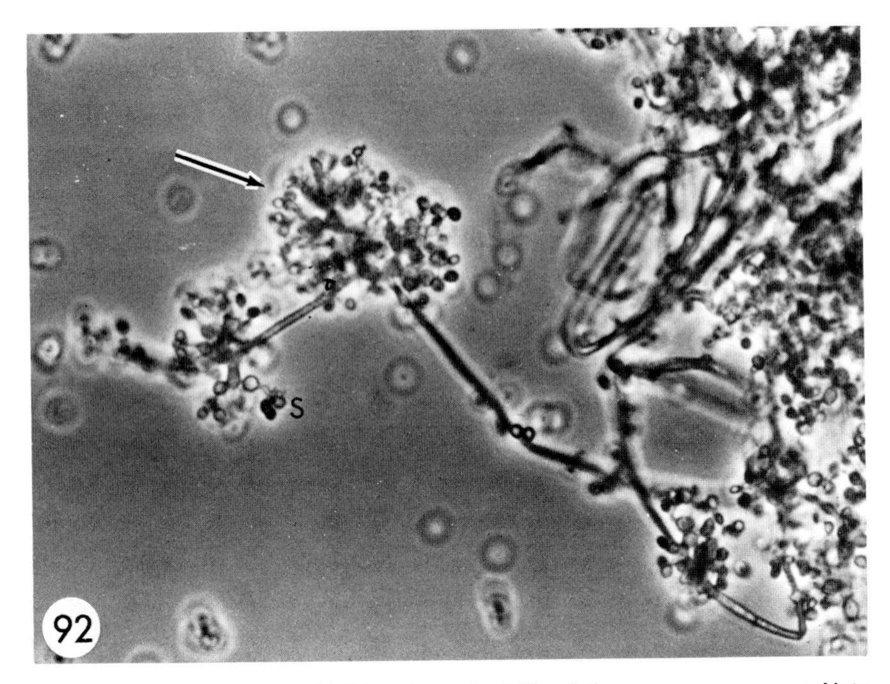

FIGURE 92. Clusters of phialides (arrow) of "*Isaria*" sp. grown on agar. Note spores (S). ×850.

FIGURE 93. *Hirsutella saussurei* on *Polistes olivaceous* (upper) (×4), and *Hirsutella* sp. on an unidentified wasp (lower) (×7.5).

FIGURE 94. An agar culture of *Hirsutella* sp. showing synnemata. ×20.

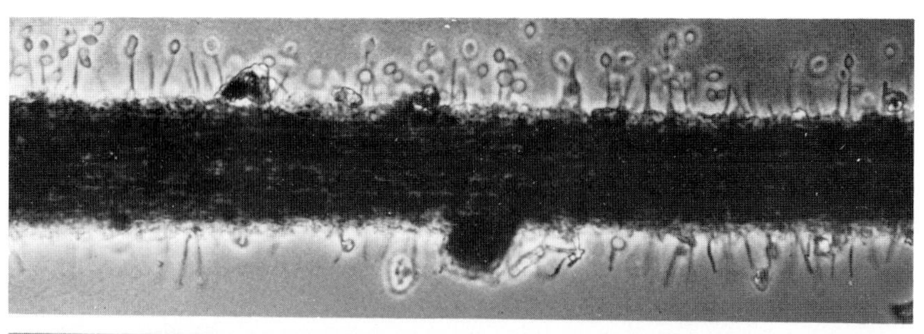

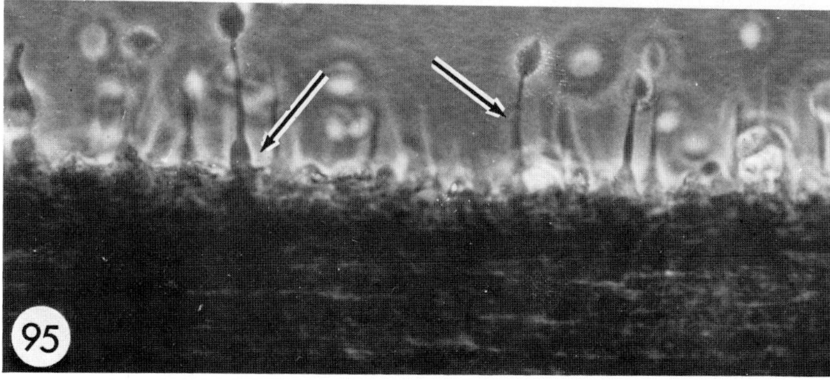

FIGURE 95. Portion of synnemata of *Hirsutella* sp. (upper) (×250), and (lower) showing phialides (arrows) (×640).

FIGURE 96. Synnemata of *Synnematium* sp. on *Panoquina* sp. ×4.

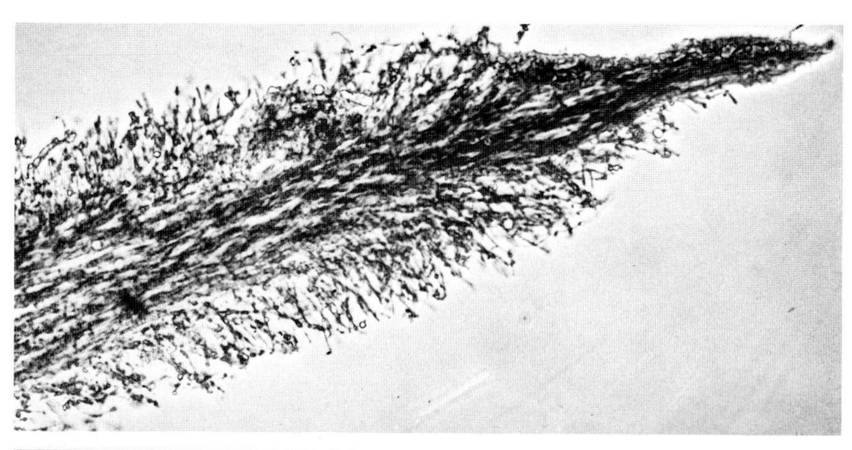

FIGURE 97. Portion of synnemata of *Synnematium* sp. (upper) (×275), and (lower) showing phialides (arrows) (×720).

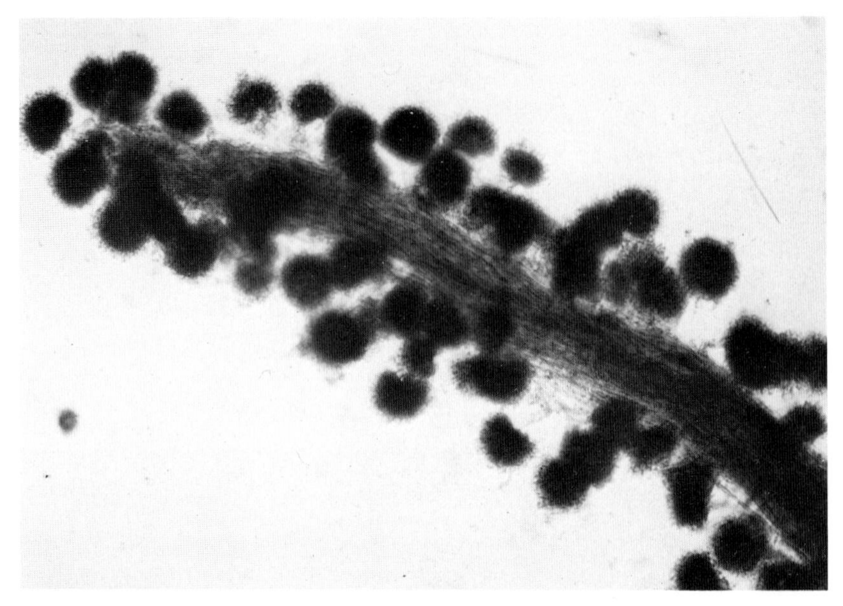

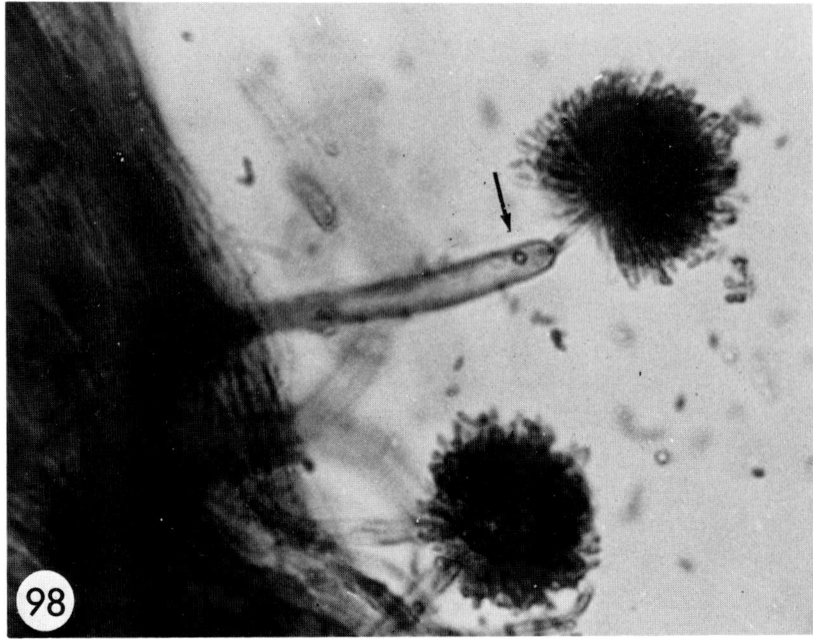

FIGURE 98. *Pseudogibellula* sp. (upper), showing a portion of the synnemata (×260), and close-up (lower), showing conidiophore with enlarged apex (arrow), bearing globose head containing prophialides, phialides, and spores (×800).

FIGURE 99. *Hymenostilbe* sp. on a dragonfly, showing the cylindrical synnemata. About 0.8× natural size.

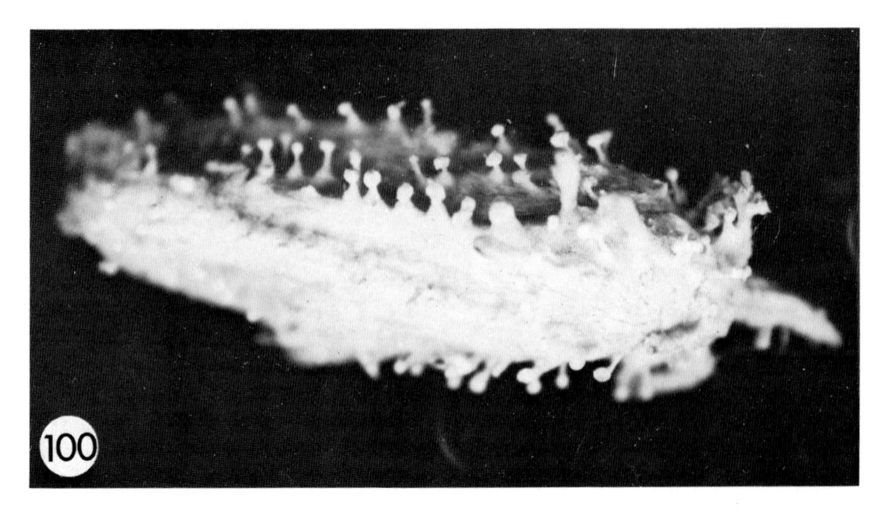

FIGURE 100. *Insecticola* sp. on a lepidopteran, showing synnemata with swollen tips. ×8.5.

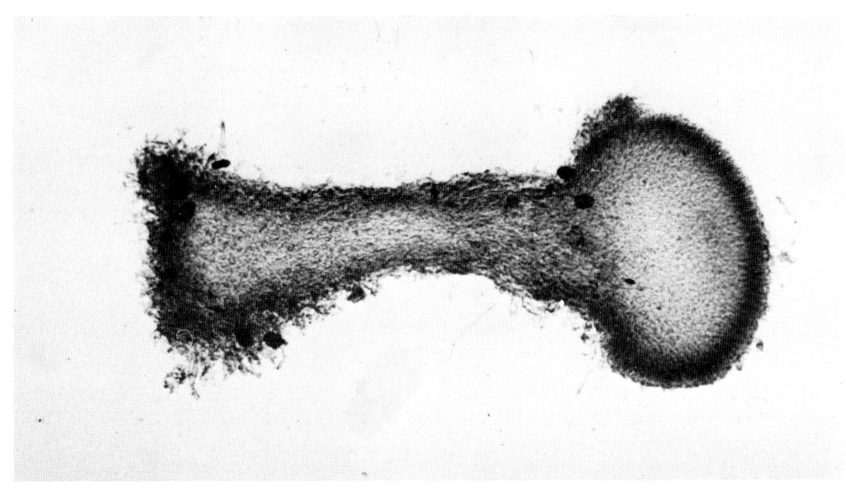

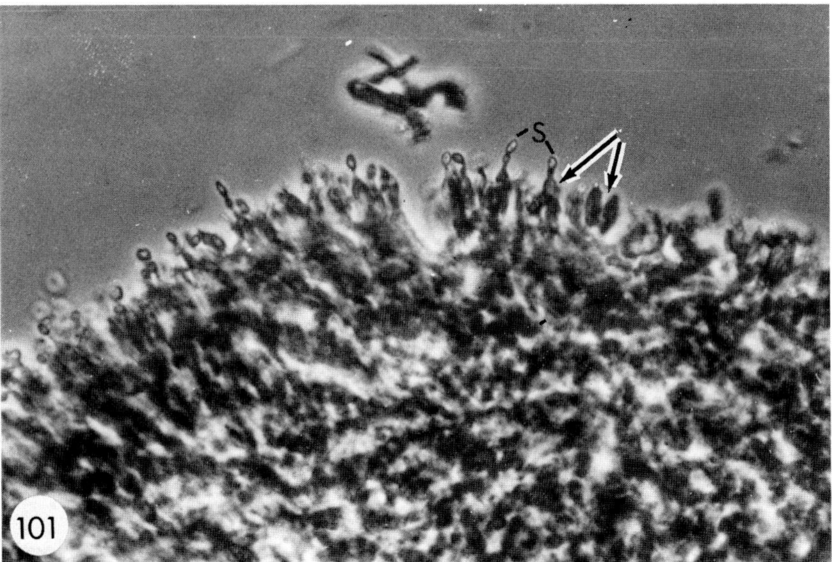

FIGURE 101. Synnematium of *Insecticola* sp. showing swollen tip (upper) (×115), and close-up of spore-bearing portion showing phialides (arrow) and spores (S) (×640).

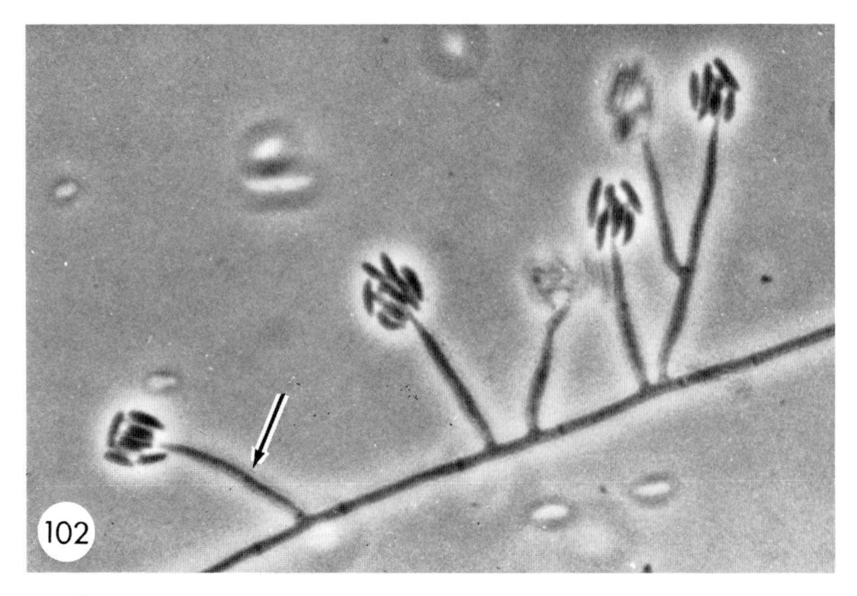

FIGURE 102. Conidia of *Fusarium* sp. borne on phialides (arrow) and held in slime drops. ×800.

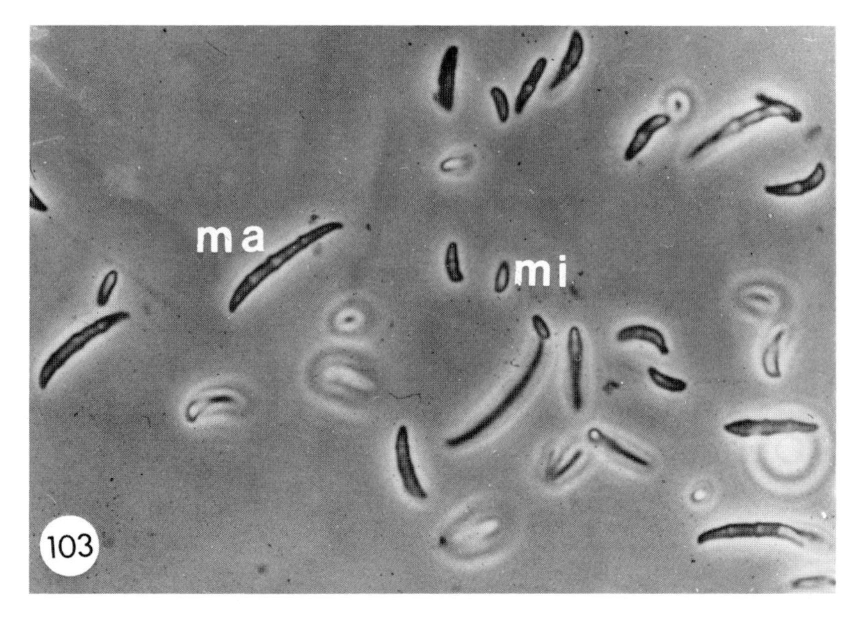

FIGURE 103. Microconidia (mi) and macroconidia (ma) of *Fusarium* sp. $\times$1024.

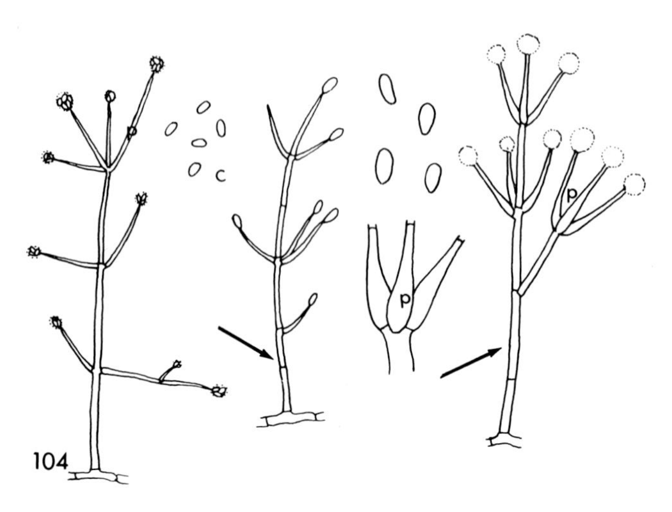

FIGURE 104. Drawing of *Verticillium* sp. showing slender conidiophores (arrow) and awl-shaped phialides (p) arranged in whorls (verticillate).

FIGURE 105. *Culicinomyces clavosporus* sporulating on the thorax of a larva of *Aedes aegypti*. Note clavate spores (arrow). (Courtesy of A. W. Sweeney.) ×300.

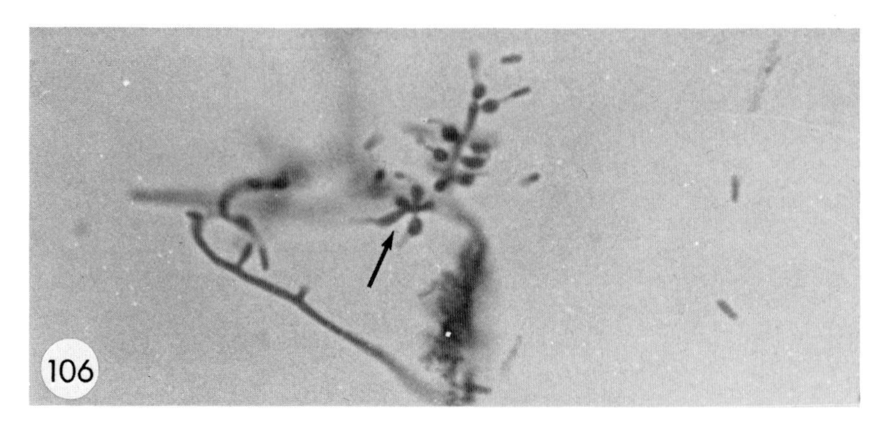

FIGURE 106. *Tolypocladium* sp. showing verticillate, flask-shaped phialides (arrow) with narrowly tapering, bent neck. ×1100.

FIGURE 107. *Beauveria bassiana*-infected weevil (upper) (×8), and tenebrionid beetle (lower) (×3).

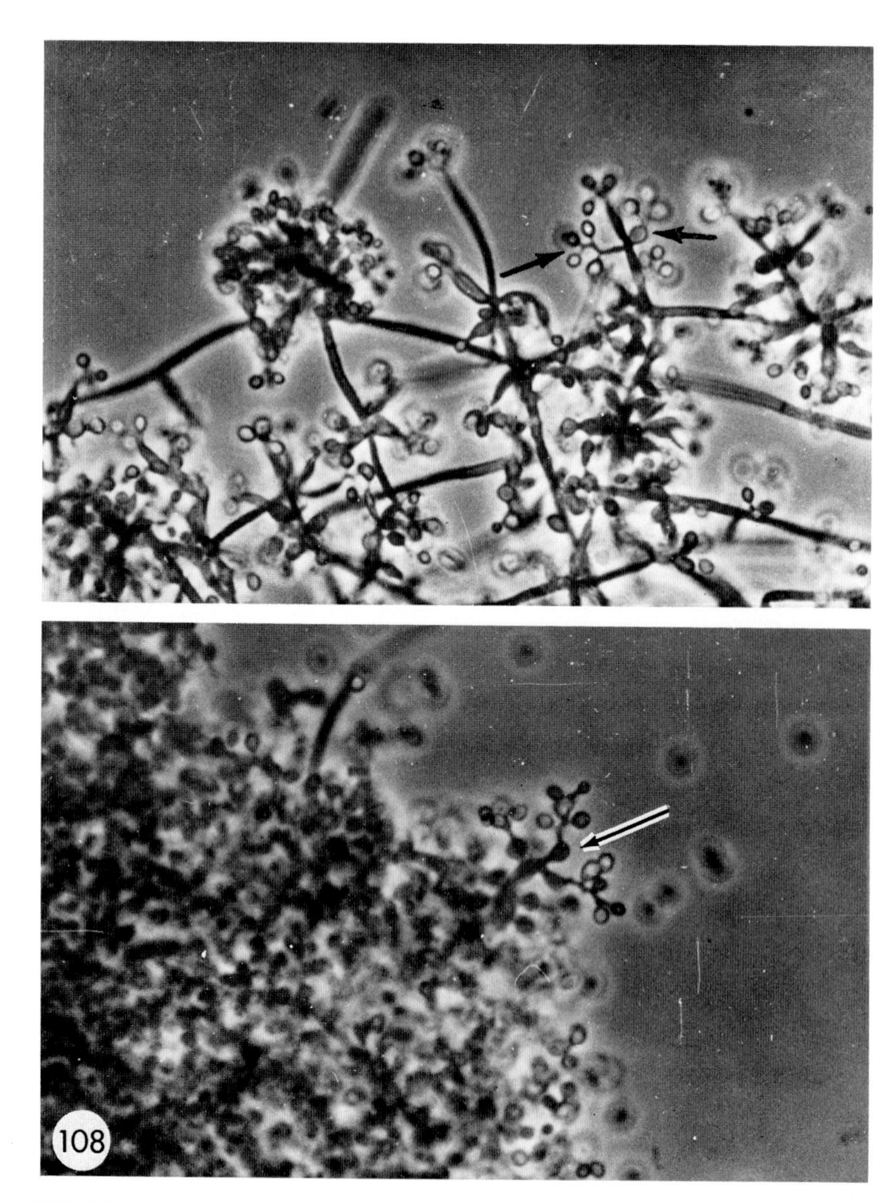

FIGURE 108. Conidia and phialides (arrows) of *Beauveria bassiana*. Upper ×800; lower ×1100.

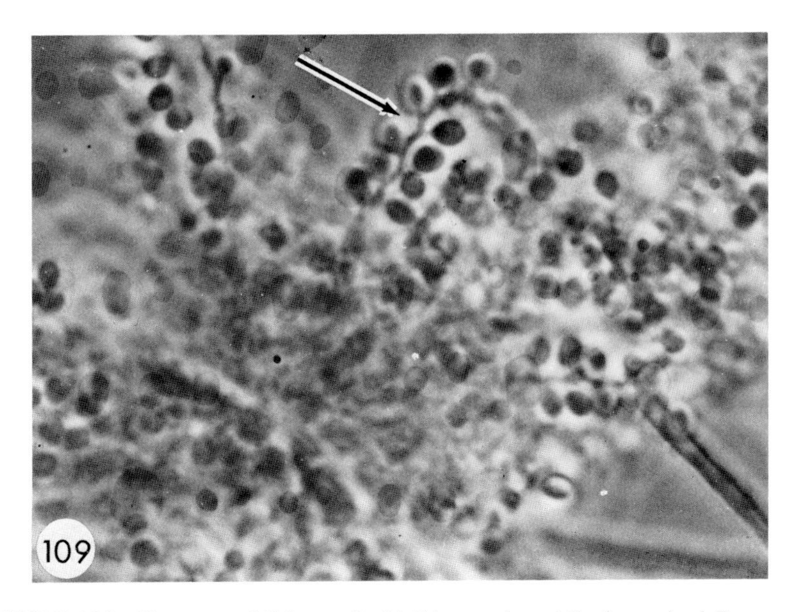

FIGURE 109. Close-up of "zig-zag" phialides and conidia (arrow) of *Beauveria bassiana.* ×1700.

FIGURE 110. *Aspergillus flavus* (arrow) on ant, *Atta texana.* ×12.

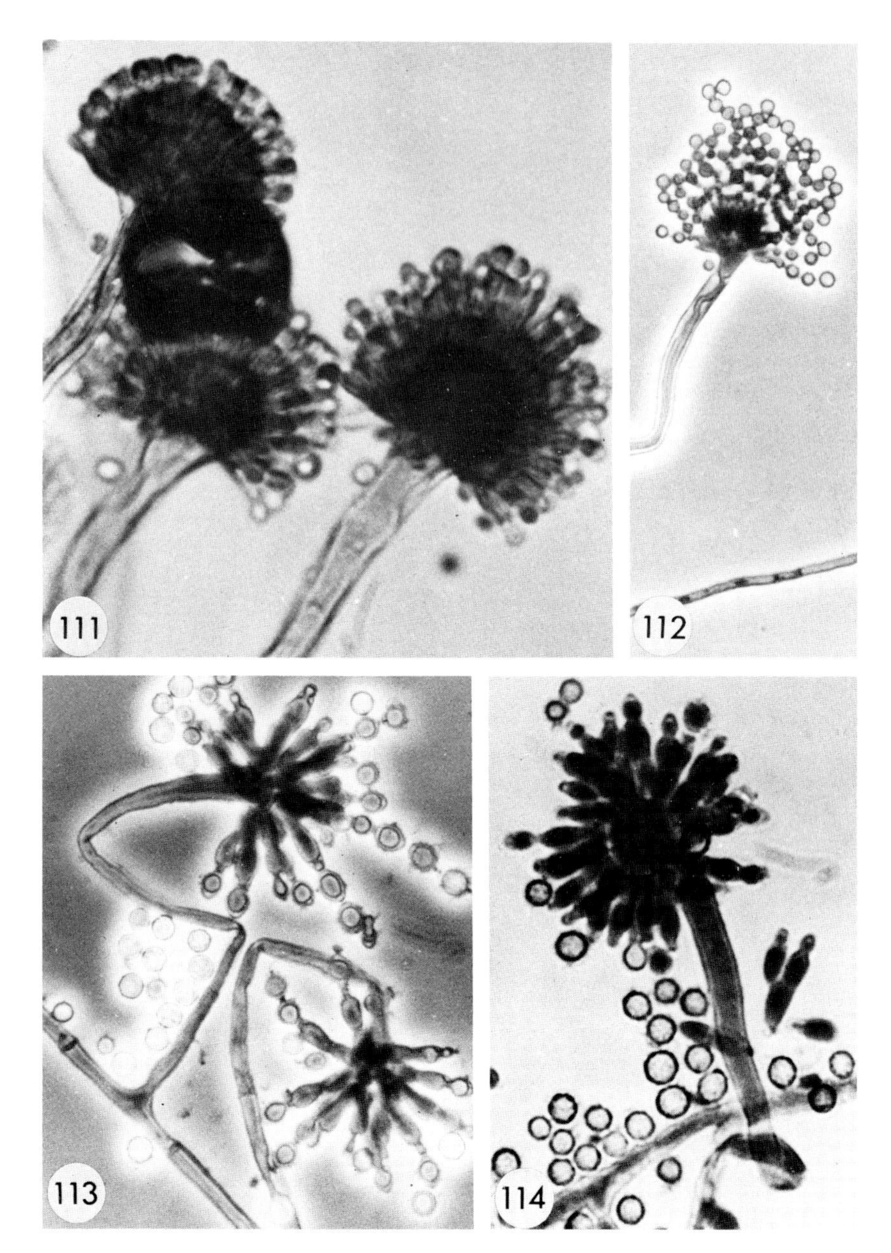

FIGURES 111–114. Phialides and spores of various species of *Aspergillus*. Fig. 111 ×800; Fig. 112 ×320; Fig. 113 ×850; Fig. 114 ×800.

FIGURE 115. Spore masses of *Metarrhizium anisopliae* on agar. ×15.

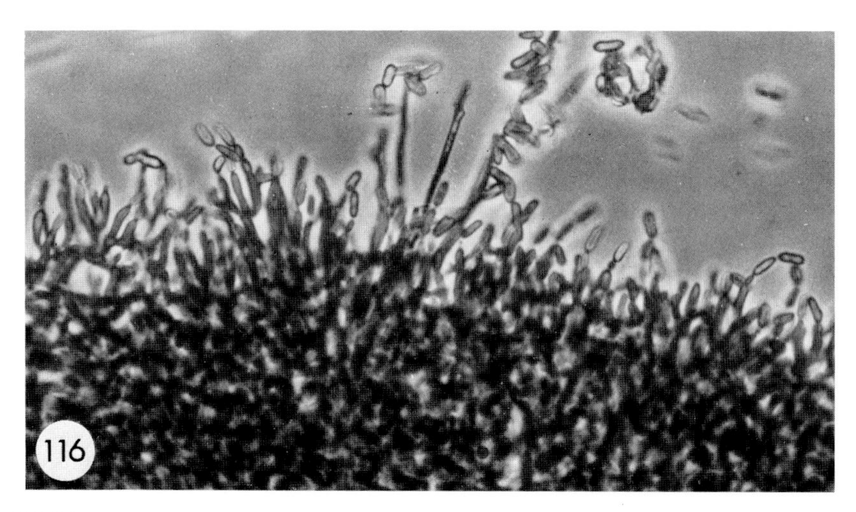

FIGURE 116. Mass of conidia and conidiophores of *Metarrhizium anisopliae*. ×640.

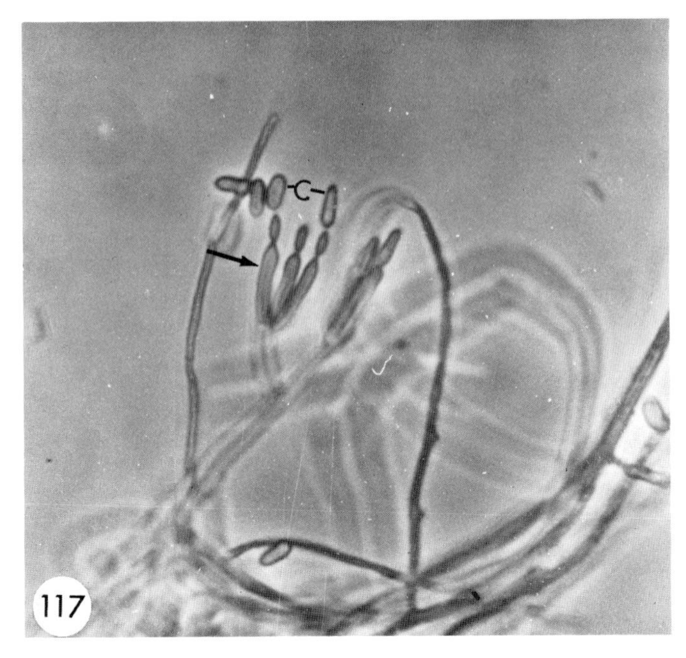

FIGURE 117. Elongate conidia (C) and conidiophores (arrow) of *Metarrhizium anisopliae*. ×1000.

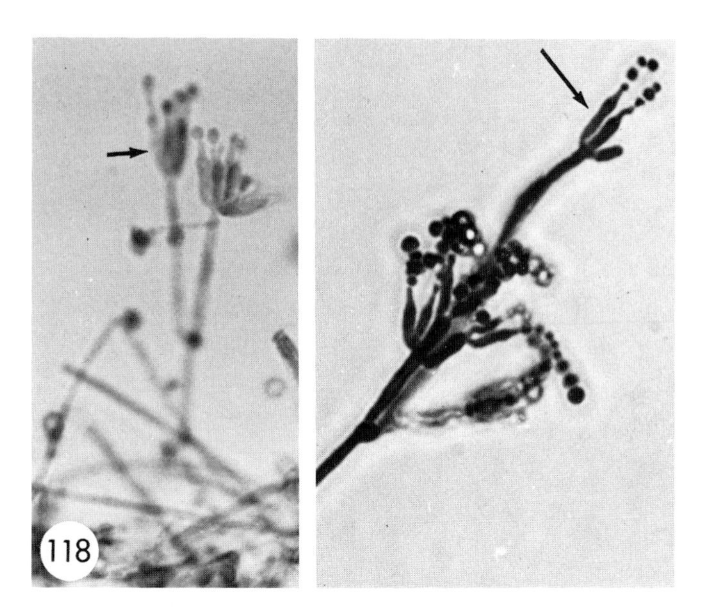

FIGURE 118. Phialides (arrows) and round spores of *Penicillium* sp. ×800.

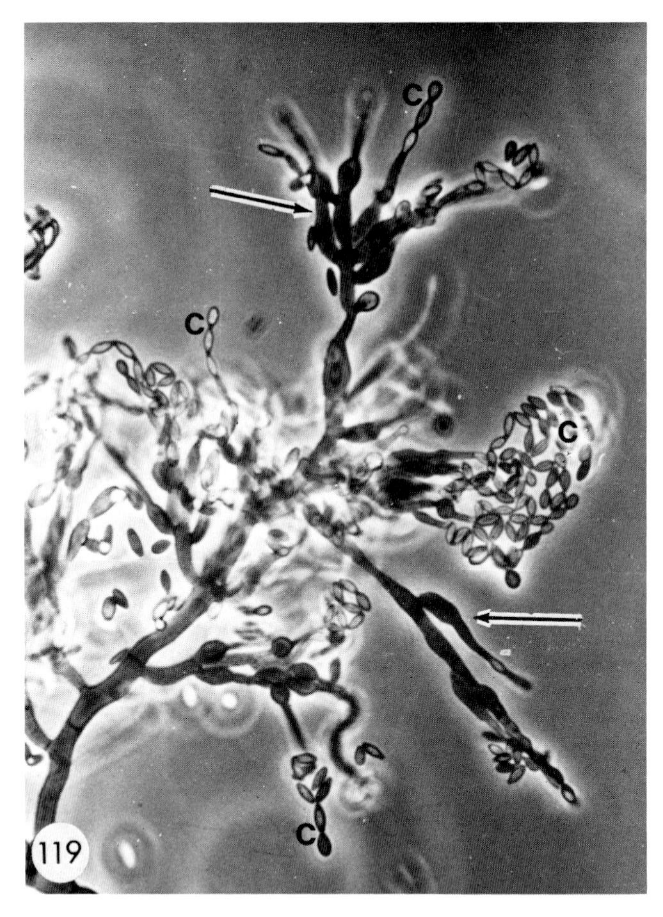

FIGURE 119. Conidia (C) and phialides (arrows) of *Paecilomyces farinosa*. ×800.

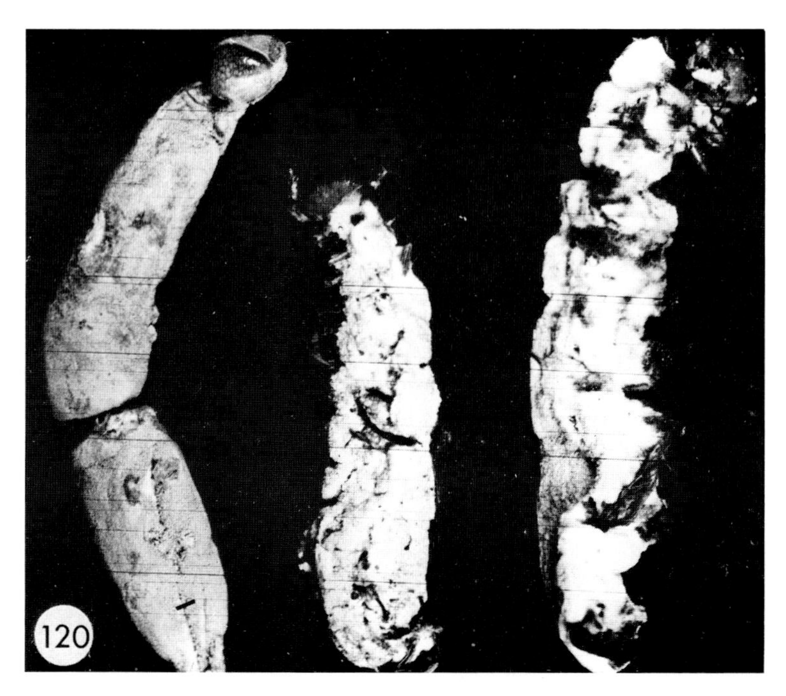

FIGURE 120. *Nomuraea rileyi* on the surface of diseased lepidopterous larvae. ×4.

FIGURE 121. Phialides (arrows) and spores (C) of *Nomuraea rileyi*. ×760.

PROTOZOA

INTRODUCTION

Representatives of many diverse groups of protozoa have associations with insects. These associations range from phoresis to obligate parasitism and in many instances insect mortality is the final result. Some symbiotic protozoa found in the gut of insects (e.g., termites, wood roaches) are important for the well-being of their bearer, while other forms seemingly innocuous for the insect carriers cause serious diseases in man and domestic animals (e.g., *Trypanosoma*).

Some groups of protozoa known to cause disease in insects also occur in other invertebrates and even vertebrates. For that reason careful tests should be conducted on host range before allowing the widespread use of certain forms as biological control agents. However, some protozoa are being used as biological control agents today. One of these, *Nosema locustae,* is commercially available as a biological control agent of grasshoppers and crickets and is distributed on particles of bran, which serve as bait. Occasionally, protozoan pathogens may attack parasites which occur in the original host. Thus, aside from infecting its normal host, *Anthonomus grandis, Mattesia grandis* also attacks a braconid parasite of the boll weevil (McLaughlin, 1965). Veremtchuk and Issi (1970) found a *Nosema* which normally infects *Pieris brassicae* L. but will also attack neoaplectanid nematodes parasitizing the host.

The minute size of many protozoa has prevented their detailed study, and the life cycles and developing stages of relatively few are known. An urgent need in this area of study remains unfilled.

TAXONOMIC STATUS

The Protozoa are generally considered to have the status of phylum. However, some have placed these organisms in a separate kingdom, the Protoctista (or Protista) (Whittaker, 1969). The latter system of classification (multikingdom) has the advantage of including organisms that resemble plants as well as animals in a broad category, since the standard definition of a protozoan as a ''unicellular animal with all body functions being performed by that cell'' leaves much to be desired.

The classification used here is based on the report by Honigberg *et al.* (1964), who split the Phylum Protozoa into several subphyla and lower categories. The majority of insect pathogenic protozoa belong to the subphyla Sporozoa and Cnidospora. Other subphyla containing insect pathogens are the Sarcomastigophora and Ciliophora. Many of the higher levels of classification are undergoing revision, and some authors propose the erection of more than one phylum for the Protozoa.

LIFE CYCLE

The infective stages (spores, cysts, etc.) of most insect pathogenic protozoa are ingested and pass into the host's alimentary tract (transovum and transovarial transmission may also occur). Entry into the hemocoel supposedly occurs through the midgut wall; however, very little is known about the initial steps of infection. Usually the first signs of infection are the presence of the developing stages in the gut epithelium, fat body, malpighian tubules, or hemolymph.

Reproduction may occur asexually by binary fission (flagellates and ciliates) or by schizogony or multiple fission (coccidians) or gamogony (gregarines). Sexual reproduction in the Sporozoa consists of the fusion of two gametes to form a zygote, which in turn undergoes repeated divisions. In some ciliates and suctorians reproduction by conjugation, autogamy, endomixis, and cytogamy may occur.

CHARACTERISTICS OF INFECTED INSECTS

There are very few specific symptoms in insects suffering from protozoan infections. Most are general symptoms which also occur in insects infected with a variety of pathogens. These include small size, morphological deformities, lethargy, difficulty in molting, reduction of feeding, loss of balance, and production of a white fecal exudate. Hosts with a transparent cuticle often have white opaque or swollen areas on the cuticle, whereas some terrestrial insects exhibit black spots on their integument. However, in many cases, collapse or mortality is the first indication of a protozoan infection, which must be confirmed by direct microscopic examination. Burkholder and Dicke (1964) devised a method of detecting insects infected with *Mattesia dispora* with ultraviolet light. This method might be extended to other protozoan infections.

METHODS OF EXAMINATION

The simplest way to detect a protozoan infection is to examine a smear mount using fat body, malpighian tubules, gut epithelium, or hemolymph under phase contrast or bright field microscopy. The mature spores are the easiest stage to distinguish. Staining with Giemsa or hematoxylin will show the developing stages more clearly (see Techniques). Both spores and developing stages are important for diagnosis.

It is sometimes difficult to differentiate protozoan spores from artifacts or resistant stages of other organisms. Most inclusion bodies of viruses dissolve in weak NaOH whereas spores of protozoa do not. Many fungal spores resemble those of protozoa and, aside from observing germination in the former, there is sometimes no sure way of separating the two. Fungal spores may be colored and are generally more uniform in texture and contents. Giemsa stain may demonstrate the fine internal structure of protozoan spores, and extrusion of the polar filiment is a simple, quick method for confirming the presence of a microsporidan. However, the presence of a protozoan in a diseased insect does not imply that it is the cause of the disease. It must be identified and compared with those forms known to cause insect disease or tested for pathogenicity. Thus, flagellates, amoebae, and ciliates may be associated with insects, but have no pathogenetic relationship.

ISOLATION AND CULTIVATION

Protozoa can best be isolated from insects by removing a sample of hemolymph or tissue from an infected insect. However, only a few pathogenic protozoa can be grown in artificial media (e.g., *Tetrahymena,*

Malpighamoeba, Plasmodium), and these methods are not practical for the general diagnostician.

IDENTIFICATION

The minute size of many protozoans makes their examination difficult. However, attempts should be made to determine the presence and type of locomotor organelles, the size and structure of the spores or vegetative stages, the number of spores in a "cyst," the number of sporozoites in a spore, and the presence or absence of a polar filament (diagnostic character for the Microsporidia). Many protozoans appear to be host and tissue specific, so it is important to identify the insect and tissue infected (e.g., fat body, gut epithelium, or malpighian tubules).

The identification of the Microsporidia—which incidentaly are considered a new phylum (Microspora) separate from the Protozoa by Sprague (1982)—is still in a state of flux. Recent studies have shown that some and possibly a fair proportion of the Microsporidia possess dimorphism in having two or even more spore types. One is usually binucleate and the other uninucleate. The former types are sometimes associated with adult insects and are carried transovarially from one host generation to the next. Using lacto-aceto-orcein stain, these double-nucleated spores can be detected in the sporoblast stage (see Techniques). Experts are now using electron microscopy to determine the number of spore wall layers, polaroplasts, polar filaments, and other internal spore structures that could be useful for classification. These characters, as well as the presence or absence of meiosis, polymorphism, gamete production, and fusion, may well be the basic characters used in subsequent classifications. For the field biologist equipped with only a light microscope, the classical methods using spore shape and size, coupled with an examination of spore nuclei and spore formation, are still the most practical way of separating these complex organisms.

TESTING FOR PATHOGENICITY

Since most entomogenous protozoa enter their host through the digestive tract, pathogenicity tests should be made by introducing protozoan spores into the mouth of the test insect. Diseased insects can be triturated in distilled water, and, after removing the debris by straining and differential centrifugation, the spore suspension can be mixed with food or sprayed on unhatched eggs or plants. It may be desirable to add a feeding stimulant. With larger insects a suspension of spores can be introduced directly into the mouth cavity with a calibrated syringe.

STORAGE

After purification, the resistant stages of some protozoans can be stored under refrigeration. For the most fastidious forms lacking resistant stages, continuous transfer from host to host may be the only method of maintenance.

LITERATURE

The text *Protozoology* by Kudo (1966) presents a general account of the Protozoa and covers most of the insect forms. The basic and advanced texts by Steinhaus (1949, 1963) also cover the insect-associated protozoa, and W. M. Brooks (1974) presented a detailed review of protozoan infections in insects. A manual for the identification of protozoa was prepared by Jahn (1949), and practical aspects of the use of protozoa for insect control have been covered by McLaughlin (1971, 1973). Classical studies of the mirosporidia have been presented by Kudo (1924) and Weiser (1961), and more recently Sprague (1977) published a systematic revision of the entire group. A discussion of the species infecting insects and mites and a review of taxonomic characters have been offered by Hazard *et al.* (1981). Various aspects of microsporidian biology have been covered by several authors (Bulla and Cheng, 1976). A review of *Nosema locustae* was given by Henry and Oma (1981) and one of *Nosema fumiferanae* by Wilson (1981). The potential for the use of *Vairimorpha necatrix,* a microsporidian with two different spore forms, was discussed by Maddox *et al.* (1981).

KEY TO THE COMMON GENERA OF ENTOMOGENOUS PROTOZOA

1. Active stages containing cilia, flagella, or pseudopodia (amoeboid movement); nonmotile resting stages not containing spores—**2**

1. Active stages absent; no stages containing cilia, flagella, or pseudopodia; resting stages represented by spores or cysts containing spores—**8**

2. Cilia and/or sucking tentacles present (Subphylum Ciliophora, Class Ciliatea)—**3**

2. Cilia and sucking tentacles absent; motile stages containing flagella or pseudopodia; resting stages may be composed of thick-walled cysts commonly found in the intestine or malpighian tubules of the host (Subphylum Sarcomastigophora)—**6**

3. Cilia present only in immature stages (usually not encountered), and absent in mature stages, which contain tentacles; usually sessile forms attached to the exoskeleton of aquatic insects by a noncontractile stalk—Subclass Suctora (Figs. 122, 123). Members of the genera *Discophrya* Lachmann, *Periacineta* Collin, *Rynchophrya* Collin, *Dactylophorya* Collin, and *Ophryodendron* Clapareda and Lachmann occur on the integument of insects in a phoretic association. For detailed descriptions of these forms, see Kudo (1966). When numerous, they may impair locomotion or respiration and thus be pathogenic.

3. Cilia present in both immature and mature stages; tentacles absent—**4**

4. Mature stages with oral cilia only; somatic cilia lacking; often stalked (Subclass Peritrichia)—**5**

4. Mature forms with cilia over their entire body—Subclass Holotrichida. The only known insect pathogens in this order belong to the genus *Tetrahymena* Furgason (Figs. 124, 125). At least three species occur in the body cavity of insects, especially diperous larvae (Fig. 126). See Corliss (1960) for a discussion of this genus.

5. Mobile forms without stalk—Suborder Mobilina. Representatives of the genus *Urceolaria* Lamarck (*Trichodina* Ehrenberg) and others occur on the integument of aquatic insects.

5. Predominantly sessile, with contractile or noncontractile stalk—Suborder Sessilina. Representatives of the genera *Vorticella* (Figs. 127, 128) (see Noland and Finley, 1931), *Epistylis* Ehrenberg (see Nenninger, 1948), and others occur on the integument of aquatic insects. Some physical impairment may occur when large numbers are present.

6. Motile stages show amoeboid movement in host cells or hemocoel during the early stages of infection (Fig. 129); resting stages are cysts in the body cavity, gut, or malpighian tubules of the host (Figs. 130, 131); flagella absent in all stages (Superclass Sarcodina, Order Amoebida)—**7.** The two species in couplet **7** are the most common entomogenous pathogens of this order. Two other species that are less frequently encountered are *Malpighiella refringens* Minchin from fleas and *Dobellina mesnili* [Keilin] from gnats. The insect-parastic amoebae are reviewed by Lipa (1963) and W. M. Brooks (1974).

6. Motile stages with one or more flagella (Fig. 132); small ovoid (leishmanial) nonmotile stages may also be present in the host's gut—Superclass Mastigophora. In general flagellates are only occasionally encountered in insects and are rarely responsible for debilitation or mortality. Members of the orders Retortamonadida Grasse, Trichomonadida Kirby, and Hypermastigida Grasse and Foa develop in the gut of insects, especially roaches and termites. Most of the parasitic genera of the order Kinetoplastida (most common in Diptera and Hemiptera) occur in the suborder Trypanosomatina and include some medically important species. The genera *Crithidia* Leger, *Blastocrithidia* Laird, *Herpetomonas* Kent, *Rhynichoidomonas* Patton, and *Leptomonas* Kent habitually live in the alimentary tract of invertebrates (see Wallace, 1966).Occasionally they break through the intestinal wall and invade the hemocoel and other internal tissues of the host. Smirnoff (1974) reported reduced viability in sawfly larvae infected with *Herpetomonas* sp. Members of *Phytomonas* Donovan are plant pathogens which are vectored by insects. The two genera *Leishmania* Ross (common in *Phlebotomus* flies in Europe, Asia, and Africa) and *Trypanosoma* Gruby (common in blood-sucking insects in Africa, South America, and tropical islands) occur in the invertebrate hemocoel or intestine (Fig. 132) and are carried by blood-sucking insects to vertebrates, often causing serious diseases. See Kudo (1966) and Wallace (1966) for accounts of the entomogenous trypanosomatids.

7. Infects the malpighian tubules of adult honeybees—*Malpighamoeba mellificae* Prell.

7. Infects the malpighian tubules and gut epithelium of locusts and grasshoppers—*Malamoeba locustae* (King and Taylor) (Figs. 130, 131). See Lipa (1963) and W. M. Brooks (1974) for a discussion of this species.

8. Spores variable in shape, usually not round or navicular; containing a polar filament (Subphyluym Cnidospora)—**9.** The polar filament is a threadlike structure coiled within the spore (Fig. 133). The filament may be used to initiate penetration into the host cells. Several methods can be used to extrude the polar filament mechanically on a microscope slide (Ishihara, 1967). One can apply pressure to the cover slip with the thumb or a blunt object or use chemicals such as acetic acid, ammonia, glycerine, hydrogen peroxide, or iodine. Rarely, the filament can be demonstrated within the spore using Giemsa stain.

8. Spores mostly round or navicular; not containing a polar filament (Subphylum Sporozoa)—**41**

9. Spores containing an elongate filament and three spherical sporoplasms (Fig. 134); filament not attached to the spore after extrusion—*Helicosporidium. H. parasiticum* Keilin (Fig. 134) was considered an ascomycete at one time, and there is still some question regarding its taxonomic position. See Kellen and Lindegren (1974) for a discussion of this pathogen, which occurs in fat body and nerve tissue of insect larvae.

9. Spores with only a single polar filament in the form of a thin, coiled tube; filament remains attached to the spore (Class Microsporidea)—**10.** These are highly specialized parasites which occur widely in insects and other animals. Identification is based mainly on spore characteristics. For a review of these groups, see Weiser (1961), Tuzet *et al.* (1971), W. M. Brooks (1974), Sprague (1977, 1982), and Hazard *et al.* (1981).

10. Spore with a short polar filament, lacking a polaroplast; many spores produced within a thick-walled sporocyst or a thin membrane or both (rare)—**11**

10. Spore with a normal polar filament, with polaroplast (Fig. 133); sporocyst (pansporoblast, sporophorous vesicle, pansporoblastic membrane) present or absent (Order Microsporidia)—**12.** This order is currently undergoing revision by several authorities, and many new taxa are being proposed. Many of the diagnostic characters separating newly proposed genera are based on the ultrastructure of the spores. The genus *Microsporidium* Balbiani is a collective genus for microsporidia that cannot be identified as belonging to a described genus.

11. Spores develop within a thick-walled envelope, parasite without any special relationship to the host nucleus; found in *Sciara* fly larvae—*Hessea* Ormières and Sprague (Family Hesseidae). See Ormières and Sprague (1973) for a discussion of this genus.

11. Spores may develop within a thick-walled cyst or a thin envelope; parasite develops adjacent to the host cell nucleus; parasites of Coleoptera—*Chytridiopsis* Schneider (Family Chytridiopsidae). See Sprague *et al.* (1972) for a review of this genus.

12. Spores from a single host are dimorphic (two different morphological types): one type is uninucleate and the other binucleate (Fig. 135)—**13**

12. Spores from a single host are all similar in shape (can be either all uninucleate or all binucleate)—**16**

13. All spores are free at maturity, not enclosed in cysts (pansporoblastic membranes); some spores are oval (uninucleate) and others are pyriform (Fig. 136) (binucleate)—*Hazardia* Weiser. See Weiser (1977) for a discussion of this genus.

13. Only some spores are free at maturity (binucleate ones), while others (uninucleate forms) are enclosed in a cyst membrane—**14**

14. Different spore types occur in different host tissues; e.g., elongate free spores (binucleate) occur in hypodermis of larval ants; broadly oval spores (uninucleate) occur in eight-spore cysts and are formed in the fat body of ant pupae—*Burenella* Jouvenaz and Hazard. See Jouvenaz and Hazard (1978) for a discussion of this genus.

14. Different spore types occur in the same host tissue—**15**

15. Free spores conical; membrane-bound spores (eight) oval; found in mosquito larvae (Culicidae)—*Culicosporella* Weiser (Figs. 135, 137). See Weiser (1977) for a discussion of this genus.

15. Free spores elongate oval; membrane-bound spores (eight) oval; occur in Lepidoptera—*Vairimorpha* Pilley. See Pilley (1976) for a discussion of this genus.

16. Spores enclosed in a pansporoblastic membrane or cyst wall—**17**

16. Spores born free, not enclosed in a cyst wall—**33**

17. Cysts containing two spores—*Issia* Weiser (found in Trichoptera), *Telomyxa* Léger and Hesse (Figs. 138, 139) (found in Coleoptera, Diptera, and Ephemeroptera), and *Glugea* Thelohan (Fig. 140). See Weiser (1961) for a discussion of these genera.

17. Cysts containing more than two spores—**18**

18. Cysts containing four spores—*Gurleya* Doflein (Fig. 141). Occurs in Diptera, Ephemeroptera, Lepidoptera, and Odonata.

18. Cysts containing eight or more spores—**19**

19. Cysts containing eight spores—**20**

19. Cysts with more than eight spores—**29**

20. Spores elongate, 5–10 times longer than wide, straight or slightly bent—**21**

20. Spores not as above—**22**

21. Spores tabular or elongate oval, at most 5–6 times longer than wide—*Octosporea* Flu (Fig. 142). See Weiser (1961) for a discussion of this genus.

21. Spores rod-shaped, usually 7–10 times longer than wide—*Mrazekia* Léger and Hesse (Figs. 143, 144). See Weiser (1961) for a discussion of this genus. *Bacillidium* Janda is considered a synonym of *Mrazekia*.

22. Spores in elongate, fusiform cysts—*Chapmanium* Hazard and Oldacre. Found in Chaoboridae and Nepidae. See Hazard and Oldacre (1975) for a discussion of this genus.

22. Not as above—**23**

23. Spores block- or barrow-shaped, with blunt ends—*Amblyospora* Hazard and Oldacre (Fig. 145). Common in mosquitos (Culicidae). See Hazard and Oldacre (1975) for a discussion of this genus.

23. Spores not as above—**24**

24. Spores pyriform-shaped—*Hyalinocysta* Hazard and Oldacre (Fig. 146). Found in mosquitos (Culicidae). See Hazard and Oldacre (1975) for a discussion of this genus.

24. Spores not pyriform-shaped—**25**

25. Spores in larvae bottle-shaped, constricted to form a neck at one end; dimorphic species with a second elongate and slightly curved spore type in adult hosts—*Parathelohania* Codreanu (Fig. 147). Found in *Anopheles* mosquitos (Culicidae). See Hazard and Anthony (1974) for a discussion of this genus.

25. Spore not bottle-shaped—**26**

26. Long, slender, U-shaped spores—*Toxoglugea* Léger and Hesse (Fig. 148). Found in Diptera, Hemiptera, Homoptera, and Plecoptera. Sprague (1977) considers the genus *Spiroglugea* a synonym of *Toxoglugea*.

26. Spores not as above—**27**

27. Spores subspherical—*Pilosporella* Hazard and Oldacre (Fig. 149). Found in mosquitos (Culicidae). See Hazard and Oldacre (1975) for a discussion of this genus.

27. Spores elliptical or oval—**28**

28. Pansporoblasts formed from plasmodia with nuclei in diplokaryotic arrangement—*Pegmatheca* Hazard and Oldacre. Found in blackflies (Simuliidae). See Hazard and Oldacre (1975) for a discussion of this genus.

28. Pansporoblasts not formed as above—*Thelohania* Henneguy (Figs. 150, 151). Occurs in Coleoptera, Diptera, Ephemeroptera, Hymenoptera, and Trichoptera. See Hazard and Oldacre (1975) for a discussion of this genus.

29. Cysts fusiform or triangular with angles prolonged into flagelliform filaments—*Mitoplistophora* Codreanu. Occurs in Ephemeroptera. See Codreanu (1966) for a discussion of this genus.

29. Cysts not as above—**30**

30. With a constant number of 16 spores in each cyst—**31**

30. With a variable number of spores in each cyst—**32**

31. Cysts unornamented—*Duboscqia* Pérez (Fig. 152). Occurs in Diptera and Isoptera. See Kudo (1942) for a discussion of this genus.

31. Cysts ornamented with three or four long spines—*Trichoduboscqia* Léger (Fig. 153). Occurs in Ephemeroptera. See Léger (1926) for an account of this genus.

32. Possessing four and eight spores in cysts—*Stempellia* Léger and Hesse. Occurs in Ephemeroptera. See Hazard and Savage (1970) for a discussion of this genus.

32. Containing more than eight spores per cyst (generally over 16)—*Pleistophora* Gurley (Fig. 154). Occurs in Diptera, Coleoptera, Dermaptera, Ephemeroptera, Isoptera, Lepidoptera, Orthoptera, and Plecoptera. See Clark and Fukuda (1971) for a discussion of this genus.

33. Spores with crests or ridges—*Weiseria* Doby and Saguez (Fig. 155). See Doby and Saguez (1964) and Jamnback (1970) for a discussion of this genus.

33. Spores not as above—**34**

34. Spores bearing a terminal tail similar in shape to sperm; lateral glatinous processes or a crest may be present—*Caudospora* Weiser (Fig. 156). Found in blackflies (Simuliidae). See Weiser (1961) for a discussion of this genus.

34. Spores not as above—**35**

35. Spores bearing terminal, short, nail-like spike—*Galbergia* Weiser. Occurs in mosquitos (Culicidae). See Weiser (1977) for a discussion of this genus.

35. Spores not as above—**36**

36. Spores conical—*Culicosporella* Weiser (Fig. 135). Occurs in mosquitos (Culicidae). See Weiser (1977) for a discussion of this genus.

36. Spores not conical—**37**

37. Spores lageniform (swollen on one end)—*Cougourdella* Hesse (Fig. 157). Found in caddisflies (Trichoptera). See Hesse (1935) for a discussion of this genus. The genus *Pyrotheca* Hesse is considered by some a synonym of *Cougourdella.*

37. Spores not lageniform—**38**

38. Spores rod-shaped, with a short caudal appendage—*Jirvecia* Weiser. Found in midges (Chironomidae). See Weiser (1977) for a discussion of this genus.

38. Spores not as above—**39**

39. Spores tooth- or wedge-shaped (coniform)—*Culicospora* Weiser (Fig. 158). Occurs in mosquitos (Culicidae). See Weiser (1977) for a discussion of this genus.

39. Spores not as above—**40**

40. Spores subspherical—*Pilosporella* Hazard and Oldacre. Occurs in mosquito larvae. See Hazard and Oldacre (1975) for a discussion of this genus.

40. Spores oval—*Nosema* Naegeli (Figs. 159, 160) [see Henry and Oma (1981) for a discussion of a *Nosema* infection which is widespread in insects] and *Tuzetia* Maurand *et al.* [see Maurand *et al.* (1971) for discussion of this genus which occurs in blackflies (Simuliidae)].

41. Spores small, nearly spherical; borne in thin-walled pansporoblasts similar to microsporidans—Order Haplosporida (Fig. 161). Representatives of the genera *Coelosporidium* Mesnil and Marchoux (see Sprague, 1940), *Haplosporidium* Caullery and Mesnil (see Sprague, 1963), *Myiobium* Swellengrebel (1919), *Myrmecisporidium* Holldobler (1930), *Mycetosporidium* Léger and Hesse (1905), and *Nephridiophaga* Ivanic (see Woolever, 1966) are found in various tissues in a range of insects. However, they are rarely encountered on routine diagnoses and usually do not cause disease.

41. Spores variable in size and shape; generally larger than those of the Microsporida and Haplosporida; borne in thin- or thick-walled cysts—**42**

42. Cysts protruding from the stomach wall of blood-sucking insects as small tumors (40–60 μm in diameter) (Fig. 162); cyst ruptures to produce elongate sporozoites that migrate to the salivary glands of the host and are infective to vertebrates; motile worm-shaped zygotes may occur in the body cavity—Suborder Haemosporina. Members of this group are heteroxenous parasites and require both invertebrate and vertebrate hosts to complete their development. The malarial parasites (*Plasmodium* spp.) (Fig. 162) occur in this suborder, along with *Haemoproteus* Kruse in pigeons and *Leucocytozoon* Danilewsky from ducks. For representatives of these genera, which all undergo partial development in blood-sucking insects, see Kudo (1966). Certain pathologies are caused in the insect host during development of these parasites, but they are rarely lethal and have been little studied.

42. Cysts not protruding from the stomach wall; monoxenous forms parasitizing invertebrates only—**43**

43. Mature trophozoites (vegetative or developing forms) large, extracellular, club-shaped, elliptical (in gut), or spherical (in body cavity); usually septate (divided into a protomerite and deutomerite); spores generally small and inconspicuous; found in the gut or body cavity (Subclass Gregarinia)—**44**

43. Mature trophozoites small; mostly intracellular; nonseptate; spores or oocysts large—**48**

44. Mature trophozoites septate; attached to gut wall or free in gut lumen—Order Eugregarinida, Suborder Cephalina (Figs. 163, 164). There are many genera parasitic in insects (especially Diptera, Orthoptera, and Coleoptera), which are treated by Kudo (1966). Since most species are considered harmless to their hosts, they will not be treated further here. For a description of a representative species in insects, see Canning (1956).

44. Mature Trophozoites aseptate; attached to gut wall or in body cavity (Order Eugregarinida, Suborder Acephalina) (Fig. 165)—**45**

45. Cysts spherical—**46**

45. Cysts sausage-shaped—*Allantocystis* Keilin. See Keilin (1920) for a description of an *Allantocystis* from the fly, *Dasyhelea obscura*.

46. Cysts found in gut lumen or malpighian tubules; gamonts elongate, motile; commonly found in mosquitos—*Lankesteria* Mingazzini (Figs. 166–169). See Sanders and Poinar (1973) for a discussion of a species in this genus.

46. Cysts found in body cavity (rarely gut lumen); gamonts spherical—**47**

47. Spores round or oval; found in Orthoptera (especially cockroaches and crickets)—*Diplocystis* Kunstler. See Weiser (1963) for an account of this genus.

47. Spores biconical (pointed at both ends); found in Coleoptera—*Monocystis* Stein (Fig. 170). See Weiser (1963) for an account of this genus. The closely related genus *Enterocystis* Zwetkow (see Codreanu, 1940) occurs in mayflys (Ephemeroptera).

48. Spores oval or navicular, containing eight sporozoites; found in gut epithelium, malpighian tubules, or fat body (Order Neogregarinida)—**49.** The three most commonly occurring genera are listed below. Others are discussed by Weiser and Briggs (1971).

48. Spores mostly spherical; with resting body; containing a variable number of sporozoites, but rarely eight; found in hemolymph, fat body, and occasionally gut epithelium (Subclass Coccidia, Order Eucoccida)—**50**

49. Cysts containing eight spores; common in *Tribolium* sp.—*Farinocystis* Weiser. See Dissanaike (1955) for an account of this genus.

49. Cysts containing one or two spores; common in stored-product insects—*Mattesia* Naville (Fig. 171) and *Ophryocystis* Schneider. See McLaughlin (1965) for a discussion of *Mattesia*. For discussion of the Schizogregarina see Weiser (1955).

50. One sporozoite in each spore; spores bivalve; gametocysts similar—*Barrouxia* Schneider. See Kudo (1966) for a discussion of this genus. *B. ornata* Schneider occurs in the gut of the bug *Nepa cinerea* L.

50. Two to six sporozoites in each spore; spores not bivalve; gametocytes dissimilar—**51**

51. Spore walls not formed; sporozoites occur within the oocyst wall—*Legerella* Mesnil. See Vincent (1927) for a discussion of this genus, which infects the malpighian tubules of fleas and the beetle *Hydroporus palustris* L. (Dytiscidae).

51. Spore walls present—**52**

52. Cysts with three spores, each spore with four to six or more sporozoites—*Chagasella* Machado. See Machado (1913) for a discussion of this genus, which is found in the intestinal epithelium of the bug *Dysdercus ruficollis* L.

52. Cysts with more than three spores—**53**

53. Cysts with four spores—*Ithania* Ludwig. *Ithania wenrichi* Ludwig occurs in the intestinal epithelium of tipulid larvae (Ludwig, 1947).

53. Cysts with usually more than four spores; each with two sporozoites—*Adelina* Hesse (Fig. 172). See Yarwood (1937) for an account of a species in this genus. They are commonly encountered in the fat body and other tissues of a variety of insects.

FIGURE 122. A representative of the class Suctora attached to a water beetle. ×34.

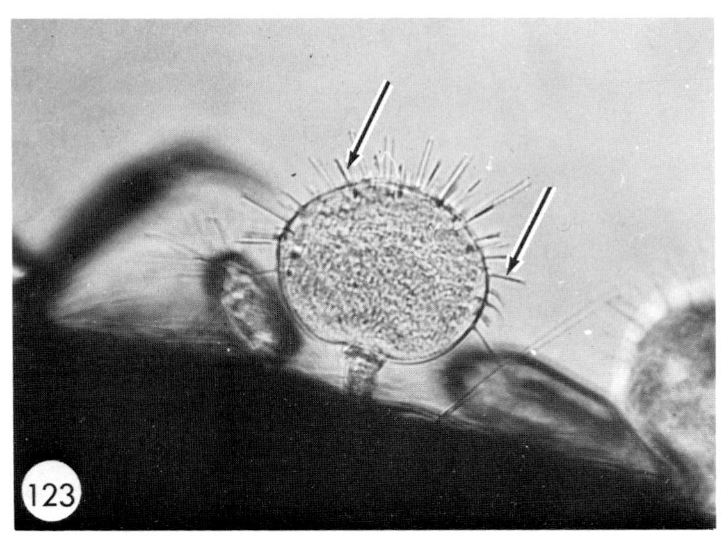

FIGURE 123. Detail of a representative of the class Suctora. Note the fine tentacles (arrows). ×340.

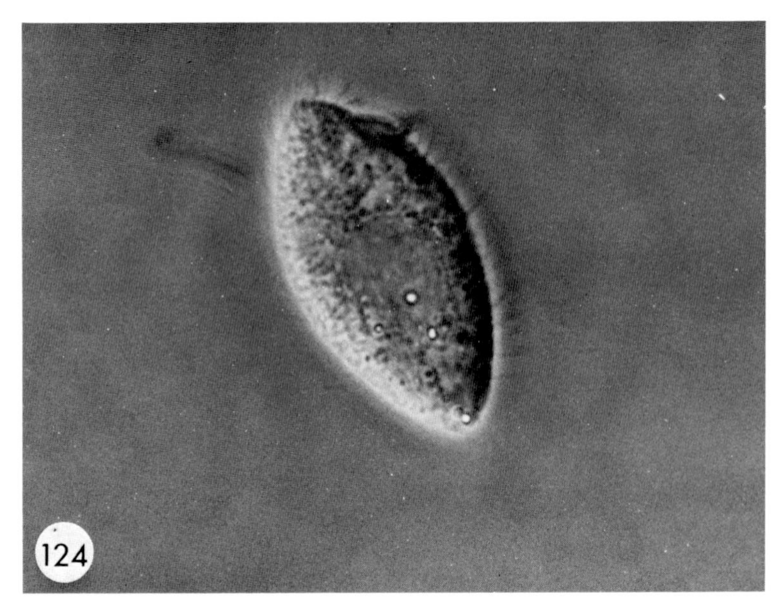

FIGURE 124. A living specimen of the ciliate *Tetrahymena pyriformis*. ×1000.

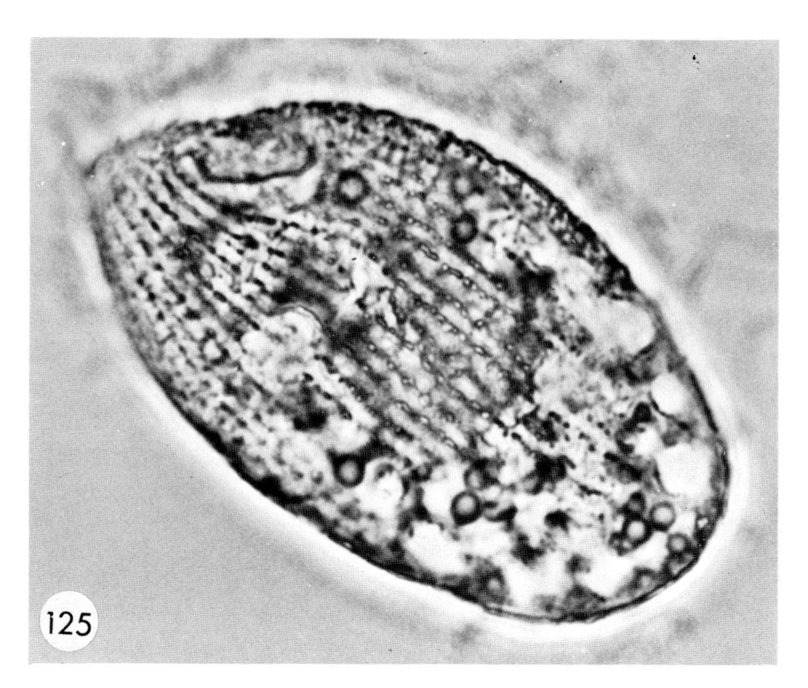

FIGURE 125. Silver nitrate stain of a fixed individual of *T. pyriformis.* ×1000.

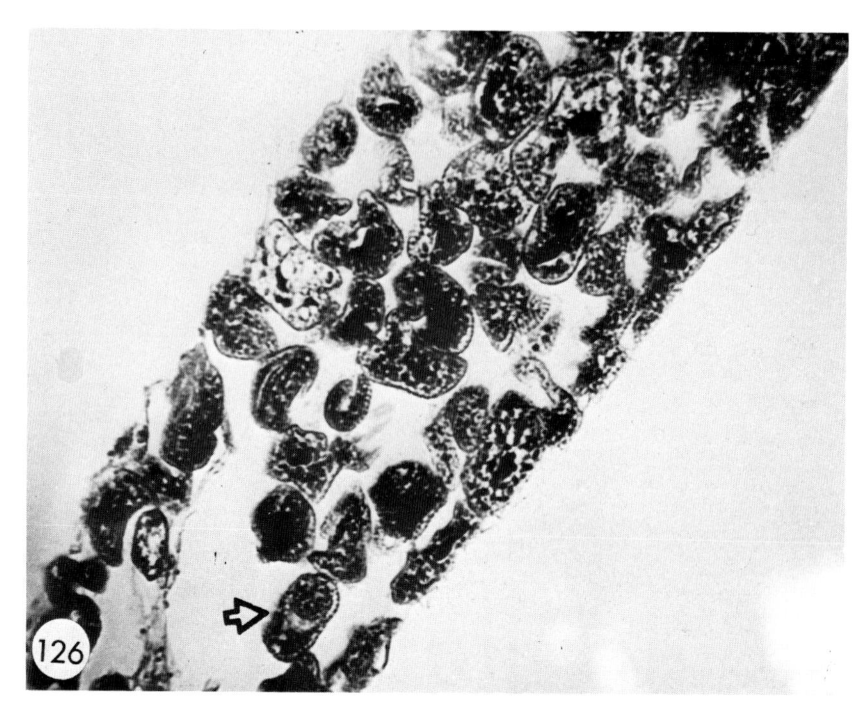

FIGURE 126. Section of a diseased mosquito larva showing body cavity filled with *Tetrahymena* sp. (arrow). (Courtesy of D. Sanders.) ×200.

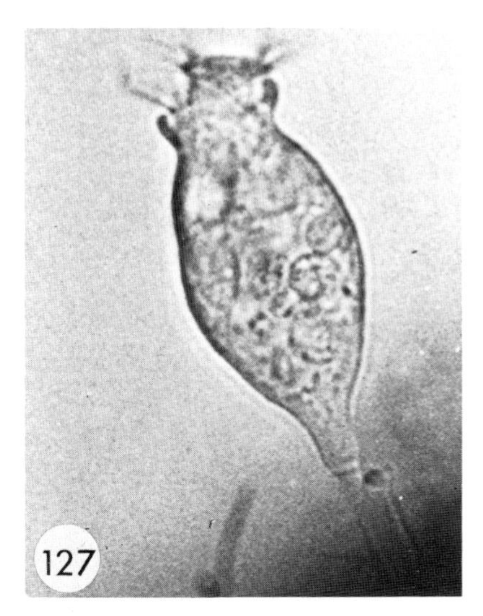

FIGURE 127. *Vorticella* sp. (Sessilina) attached to the cuticle of a chironomid larva. Note expanded cilia around mouth opening of an actively feeding form. ×970.

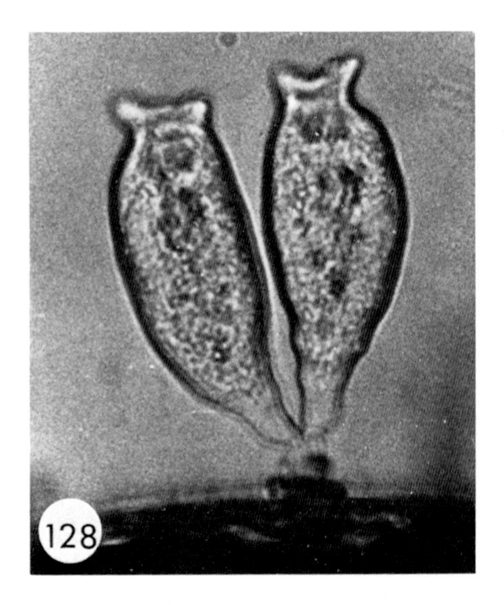

FIGURE 128. Two nonfeeding, quiescent forms of *Vorticella* sp. ×700.

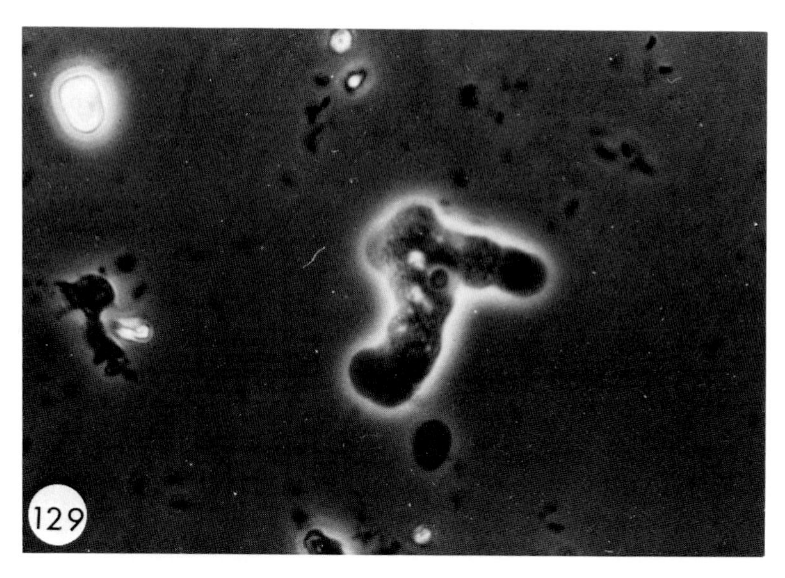

FIGURE 129. Motile amoeboid stage of a representative of the order Amoebida in the hemocoel of an insect. ×1000.

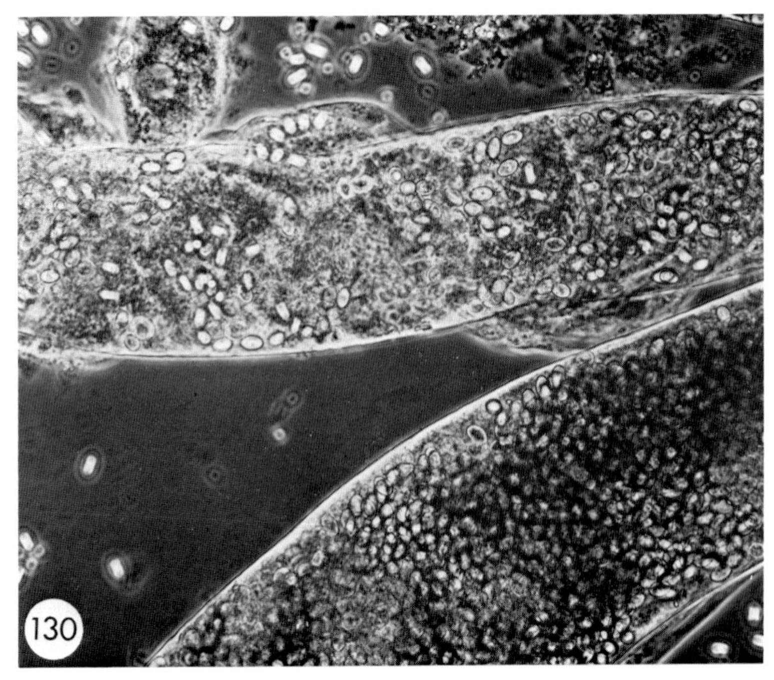

FIGURE 130. Spores of *Malamoeba locustae* in the malpighian tubules of a grasshopper. (Courtesy of J. Evans.) ×260.

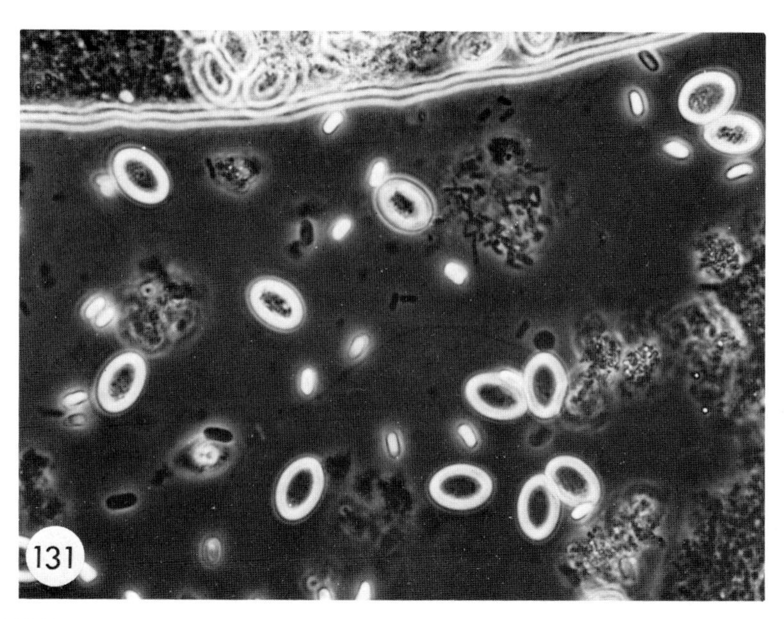

FIGURE 131. Larger spores of *Malamoeba locustae* together with the smaller spores of *Nosema locustae* in a grasshopper with a double infection. (Courtesy of J. Evans.) ×640.

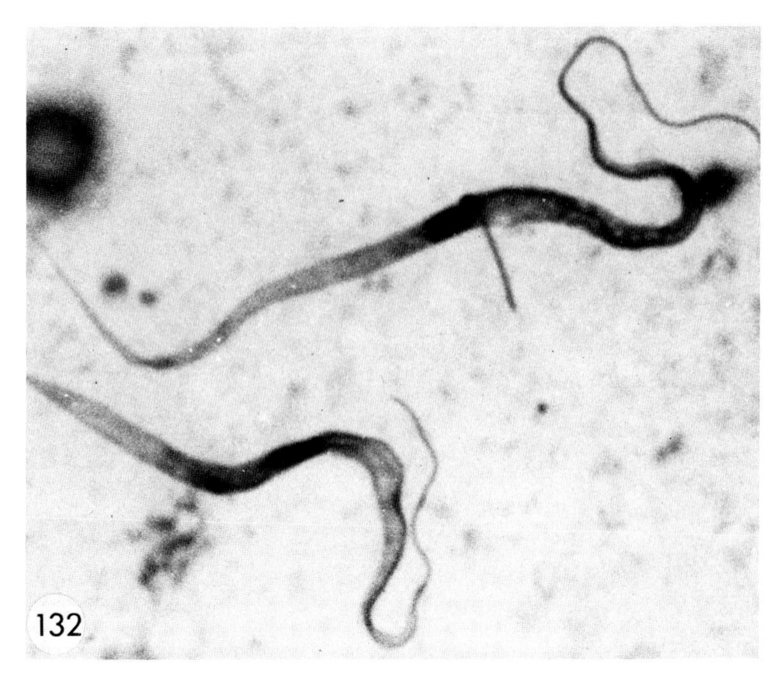

FIGURE 132. Two specimens of *Trypanosoma rangeli* from the gut of *Rhodnius prolixus*. (Courtesy of R. Zarate.) ×2000.

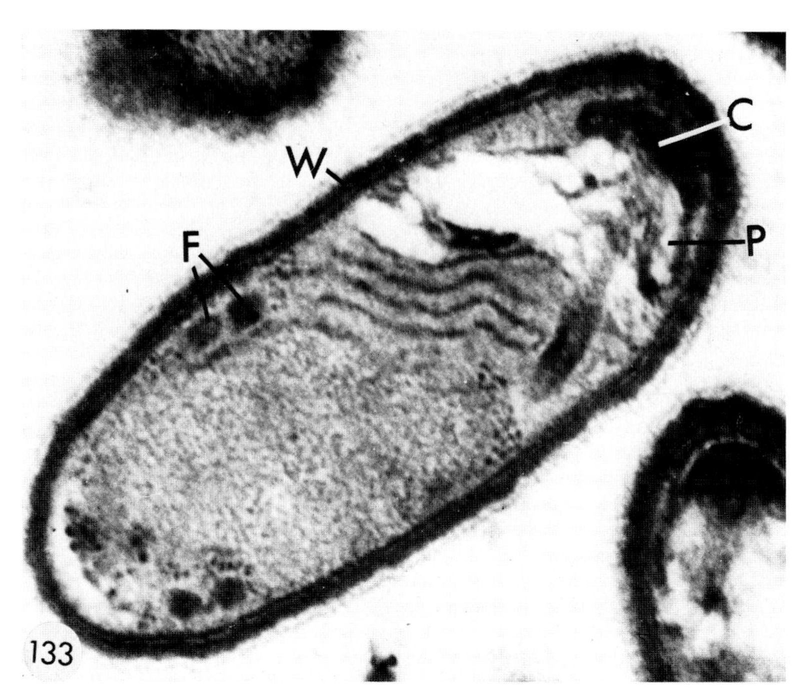

FIGURE 133. Structure of a microsporidan spore (*Pleistophora* sp.), showing the polar cap (C), polar filament (F), polaroplast (P), and spore wall (W). (Courtesy of D. Sanders.) ×84,900.

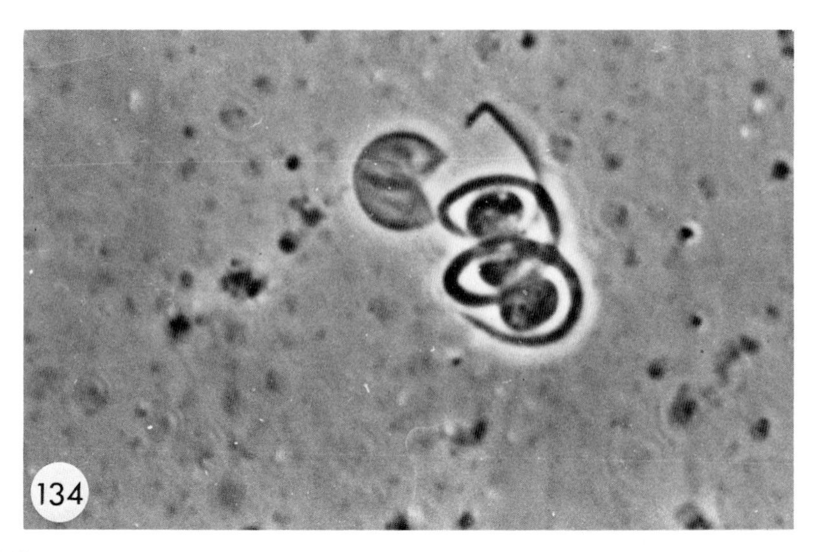

FIGURE 134. Elongate coiled filament and three sporoplasms of *Helicosporidium parasiticum.* (Courtesy of W. Kellen.) ×1700.

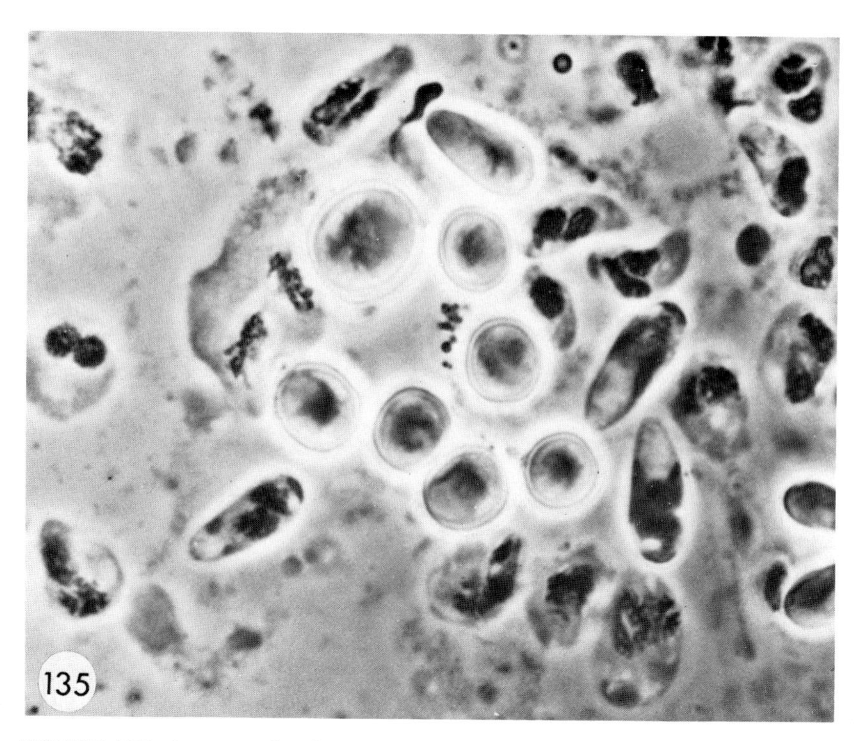

FIGURE 135. An example of dimorphic spores in *Culicosporella lunata* from the mosquito, *Culex pilosus.* The elongated, pyriform spores are binucleate and the oval spores are uninucleate. Lacto-aceto-orcein stain. (Courtesy of T. Fukuda and E. I. Hazard.) ×2000.

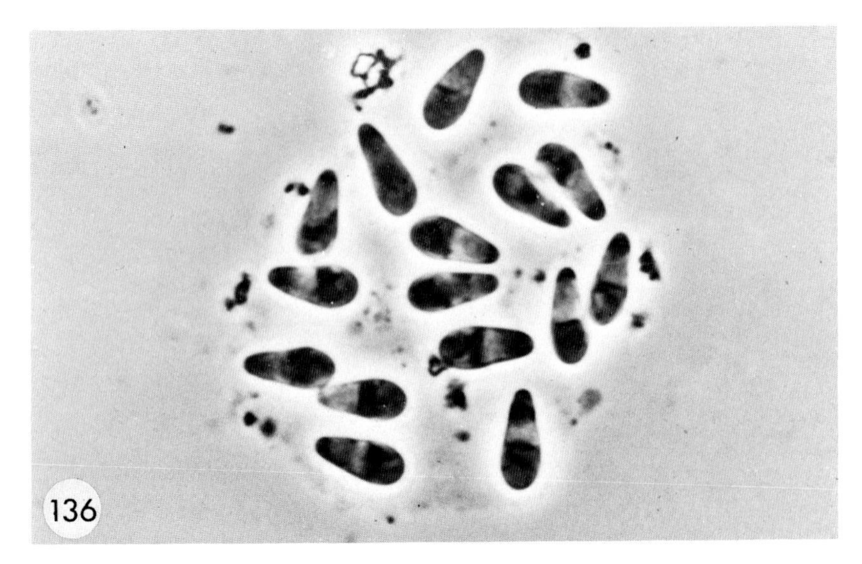

FIGURE 136. Spores of *Hazardia milleri* from the mosquito *Culex pipiens quinquefasciatus.* Lacto-aceto-orcein stain. (Courtesy of T. Fukuda and E. I. Hazard.) $\times 2000$.

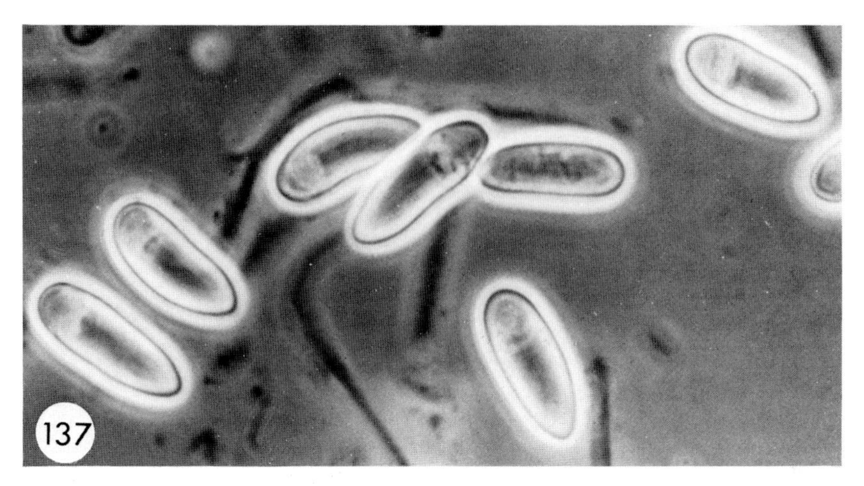

FIGURE 137. Binucleate spores of *Culicosporella lunata* from the mosquito *Culex pilosus*. (Courtesy of T. Fukada and E. I. Hazard.) ×2000.

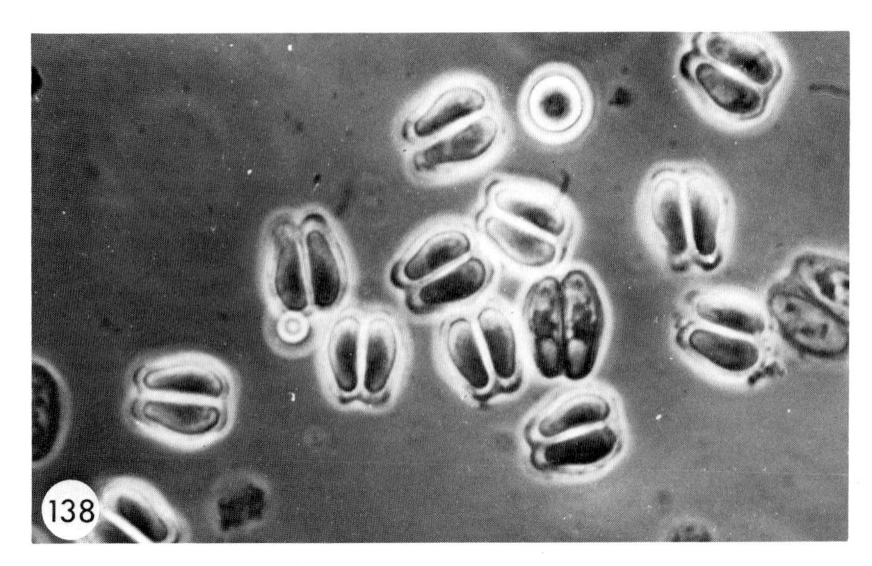

FIGURE 138. Diplospores of *Telomyxa* from a helodid beetle larva. (Courtesy of E. Hazard.) ×1370.

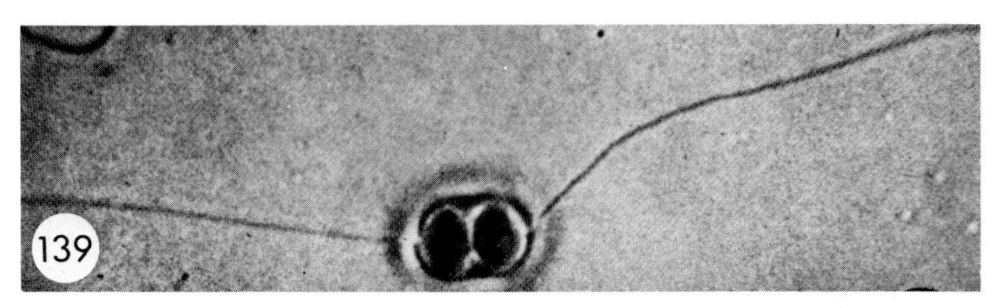

FIGURE 139. A diplospore of *Telomyxa* with extruded polar filaments. (After Codreanu and Vavra, 1970.) ×2000.

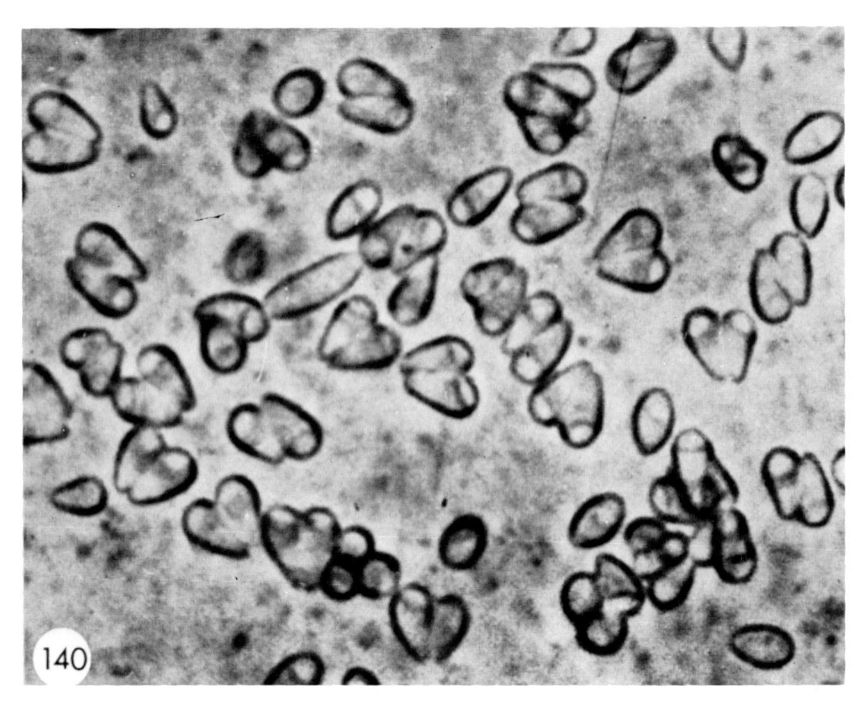

FIGURE 140. Pansporoblasts of *Glugea trichopterae* containing paired spores. (Courtesy of J. Weiser.) ×2000.

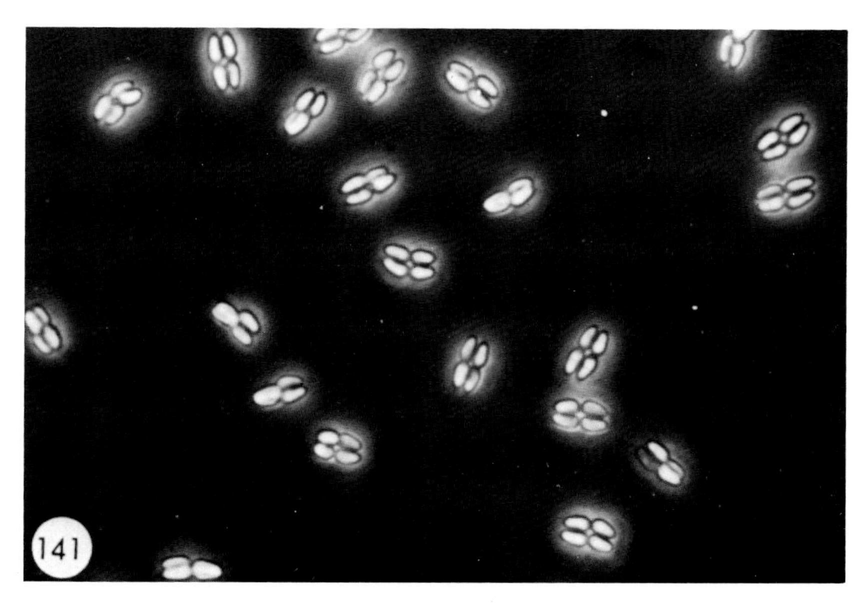

FIGURE 141. Four-spored pansporoblasts of *Gurleya* sp. from a mayfly (Ephemeroptera) nymph. ×750.

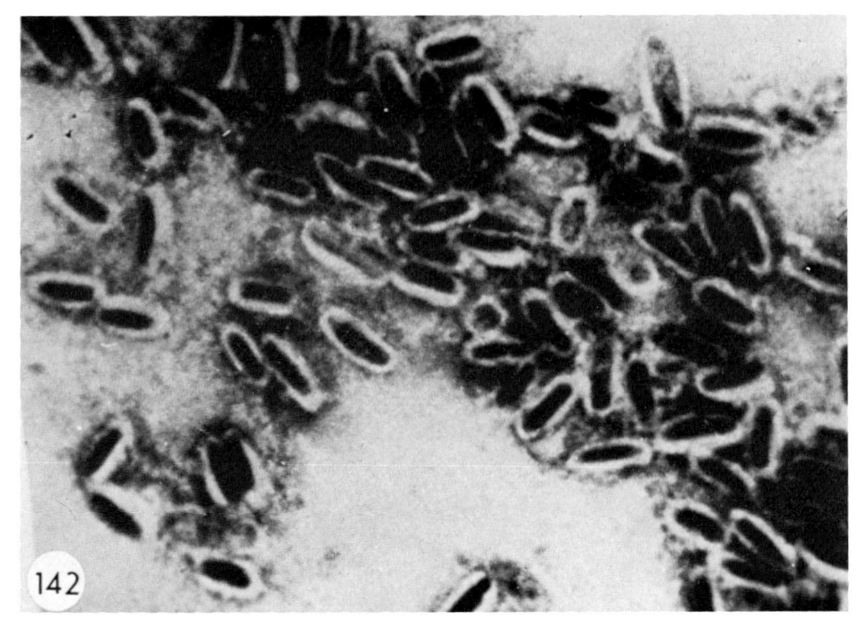

FIGURE 142. Spores of *Octosporea vividanae.* (Courtesy of J. Weiser.) ×1000.

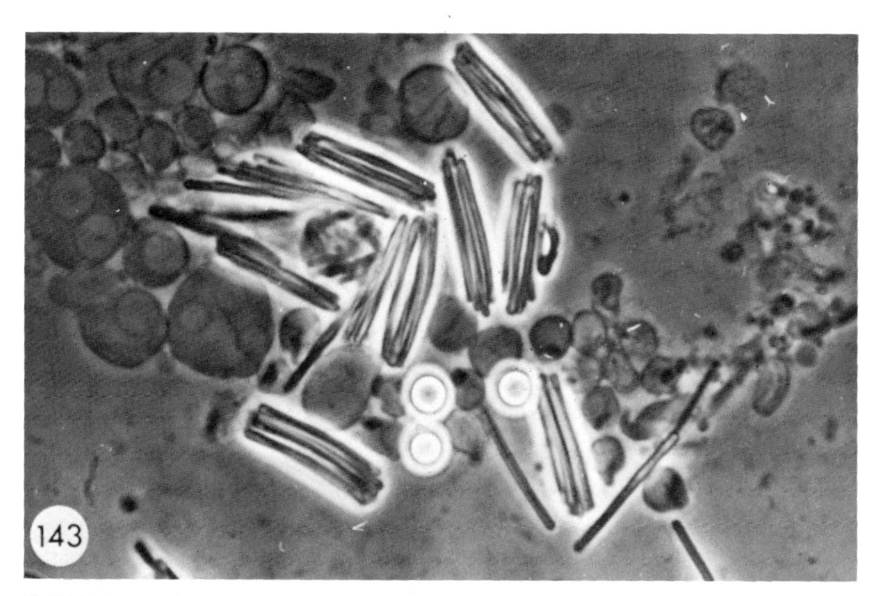

FIGURE 143. Spores of *Mrazekia* sp. from *Clodotanytarsus* sp. (Courtesy of E. Hazard.) ×1320.

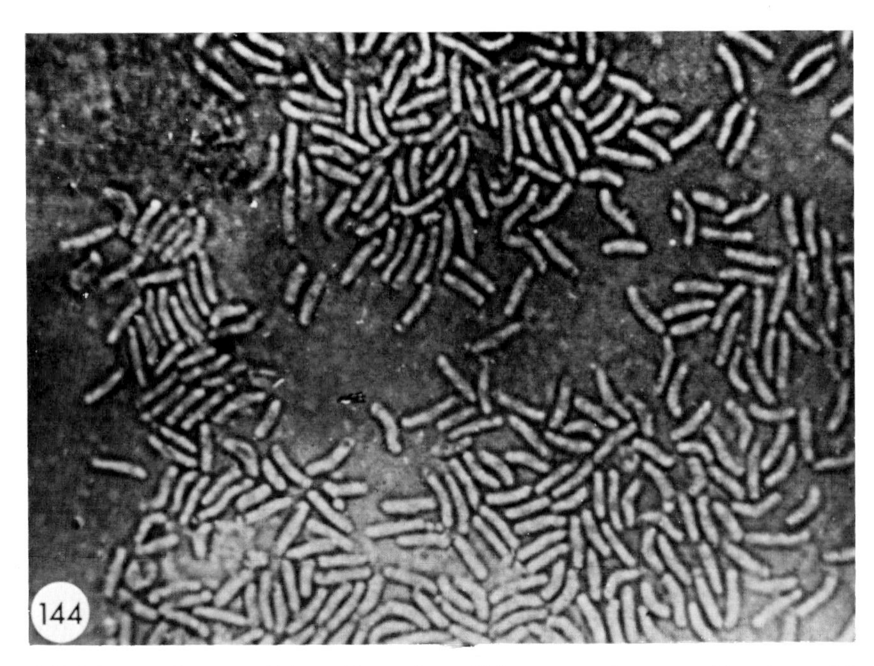

FIGURE 144. Spores of *Mrazekia* sp. from *Chironomus plumosus.* (Courtesy of J. Weiser.) $\times 600$.

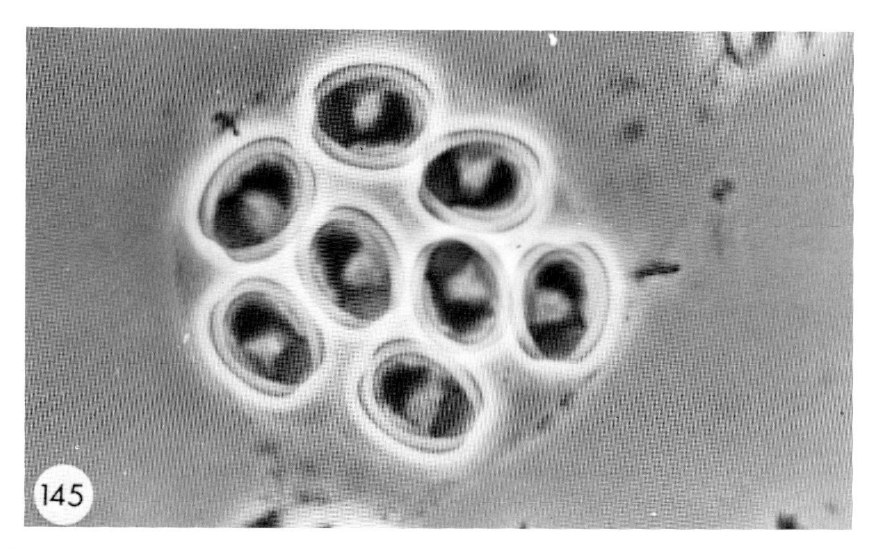

FIGURE 145. Haploid spores of *Amblyospora* sp. from the mosquito *Culex solinarius*. Lacto-aceto-orcein stain. (Courtesy of T. Fukuda and E. I. Hazard.) ×2000.

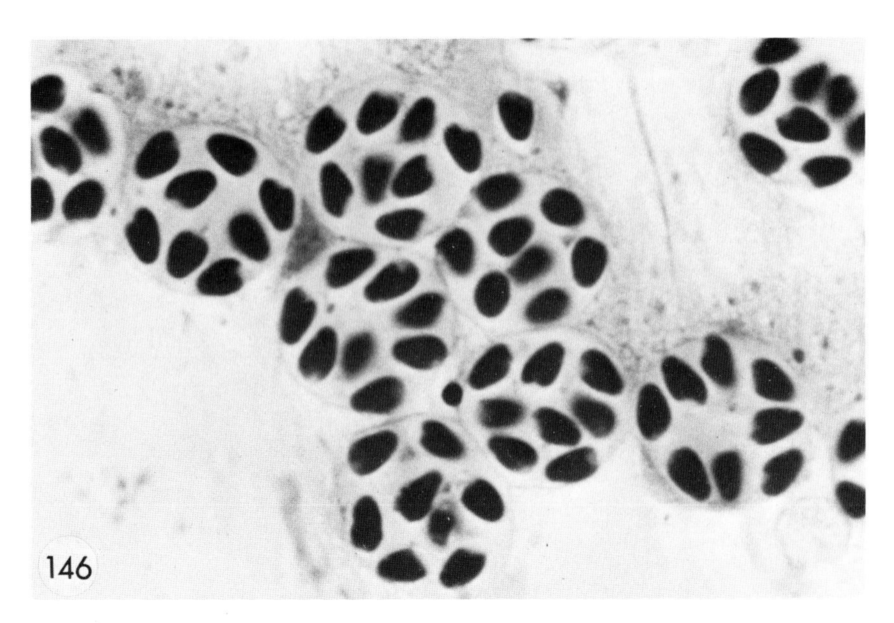

FIGURE 146. Spores of *Hyalinocysta chapmani* from the mosquito *Culiseta melanura*. Heidenhain's hematoxylin stain. (Courtesy of T. Fukuda and E. I. Hazard.) ×2000.

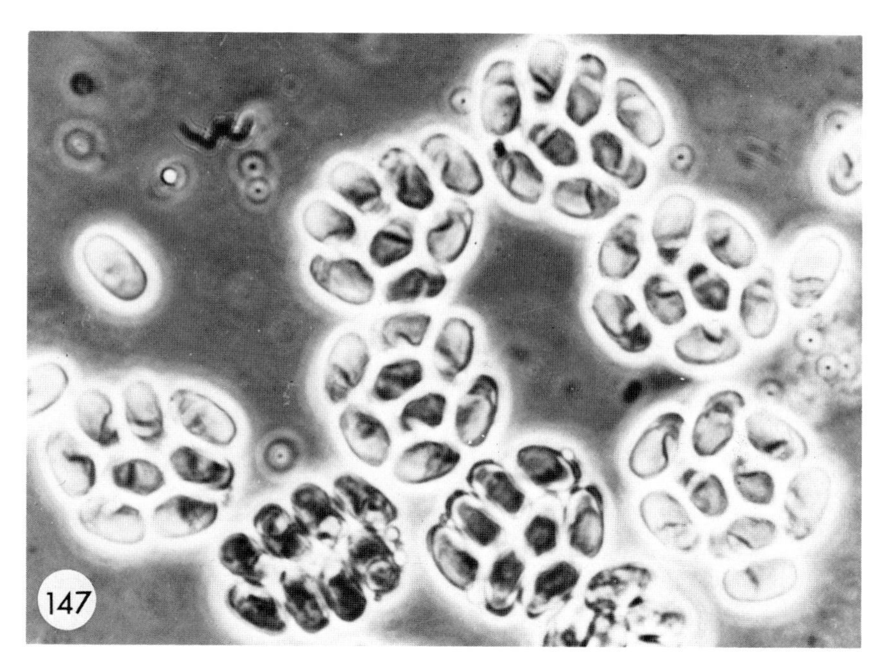

FIGURE 147. Haploid spores of *Parathelohania* sp. from the mosquito *Anopheles bradleyi*. (Courtesy of T. Fukuda and E. I. Hazard.) ×2000.

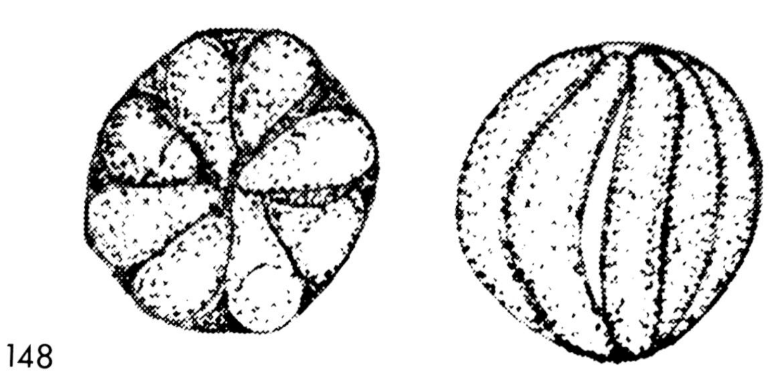

FIGURE 148. Eight-spored pansporoblast of *Toxoglugea* sp. (After Weiser, 1961.) Diagrammatic, about ×2100.

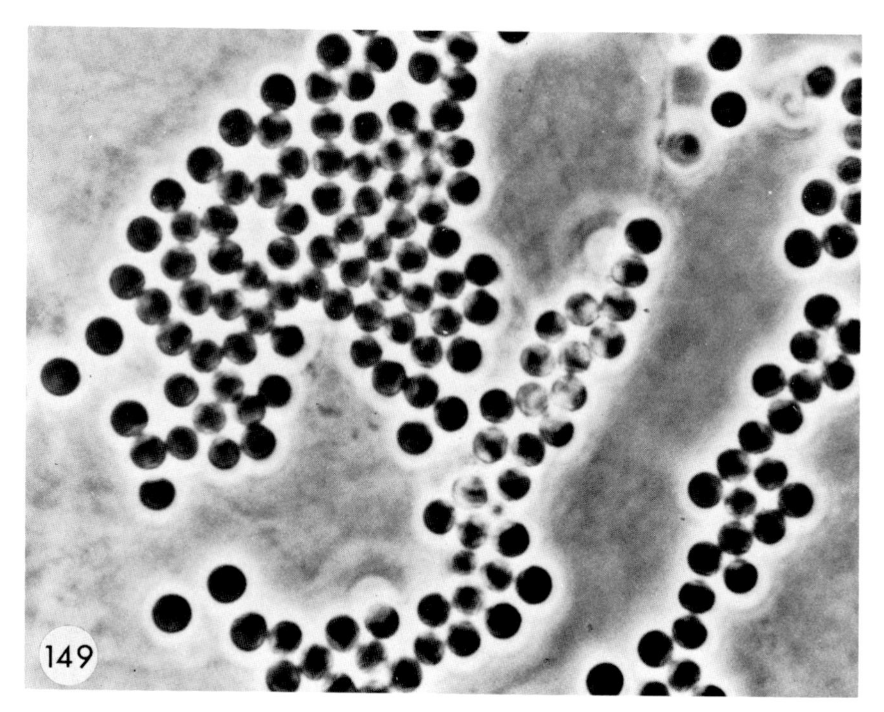

FIGURE 149. Spores of *Pilosporella* sp. from the mosquito *Aedes triseriatus*. Lacto-aceto-orcein stain. (Courtesy of T. Fukuda and E. I. Hazard.) ×2000.

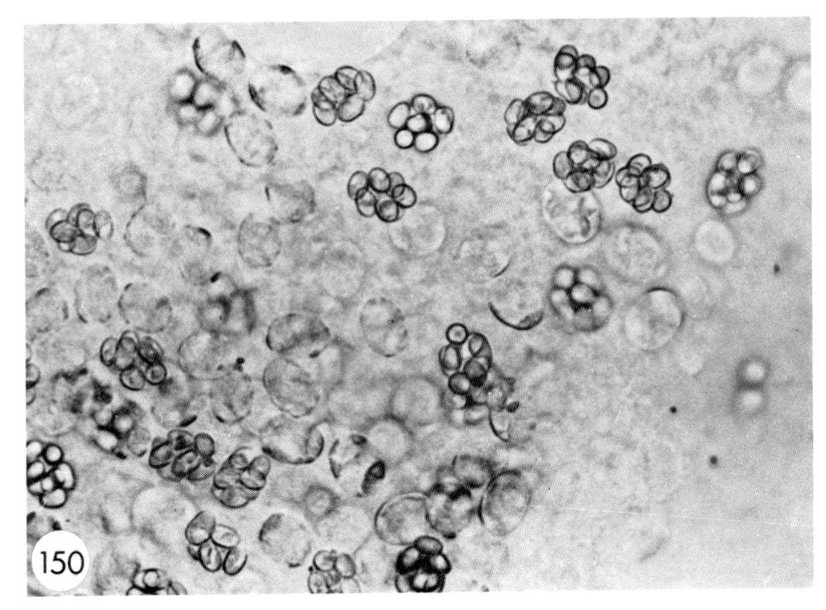

FIGURE 150. Pansporoblasts of *Thelohania* sp. ×1200.

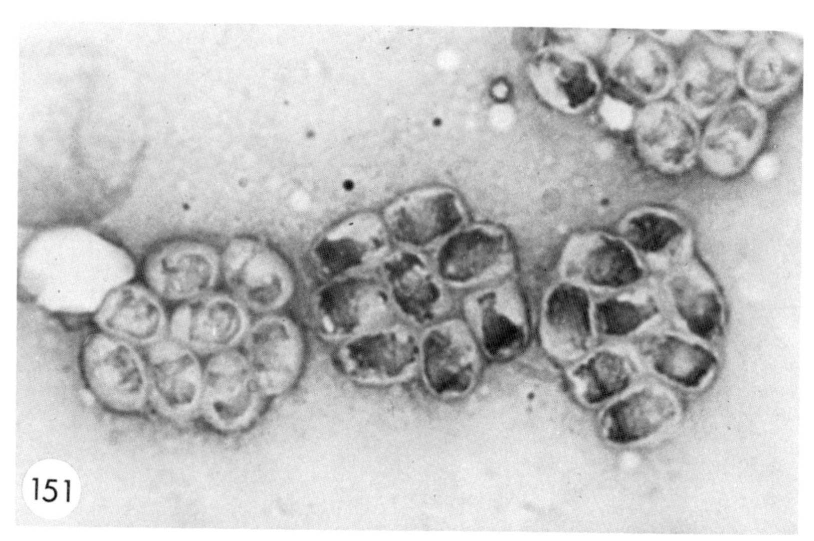

FIGURE 151. Eight-spored pansporoblasts of *Thelohania californica* from a larva of the mosquito *Culex tarsalis*. (Courtesy of D. Sanders.) ×1600.

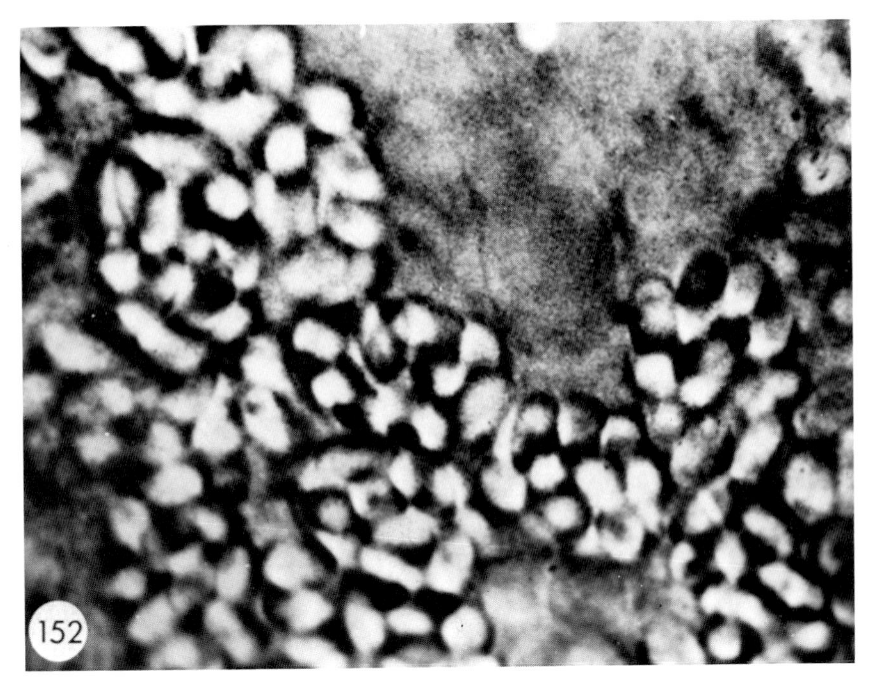

FIGURE 152. Pansporoblasts of *Duboscqia legeri* (after Kudo, 1942.) ×2400.

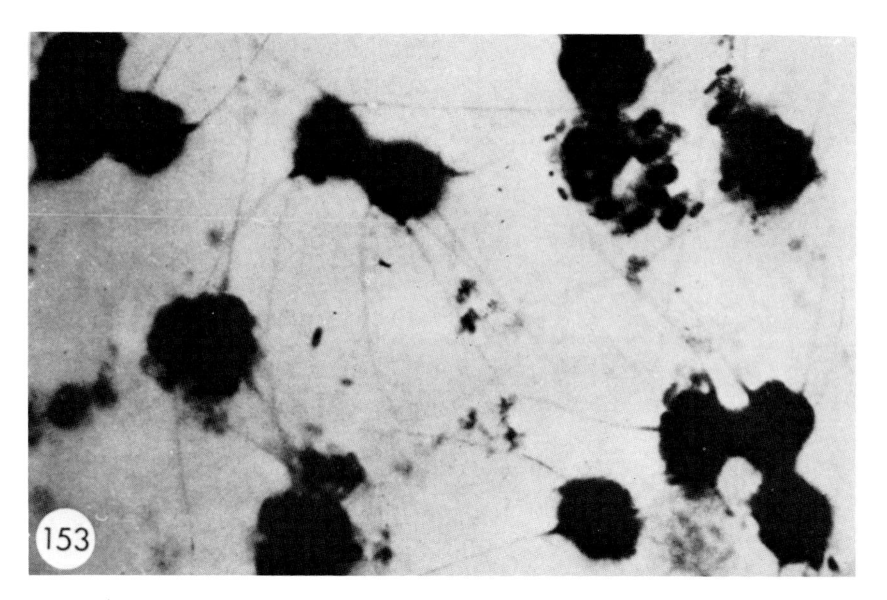

FIGURE 153. Pansporoblasts of *Trichoduboscqia epeori.* (Courtesy of J. Weiser.) ×500.

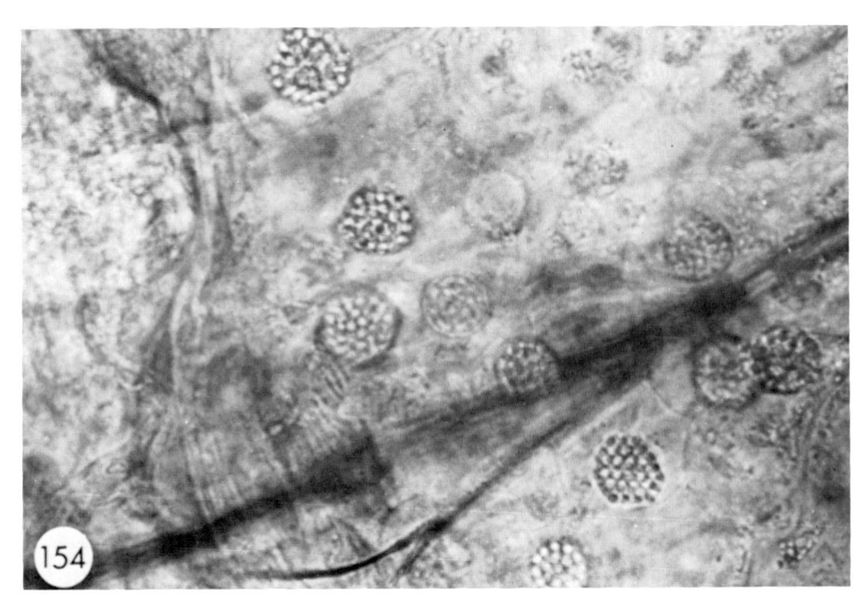

FIGURE 154. Pansporoblasts of *Pleistophora* sp. from a mosquito larva. (Courtesy of D. Sanders.) ×1200.

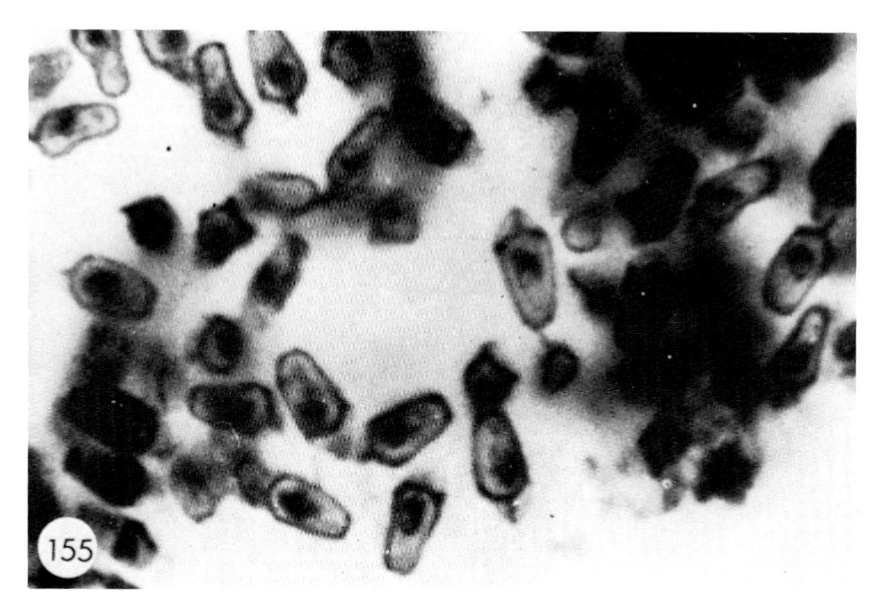

FIGURE 155. Spores of *Weiseria laurenti*. (Courtesy of J. Weiser.) ×1000.

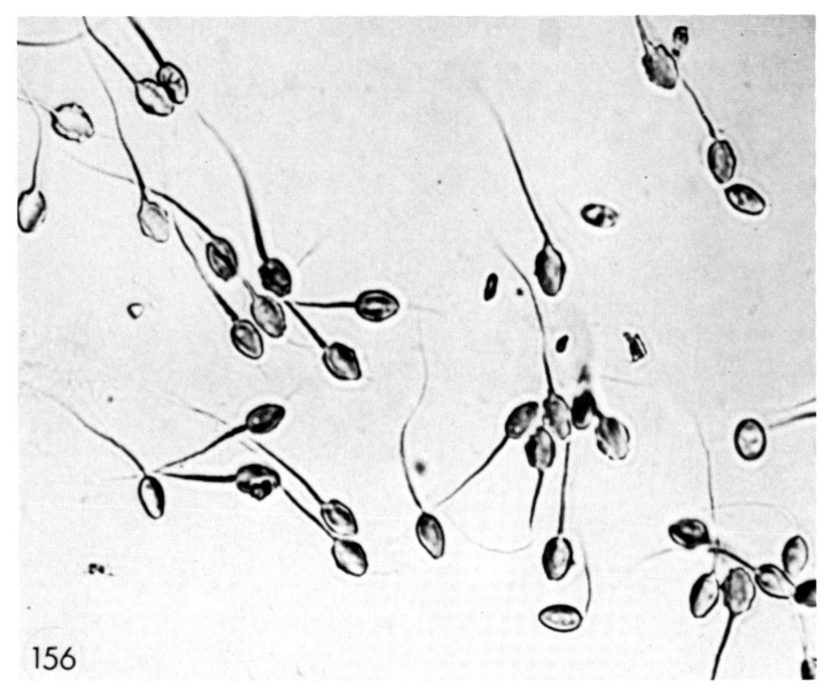

FIGURE 156. Spores of *Caudospora simulii*. (Courtesy of J. Weiser.) ×1000.

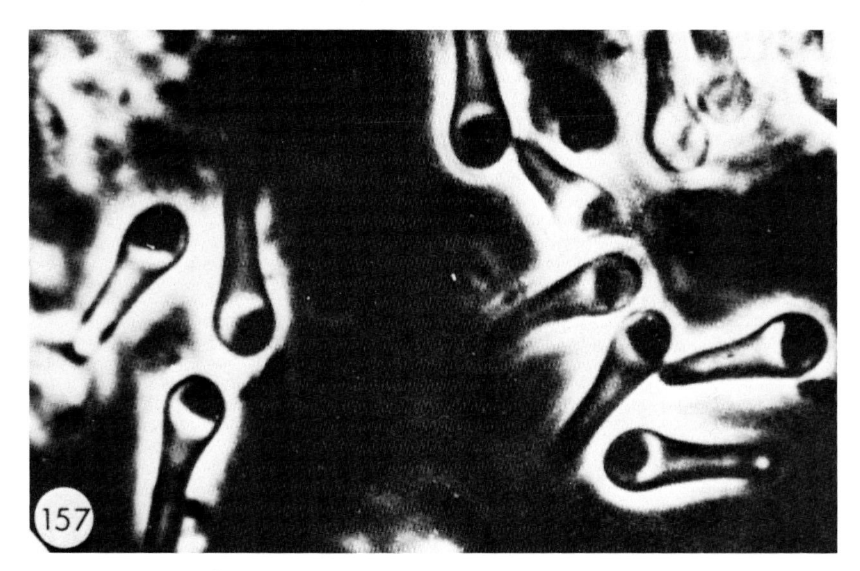

FIGURE 157. Spores of *Cougourdella rhyacophilae.* (After Baudoin, 1969.) ×1600.

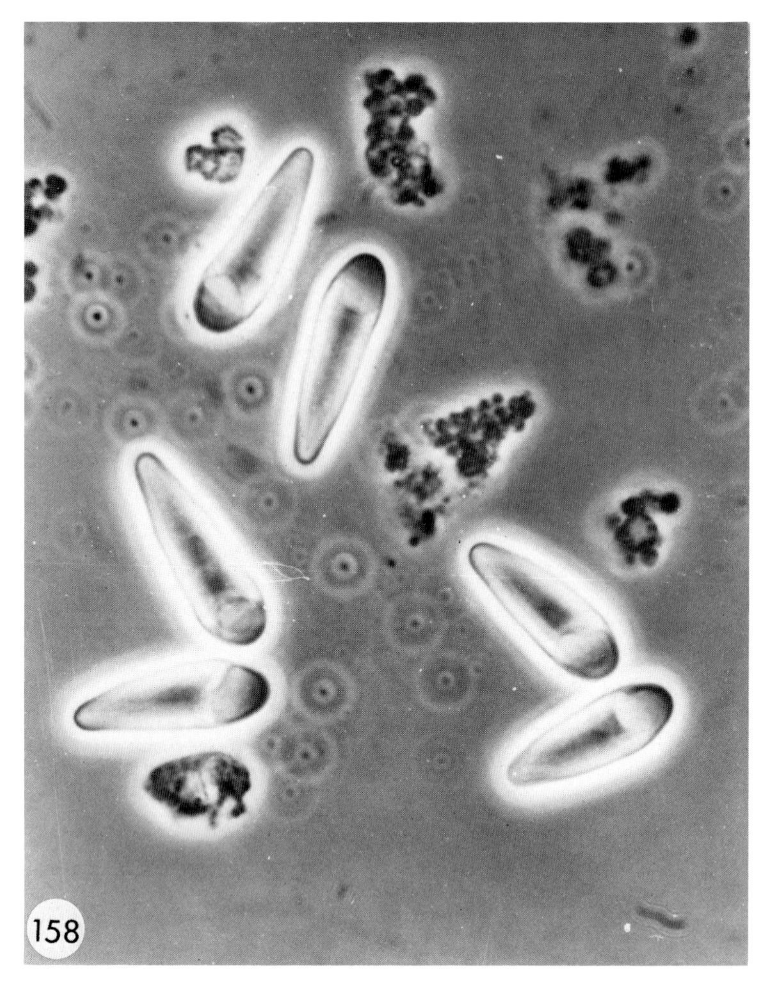

FIGURE 158. Spores of *Culicospora magra* from the mosquito *Culex restuans*. (Courtesy of T. Fukuda and E. I. Hazard.) ×2000.

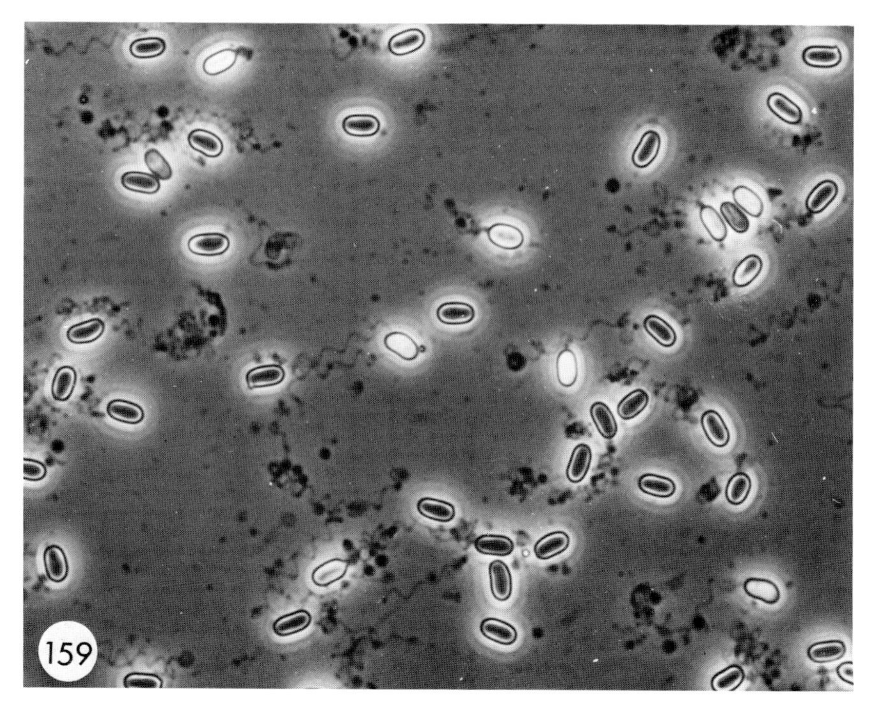

FIGURE 159. Spores of *Nosema locustae.* (Courtesy of J. Evans.) ×850.

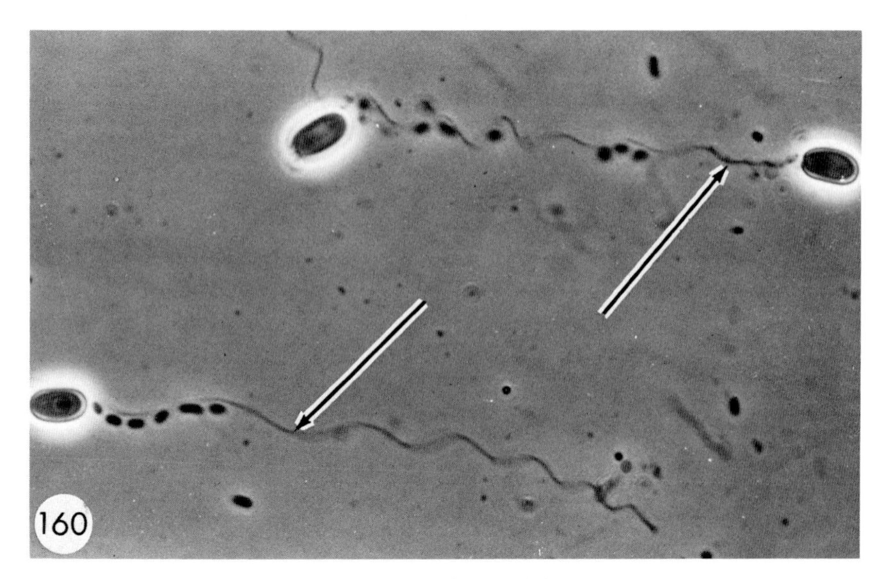

FIGURE 160. Extruded polar filaments (arrows) from *Nosema locustae* spores. ×1600.

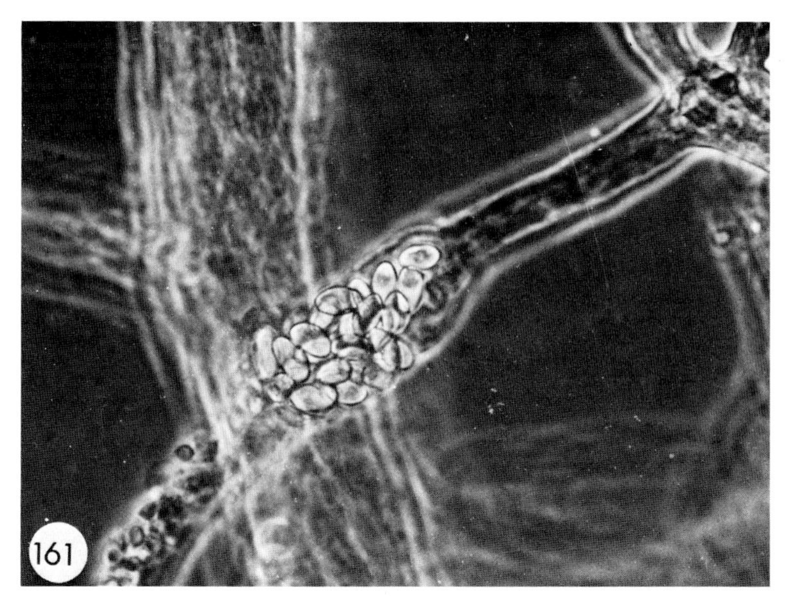

FIGURE 161. Spores of a Haplosporida in the ganglion of the beetle *Tenebrio molitor.* (Courtesy of W. Kellen.) ×900.

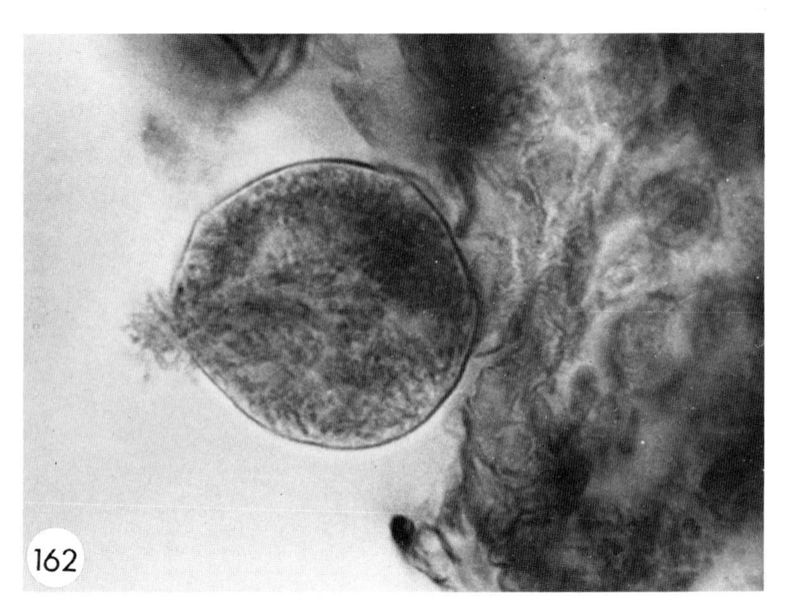

FIGURE 162. Rupturing oocyst of *Plasmodium circumflexum* attached to the gut wall of *Culiseta morsitans*. (Courtesy of M. Laid and E. Greiner.) ×900.

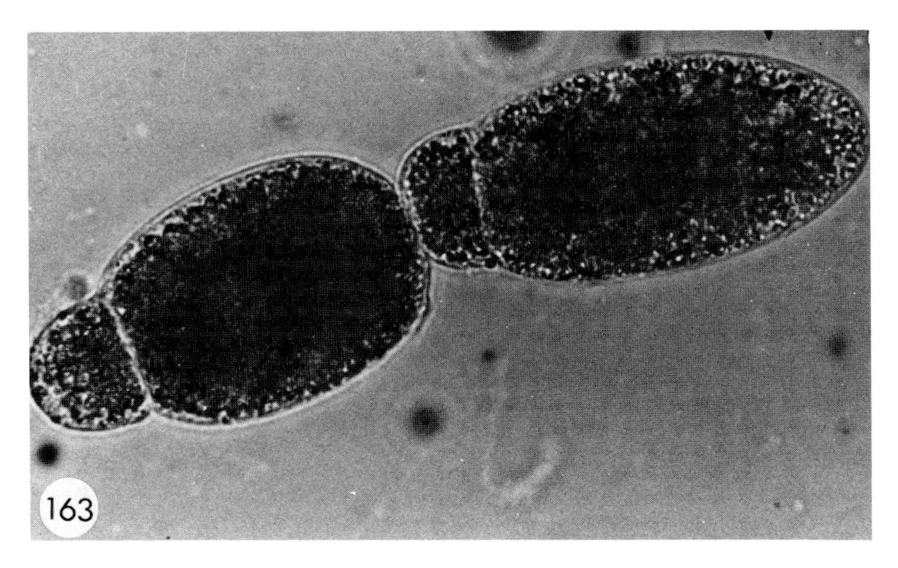

FIGURE 163. Mature trophozoites of a cephalinid gregarine removed from the gut of the beetle *Carpophilus* sp. ×640.

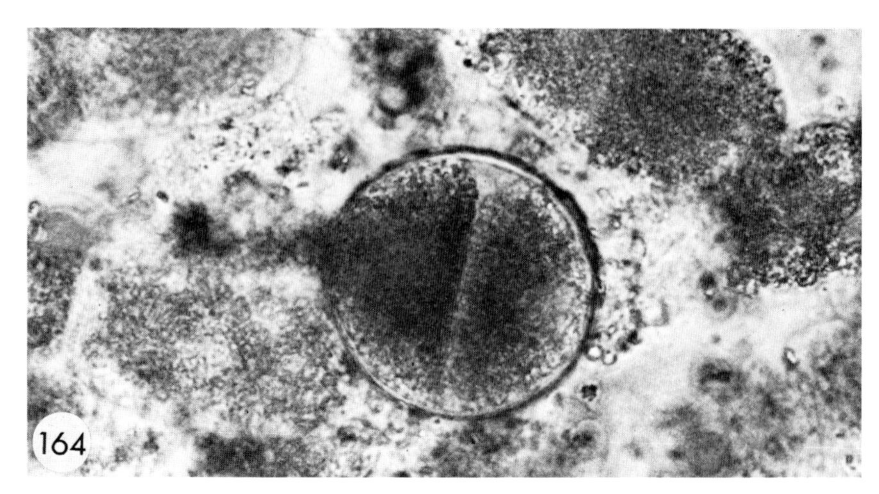

FIGURE 164. Cyst of a cephalinid gregarine removed from the gut of the beetle *Carpophilus* sp. ×640.

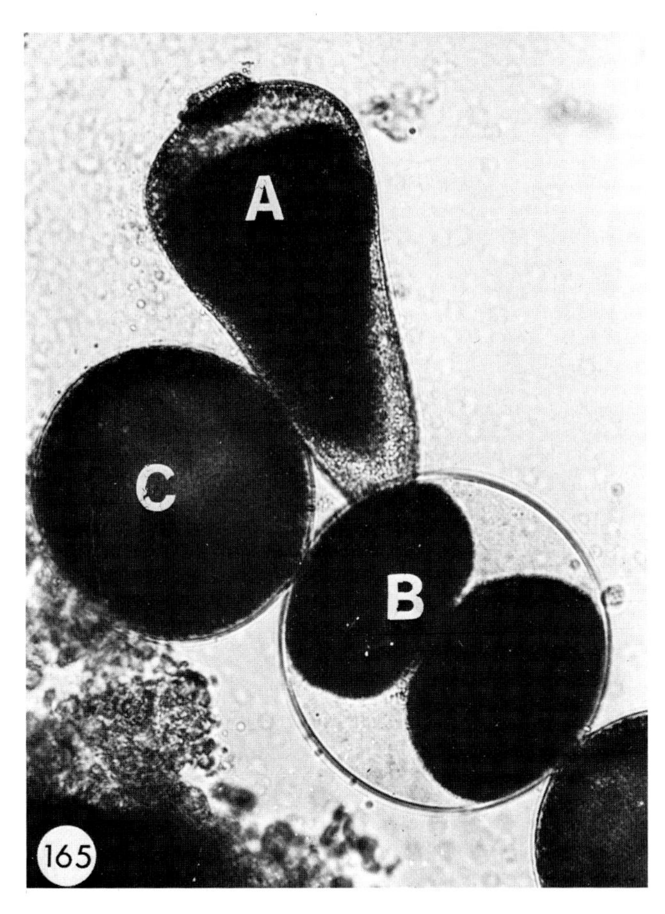

FIGURE 165. Mature trophozoite (A), encysted gamonts (B), and gamontocyst (C) of an acephalinid gregarine removed from the body cavity of a flea. (Courtesy of B. Nelson.) ×640.

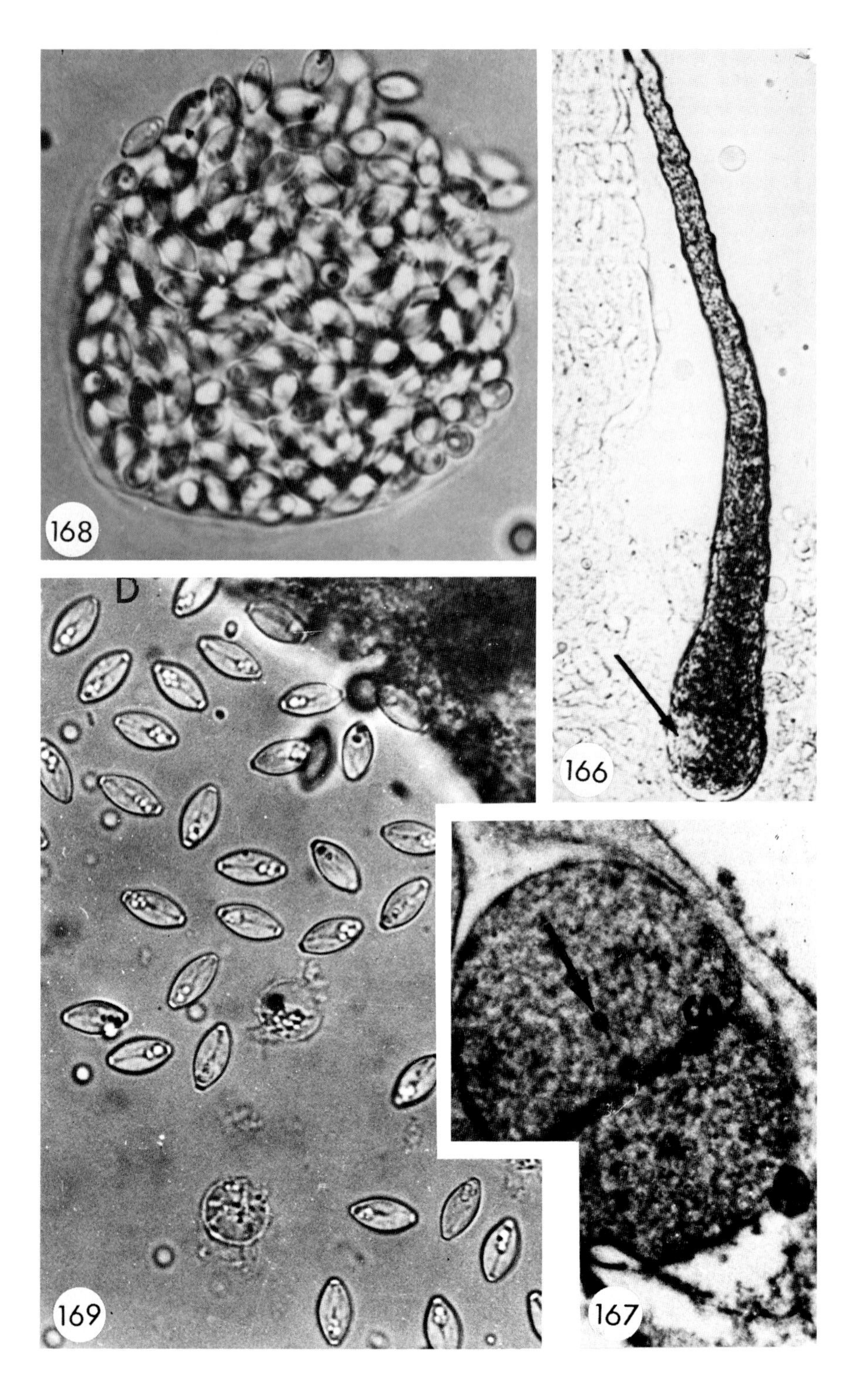
168
166
169
167

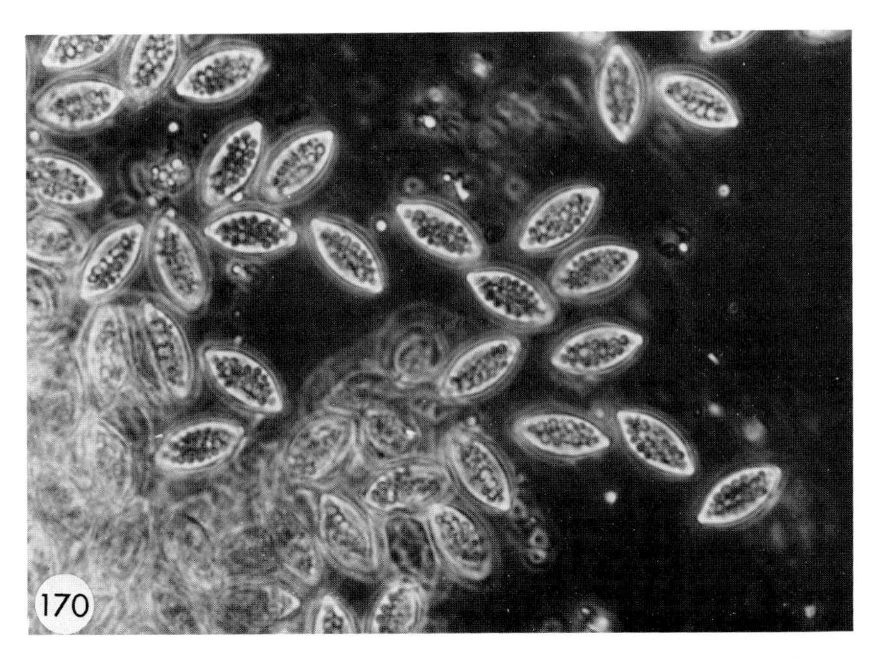

FIGURE 170. Mature spores of the eugregarinid *Monocystis* sp. from the hemocoel of a scarabaeid larva. ×700.

FIGURE 166. Gamont of *Lankesteria clarki,* an acephalinid gregarine, in the midgut of the mosquito *Aedes sierrensis.* Arrow indicates nucleus. (Courtesy of D. Sanders.) ×800.

FIGURE 167. Young gamontocyst of *Lankesteria clarki* in the malpighian tubule of *Aedes sierrensis.* Arrow shows possible gamete. (Courtesy of D. Sanders.) ×500.

FIGURE 168. Mature gamontocyst of *Lankesteria clarki* containing spores. (Courtesy of D. Sanders.) ×650.

FIGURE 169. Mature spores of *Lankesteria clarki* released from the gamontocyst. (Courtesy of D. Sanders.) ×1000.

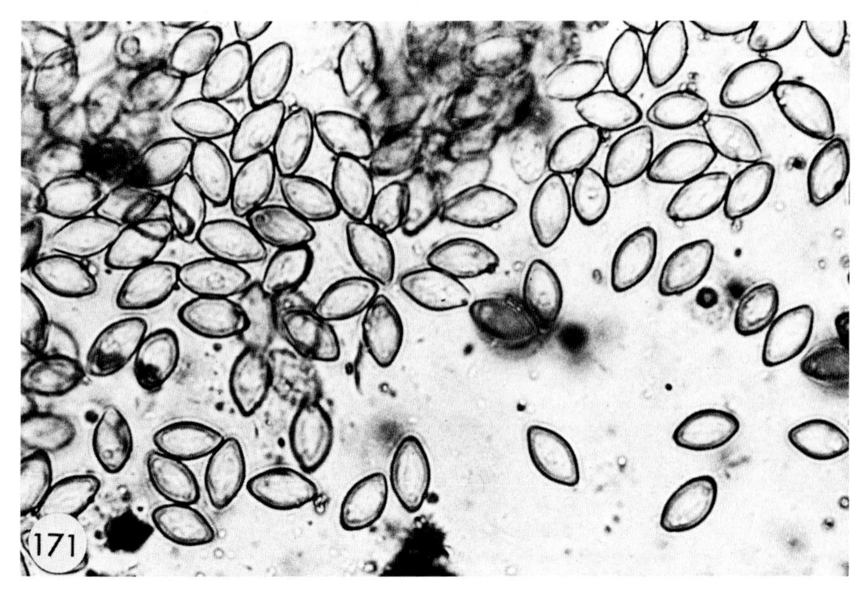

FIGURE 171. Mature spores of the neogregarinid *Mattesia* sp. from the wasp *Bathyplectes* sp. ×680.

FIGURE 172. Cysts of the coccid *Adelina* sp. releasing spores in the hemocoel of a beetle, *Trogoderma* sp. ×1000.

NEMATODES

INTRODUCTION

The literature includes over 3100 natural associations between insects and nematodes, involving 11 orders of nematodes and 19 orders of insects. If we eliminate phoretic nematodes, which use insects for transport, then we are left with eight orders of nematodes with representatives capable of parasitizing insects. The great majority of these parasites are found attacking beetles (Coleoptera) and flies (Diptera), although representatives in 17 other orders are also stricken. Nematodes attack both terrestrial and aquatic insects and occur in all hosts stages except the egg.

Insect parasitism has arisen independently in three separate nematode lines. The first, and most primitive from the evolutionary standpoint, comprises the bacterial-feeding microbotrophic rhabditoids, which gave rise to the gut-inhabiting Oxyurida and bacterial-carrying forms like *Neoaplectana* and *Heterorhabditis*.

The second line of insect parasites evolved from plant parasites in the orders Aphelenchida and Tylenchida. These gave rise to groups such as the Entaphelenchidae and Allantonematidae, which now use their stylets for entering insect hosts.

The third line evolved from predaceous members of the Dorylaimida and are represented by the Mermithidae and Tetradonematidae.

TAXONOMIC STATUS

The present trend regards the Nematoda as a separate phylum and divides this taxon into two classes, the Adenophorea and the Secernentea.

Roughly speaking, members of the Adenophorea are mostly free living, microbotrophic aquatic forms which are commonly encountered in marine and freshwater habitates. Representatives do occur in soil and other locations and there are a few parasitic groups. One order, the Mermithida, is strictly parasitic in invertebrates, including a wide range of insects.

In contrast, the class Secernentea has demonstrated an amazing genetic plasticity, which has resulted in the origin of many polyphyletic parasitic groups, involving not only insects, but plant and vertebrate hosts as well. There are relatively few aquatic representatives in this class, the majority of the free-living microbotrophic forms being terrestrial in habitat.

The following presentation of the higher insect nematode categories as discussed in this guide gives some idea of the relationship of one group to another:

- Class Adenophorea
  - Order Mermithida
    - Suborder Mermithina
      - Superfamily Mermithoidea
        - Family Mermithidae
        - Family Tetradonematidae

- Class Secernentea
  - Order Rhabditida
    - Suborder Rhabditina
      - Superfamily Diplogasteroidea
        - Family Diplogasteridae
      - Superfamily Rhabditoidea
        - Family Rhabditidae
        - Family Steinernematidae
        - Family Heterorhabditidae
  - Order Tylenchida
    - Suborder Hexatylina
      - Superfamily Neotylenchoidea
        - Family Neotylenchidae
      - Superfamily Allantonematoidea
        - Family Allantonematidae
        - Family Contortylenchidae
      - Superfamily Sphaerularioidea
        - Family Sphaerulariidae

Order Aphelenchida
Suborder Aphelenchina
Superfamily Aphelenchoididea
Family Aphelenchoididae
Family Entaphelenchidae
Family Parasitaphelenchidae
Order Strongylida
Order Ascaridida
Order Spirurida
Order Oxyurida

LIFE CYCLE

Although diverse in form and biology, the various groups of insect-parasitic nematodes are similar in regard to such characteristics as basic patterns of infection and growth stages.

The Mermithida constitute the only group of obligate monoxenous insect parasites in which oviposition occurs in an environment outside the host. In other groups such as the Entaphelenchidae, Allantonematidae, Neotylenchidae, Sphaerulariidae, and insect-parasitic Rhabditida, oviposition occurs inside the insect's body cavity.

Regarding the mode of infection, juveniles constitute the infective stage for mermithids, heterorhabditids, steinernematids, aphelenchoidids, and some oxyurids, even though these groups are for the most part widely separated taxonomically. In contrast, for the sphaerulariids, allantonematids, neotylenchids, and entaphelenchids, it is the fertilized female that enters the host. Aside from some highly specialized representatives of the Mermithida, only members of the Oxyurida initiate infection with the embryonated egg (in some oxyurids, the egg may first hatch and the second stage juvenile is the infective stage).

Most nematode groups are similar in releasing juveniles from the diseased host when the parasites are mature and prepared to enter the environment. This applies to entomogenous forms in the Rhabditida, Tylenchida, Aphelenchida, and Mermithida. The Oxyurida are unique in having the egg stage as both the release stage and the infective stage (in most cases).

In summary, the life cycle of insect-parasitic nematodes involves a growth phase inside the host and a developmental stage in the environment. The growth phase may last for only a relatively short period and involve only two juvenile stages (i.e., the Mermithida) or it may include oviposition and growth of the newly hatched juveniles to the fourth stage (i.e., the Allantonematidae). In some nematodes, there are two separate cycles inside the insect. Both cycles may be amphimictic, as in the case of *Neoaplectana,* or one cycle may be hermaphroditic and the other amphimictic, as in the case of *Heterorhabditis* and *Psyllotylenchus.*

CHARACTERISTICS OF INFECTED INSECTS

The evidence of insect parasitism by nematodes varies depending on the degree of damage inflicted by the parasite. In the most severe cases, involving *Neoaplectana* and *Heterorhabditis,* in which the insect dies after 48 hr, death is accompanied by a characteristic color imparted by the symbiotic bacteria. This may be a cream to gray color (*Neoaplectana* spp.) or a reddish color (*Heterorhabditis* spp.) (color plate, A). Less striking, but also indicative of an infection, are instances, most commonly noticed with the long-bodied Mermithidae, in which the nematodes either are seen through the body wall of transparent hosts or distort the body to a noticeable degree. Mermithids which infect ant and midge larvae and are carried over into the adult stage of the insect may cause external morphological modifications. Parasitized midges that are genetically males may have female antennae, wing venation, and front legs. Instead of this intersex effect, parasitized ants may exhibit intercastes, or individuals with characters typical of several castes. Often modified in ants as a result of mermithid parasitism are the sizes of the thorax, abdomen, head, eyes, and wings. The degree of modification depends on the time when infection occurs in the larval stage. Normally, the longer the parasite stays in the host, the greater the degree of modification.

The remaining groups of nematode parasites produce little, if any, outward sign of infection. If the behavior of the host insect is well known, then aberrant activity can be recognized as the result of nematode activity. Examples of this are differences in the swarming behavior of adult midges parasitized by mermithids, the loss of a nesting instinct in queen bumblebees infected with *Sphaerularia bombi,* and a difference in gallery structure in bark beetles attacked by allantonematids.

However, the behavior of host insects may not be known or may not be observed, and in these and many other cases dissection or the examination of feces may be the only methods of detection.

METHODS OF EXAMINATION

All insects suspected of harboring nematodes should be dissected in physiological saline (Ringer's solution or 0.9% NaCl) to avoid rupturing the nematode as a result of osmotic imbalance. The body contents should be examined under a dissecting microscope with transmitted light, which shows the nematodes better than direct reflected light. Various tissues should be examined, including the blood, alimentary tract, malpighian tubules, muscles, fat body, head glands, colleterial glands, and reproductive system. The free-living stages of entomogenous nematodes can be placed in water. Although it is often useful to examine freshly dissected material directly under the microscope by making temporary mounts in physiological saline, permanent mounts should be made for detailed examinations.

Specimens should be properly prepared for permanent mounts by carefully killing and fixing the material. To avoid distortion brought about by placing living specimens directly in fixative, the nematodes should be heat killed, a process which leaves them in a natural position. This is best done by adding hot (60°C) physiological saline to a small container with living specimens. After death, the nematodes should be transferred immediately to a fixative. We recommend TAF (97 ml 40% formalin, 2 ml triethanolamine, and 91 ml distilled water). If TAF is not available, then 3% formalin is adequate. Alcohol (70%) can be used in the absence of formalin. After a week, the specimens should be removed from the fixative and placed in a small petri dish, a square observation dish, or an equivalent vessel containing a solution of 70 ml 95% ethanol, 5 ml glycerin, and 25 ml water. This vessel is left partially covered at room temperature for one to two weeks depending on the humidity present in the room. This simple method involves the removal of water and alcohol by evaporation and the nematodes are left in glycerin. The specimens can now be mounted directly on microscope slides according to the following procedure:

1. Transfer the nematodes from the vessel to a drop of glycerin placed in the center of a microscope slide (use a small pointed instrument such as an insect pin or a dental pulp canal file).
2. Add three small supports (glass, wire, etc.) with a thickness equal to or slightly larger than that of the nematodes (to keep the cover slip from flattening the specimens).
3. Push the nematodes to the bottom of the glycerin drop and ar-

range them in the form of spokes radiating out from the center of the drop.
4. Add a cover slip and then carefully seal the outer rim with a ringing compound such as nail polish, wax, or a synthetic resin.
5. After 48 hr, examine for leaks and reseal if necessary.

ISOLATION AND CULTIVATION

Many of the facultative parasites, which have the ability to obtain nourishment in the host's environment, can be cultivated outside their host. The obligate parasites can normally only be "maintained" unless special axenic culture media are employed. The use of these specialized media is beyond the scope of most insect pathology laboratories.

Most facultative members of the Rhabditida feed on bacteria and can often be cultured on standard agar media seeded with bacteria. Sometimes the choice of bacteria is important, since certain forms are not acceptable to all nematodes. Using this method, continuous developmental cycles can be established, which incidentally should be transferred every two to three weeks. The case of bacterial feeding in *Neoaplectana* and *Heterorhabditis* nematodes is highly specialized, since in their insect hosts they feed on a breakdown product of the host's tissues together with cells of symbiotic bacteria of the genus *Xenorhabdus* which they carry into the insect. Some nematode development can be achieved by simply placing the *Xenorhabdus* bacteria on the surface of a nutrient agar plate. Continuous development can be achieved when a richer medium is used.

Facultative members of the Neotylenchidae (*Deladenus* spp.) and Aphelenchoididae (*Parasitaphelenchus* spp.) feed on fungi in the environment and can be maintained on agar plates seeded with fungi. In the case of *Deladenus* spp., the preferred fungus is a specialized form that is symbiotically associated with the insect host and commonly encountered in the host's environment. It is possible that other entomogenous nematodes require specialized fungi as well to continue their development.

IDENTIFICATION

Groups of entomogenous nematodes are always fairly specific in regard to their location in the host. Neoaplectanids, allantonematids, and mermithids always develop in the insect's body cavity. Oxyurids always develop in the host's alimentary tract. The location of the developmental stages can be used as an aid in identification. However, it should be remembered that in some groups (i.e., Allantonematidae, Neotylenchidae), whereas development occurs in the body cavity, juveniles often appear in the intestine or reproductive system as they leave the host. Also, in many forms (e.g., *Heterorhabditis, Neoaplectana,* Oxyurida) the males are usually much smaller than the females and may be overlooked.

The presence of bacteria belonging to the genus *Xenorhabdus* always implies an association with nematodes of the genera *Neoaplectana* or *Heterorhabditis,* and various forms can usually be found inside the insect host. Occasionally, the *Xenorhabdus* bacteria will be present without a trace of a living nematode. In these cases, it is probable that the nematode died soon after entering and liberating the bacteria.

The general size of a nematode is also a clue to its identity. Long, whitish nematodes that are several times the length of their host are probably mermithids (horsehair worms are also long and slender, but they are dark in color and belong to another phylum). Swollen, ovipositing females in the hemocoel of living insects are probably allantonematids. To determine whether the nematode is a juvenile or an adult, it is necessary to look for mature sexual characters, which are the vulval opening and ovaries (often with eggs) in the female and spicules (exsertable sclerotized structures) in the male.

Those nematodes enclosed in host tissue, cells, or capsules are for the most part heteroxenous parasites that utilize insects as intermediate hosts and vertebrates as definitive hosts. Insects ingest the eggs or first-stage juvenile nematodes and the parasites then develop to the third stage, which is infective for the vertebrate. Under normal conditions, these small parasites have little detrimental effect on the insect. A few specialized female nematodes evert their reproductive system through their vulva and use the inverted uterine cells to absorb nutrients directly from the host's hemolymph. There are various degrees of uterine development, and the height of this behavior is reached in *Sphaerularia bombi,* where the enlarged uterine sac dwarfs the original female nematode.

Members of the Tylenchida and Aphelenchida possess a stylet that is usually clearly developed in certain stages. In most of these groups, the stylet is very noticeable in the infective-stage female and less distinct in the juveniles.

One character that is rather difficult to observe is the presence of a metacorpus containing the openings of the ventral and dorsal pharyngeal glands. Since these openings are very small, oil immersion is required to verify their presence.

Insect-parasitic members of the Neotylenchidae undergo a plant-feeding cycle that is initiated by females morphologically different from those destined to attack insects. Most species belong to the genus *Deladenus* and are commonly found in members of the Siricidae. It may be difficult to distinguish between the swollen females of *Deladenus* and some of the Allantonematidae on the basis of morphology alone.

TESTING FOR INFECTIVITY

Nematodes are frequently associated with dead and dying insects and it is necessary to determine whether the nematodes are pathogenic. Koch's postulates should be followed to determine this. Similarly, it may be necessary to explore the host range of a given parasite. In both cases, the following procedures for attempting artificial infections can be followed.

If it is not known how the parasites enter their host in nature, then it is best to attempt infection by placing the insects and nematodes together in an infection chamber. One should simulate the natural surroundings as much as possible inside the chamber, although sometimes infection can be achieved by placing the organisms between two pieces of moist filter paper.

If it is known that infection is initiated by an oral route, then the nematodes can be fed or directly introduced into the oral cavity of the potential host. This method of microfeeding or per os inoculation is beneficial when only a few nematodes are available and care is needed to insure a high degree of survival.

A microinjector is required to hold the syringe for per os injections and a glass-tipped needle is best used since there is less chance of piercing the host's alimentary tract. A predetermined amount of inoculum can then be introduced into each insect for detailed studies.

Nematodes can also be directly inoculated into the body cavity of potential hosts. This is a good way to determine what, if any, host reaction may limit nematode development. When this method is employed, care should be taken to have a clean, preferably axenic inoculum in order that associated microbial agents do not interfere with the host's natural defense system or even overwhelm and destroy the host.

STORAGE

Unfortunately, nematodes do not form spores or long-lasting cysts that can retain their viability for extended periods. In general, however, in every life cycle of an entomogenous nematode there is one stage that is more durable and lasting then all other stages, and this is the logical stage to store. It usually is the free-living stage which is most resistant to external conditions. Thus, it is the adult or egg stage for mermithids, the third-stage juvenile for neoaplectanids and heterorhabditids, and the adult female for allantonematids and sphaerulariids.

In general, a longer shelf life is achieved when the nematodes are kept moist and maintained at a temperature of 5–15°C.

LITERATURE

The field of insect nematology is relatively recent. Some of the earlier publications in this field include the classical works of Hagmeier (1912), Bovien (1937), Glaser (1932), and later Christie (1950) and Rühm (1956).

More recently, there has been a renewed interest in entomogenous nematodes, initiated by the studies of Welch (1963, 1965). Host lists of natural associations between insects and nematodes were initiated by Rudolphi (1809), followed by Hope (1840), Diesing (1851), Assmus (1858), Van Zwaluwenberg (1928), La Rivers (1949), and Poinar (1975).

Keys for the identification of entomogenous nematodes have been provided by Poinar (1975, 1979). Many of these proceed to nematode genus, and the reader may wish to consult them after the group has been determined using the key in this guide.

Bibliographies of entomogenous nematodes have been compiled by Shepard (1974) and Poinar (1975). General accounts of the life patterns of various groups of insect nematodes have been presented by Poinar (1975, 1979, 1983).

KEY TO THE COMMON GROUPS OF NEMATODE PARASITES OF INSECTS

1. Nematodes occurring in the body cavity—**2**
1. Nematodes occurring in the alimentary tract—**13**

2. Adults and juveniles developing inside dead insects (Fig. 173); symbiotic bacteria (*Xenorhabdus* spp.—see Fig. 62 in Bacteria) present in host's hemolymph—**3**
2. Adults and/or juveniles developing inside living insects; symbiotic bacteria of the genus *Xenorhabdus* absent from the host's hemocoel—**4**

3. First generation composed of hermaphroditic females only (Fig. 174); following parasitic generation with both males (Fig. 175) and females; males with a bursa (Fig. 176); female tail narrows to a point (Fig. 177); insect hemolymph takes on a reddish gummy consistency owing to symbiotic bacteria (*Xenorhabdus luminescens*)—*Heterorhabditis* (Heterorhabdidae).
3. All generations with both sexes present (Fig. 178); males lacking a bursa (Figs. 179, 180); female tail blunt with a point at the tip (Figs. 181, 182); insect hemolymph takes on a creamy pasty consistency owing to symbiotic bacteria (*Xenorhabdus nematophilus*)—*Neoaplectana* (Steinernematidae).

4. Nematodes enclosed in host tissue (capsules) (Figs. 183, 184) or within host cells (malpighian tubule, muscle, or fat). With rare exception, development is halted when the third stage juvenile is reached—Representatives of the Spirurida, Ascaridida, and Strongylida (the former group is most frequently encountered in insects).
4. Nematodes found free in the hemocoel, not enclosed in host tissue—**5**

5. Nematodes (at the end of their growth phase in the insect) long and slender (Figs. 185–188) (normally over 3 mm in length and coiled up inside the host), white (rarely pink or green); usually less than ten per host; normally only juveniles inside the insect; insect dies soon after the nematodes exit and enter the environment—**6**
5. Nematodes usually less than 3 mm in length; usually over ten individuals in each host; insects remain living after nematodes leave and enter the environment—**7**

6. Nematodes usually greater than 1 cm in length; juvenile stage and rarely adult stage in insect occur in a range of aquatic and terrestrial insects—Mermithidae (Figs. 185–188) (consists of some 30 genera).
6. Nematodes less than 1 cm in length; adult forms develop in the host; mating and oviposition may occur but the eggs do not hatch inside the insect; common in lower Diptera (Sciaridae) and rarely in Coleoptera (Nitidulidae)—Tetradonematidae (consists of 5 genera).

7. Juveniles only present in living insects; usually less than 20 per host; little noticeable effect on host—**8**
7. Juveniles and swollen adult females present in the hemocoel of living insects (Fig. 201); usually more than 20 per host; may kill or debilitate hosts—**9**

8. Stylet absent; pharynx without a distinct median bulb (Fig. 189); occur in dung beetles and rarely bark beetles; when placed on nutrient agar, develop on bacteria (microbotrophic nematodes)—Rhabditidae (Figs. 189, 190) and Diplogasteridae (Fig. 191).
8. Stylet present (may be faint); pharynx with a distinct median bulb (Fig. 192); occur in bark beetles; when placed on nurtient agar, will develop on fungae—Aphelenchoididae (*Parasitaphelenchus*) (Figs. 192, 193).

9. Female nematode with an extruded reproductive system (Figs. 194, 195)—**10**
9. Female nematode without an extruded reproductive system—**11**

10. Extruded reproductive system forms a uterine sac that grows and may eventually dwarf the original nematode; found in queen bumblebees—*Sphaerularia* (Sphaerulariidae) (Figs. 194, 195) or bark beetles, *Sphaerulariopsis* (Allantonematidae).
10. Reproductive system only partially extruded or if totally extruded never greater in size than the original nematode; found in Diptera—*Tripius* (Sphaerulariidae) (Fig. 196).

11. Nematodes are facultative parasites with one stage capable of initiating a continuous fungal feeding cycle. Two morphologically different females present; commonly found in *Sirex* woodwasps—*Deladenus* (Neotylenchidae) (Figs. 197, 198).
11. Nematodes not facultative parasites, unable to maintain a separate cycle apart from the insect host; only one type of female present; found in a variety of insects, expecially members of the Coleoptera and Diptera—**12**

12. All stages, including parasitic females, with a median bulb containing the openings of the ventral and dorsal pharyngeal glands (Fig. 199); relatively rare—Entaphelenchidae (Figs. 199, 200) (consists of four genera).

12. Metacorpus lacking; dorsal pharyngeal gland opening posterior to the stylet; relatively common—Allantonematidae (Figs. 201–204) (consists of some 17 genera).

13. Only juveniles present, often exhibiting rapid movement; common in bark beetles (Scolytidae)—*Parasitorhabditis* (Rhabditidae) (Fig. 205).

13. Both juveniles and adults present; tail of female usually drawn out to form a long spine—Oxyurida (Fig. 206) (some 30 genera occur in insects).

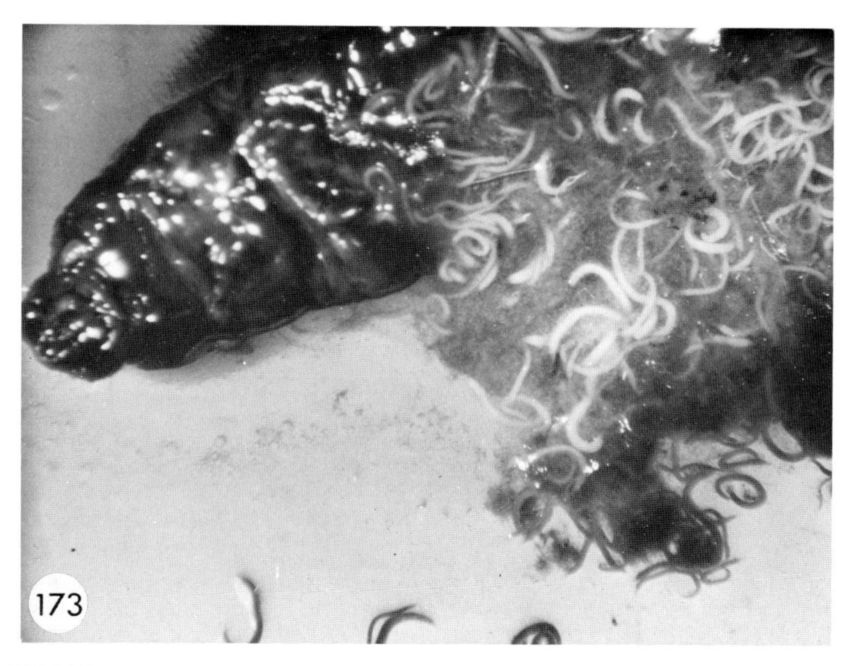

FIGURE 173. A dead wax moth larva (*Galleria mellonella*) partially opened to expose the males and females of the amphimicitic generation of *Heterorhabditis bacteriophora*.

FIGURE 174. A hermaphroditic female of *Heterorhabditis* sp.

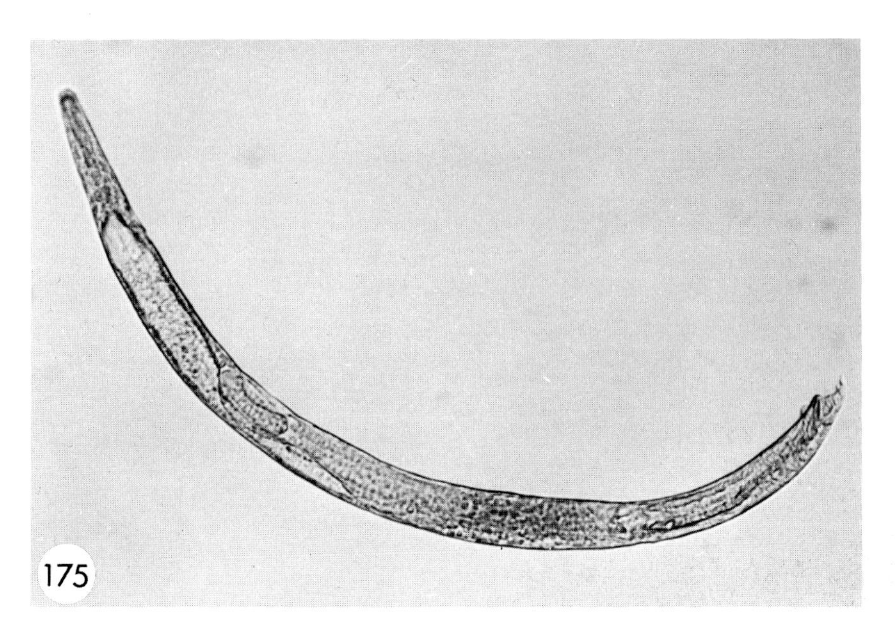

FIGURE 175. A male of *Heterorhabditis* sp.

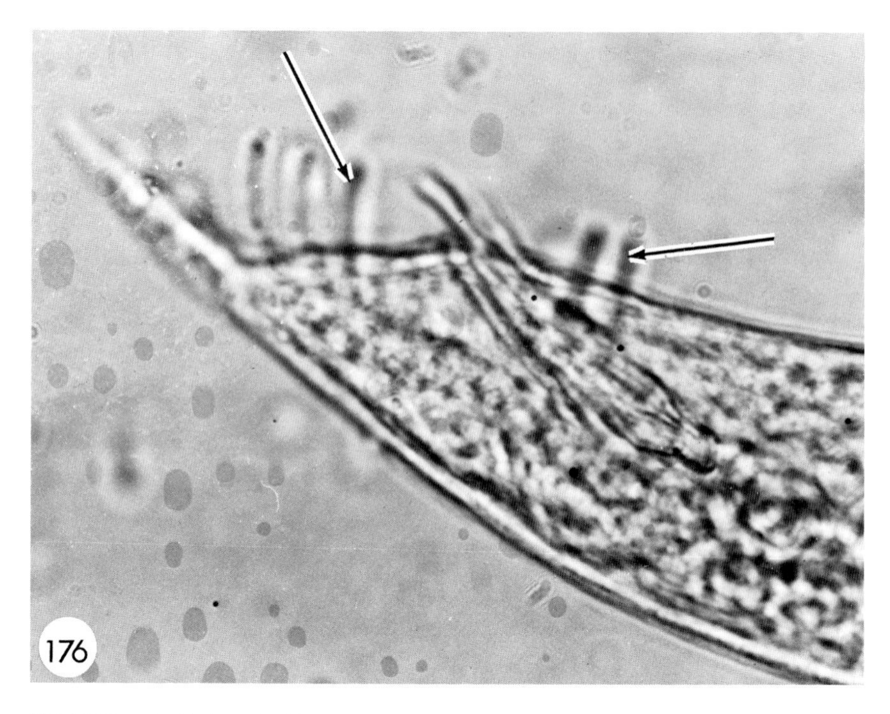

FIGURE 176. Male tail of *Heterorhabditis* sp. Note the straight spicules and bursal rays (arrows).

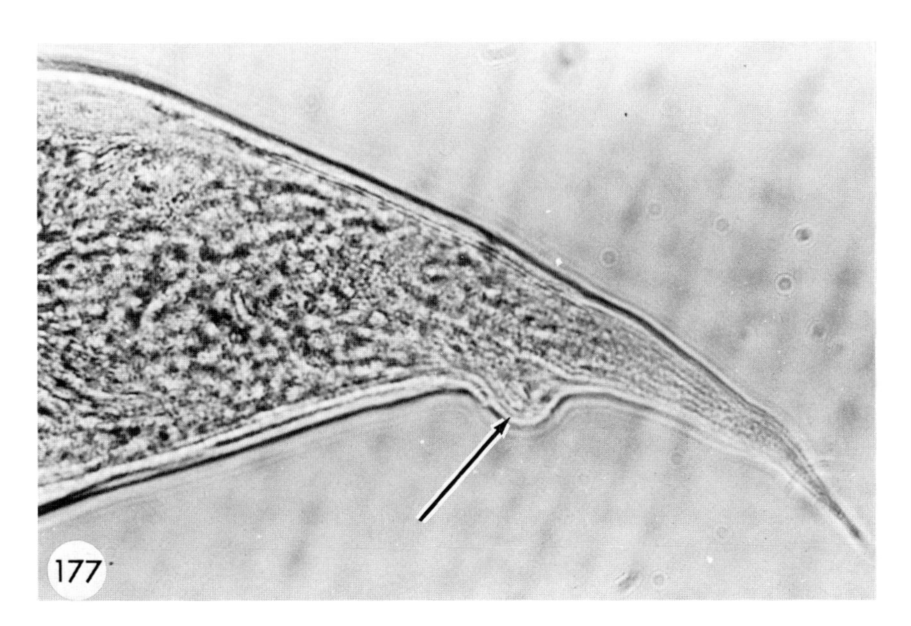

FIGURE 177. Female tail of *Heterorhabditis* sp. Note the anal swelling (arrow) and the attenuated tail.

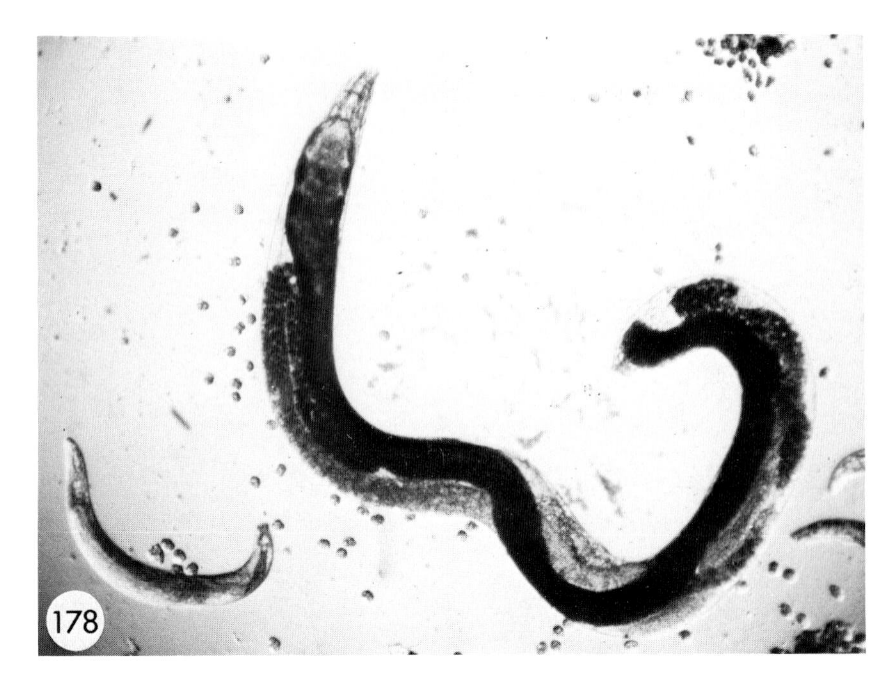

FIGURE 178. Larger female and smaller male of *Neoaplectana* sp. Note eggs in the background and in the female uterus.

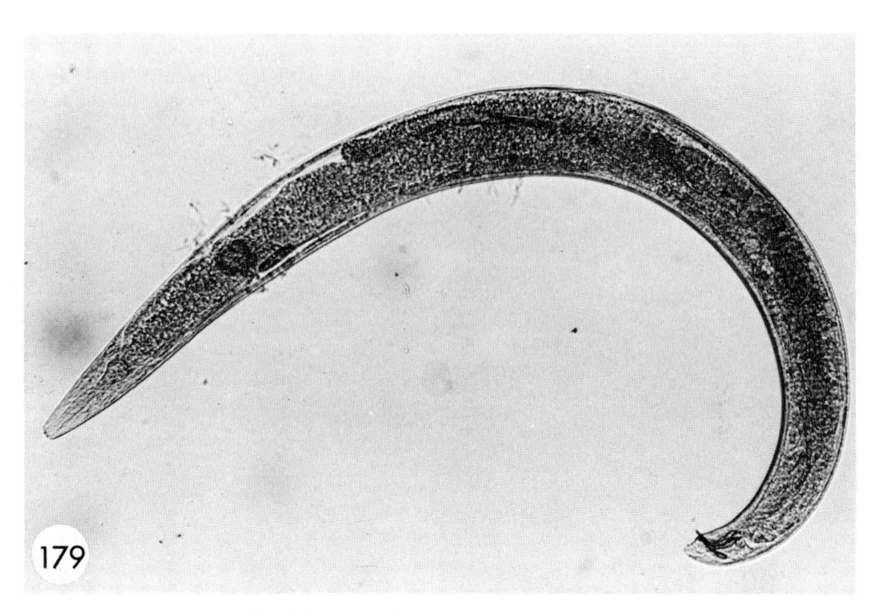

FIGURE 179. A male *Neoaplectana* sp.

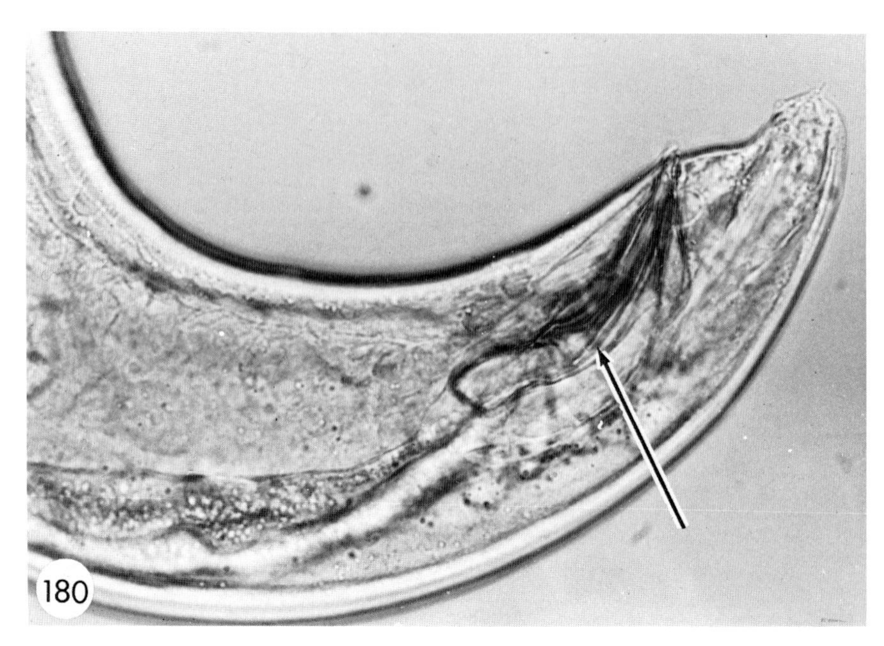

FIGURE 180. Male tail of *Neoaplectana* sp. showing curved spicules (arrow) and absence of bursal rays.

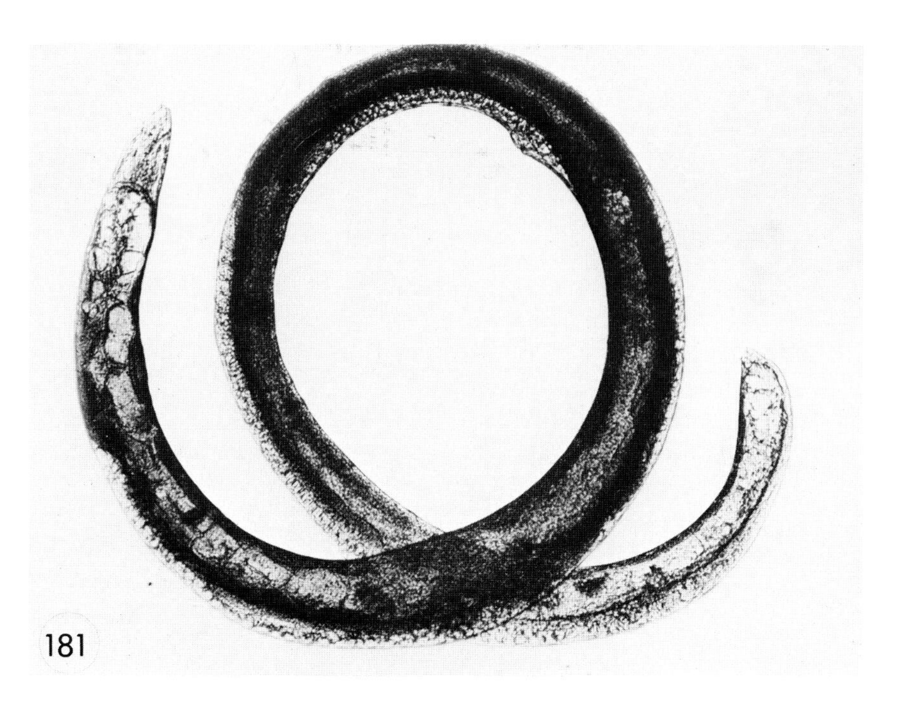

FIGURE 181. A giant first-generation female of *Neoaplectana* sp.

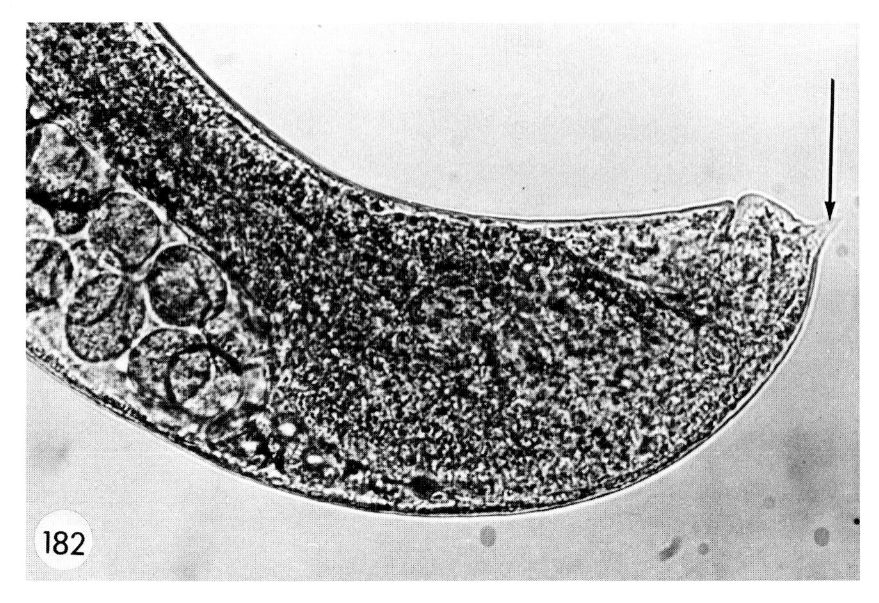

FIGURE 182. Female tail of *Neoaplectana* sp. Note pointed tail projection (arrow).

FIGURE 183. An encapsulated spirurid from the body cavity of a grasshopper.

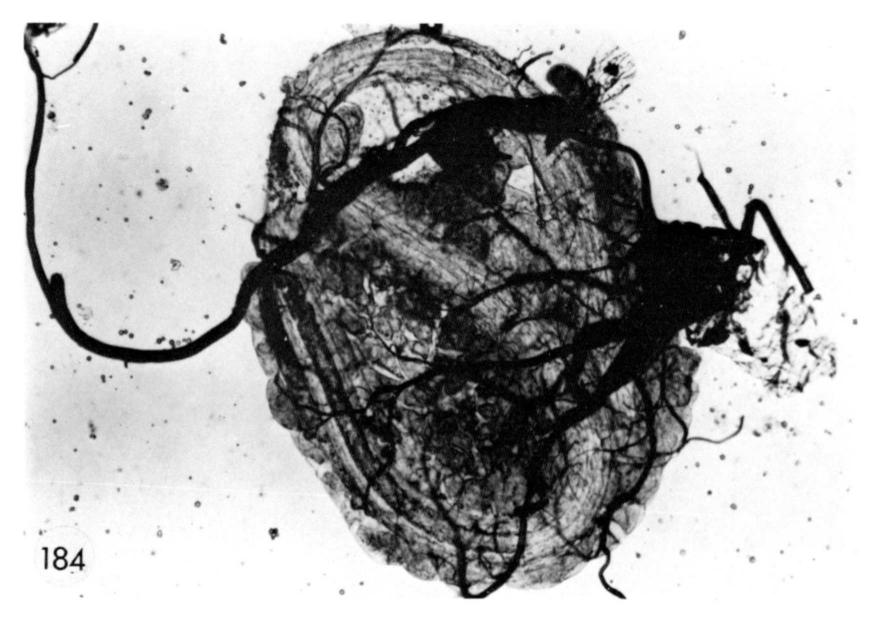

FIGURE 184. *Habronema muscae* (Spirurida) developing in the adipose tissue of a fly (*Musca* sp.). (Specimen from R. Merritt.)

FIGURE 185. A mermithid nematode (Mermithidae) emerging from a cricket.

FIGURE 186. A mermithid nematode (Mermithidae) emerging from a grasshopper. (Specimen from J. Evans.)

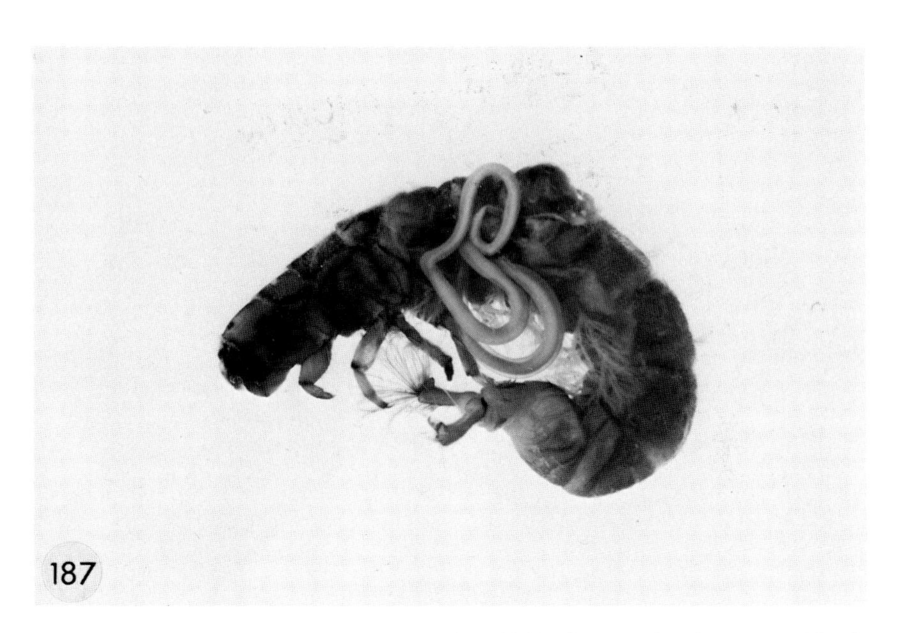

FIGURE 187. A mermithid nematode (Mermithidae) emerging from a caddis fly larva (Trichoptera). (Specimen from R. Merritt.)

FIGURE 188. An ant with a coiled-up mermithid nematode in its abdomen.

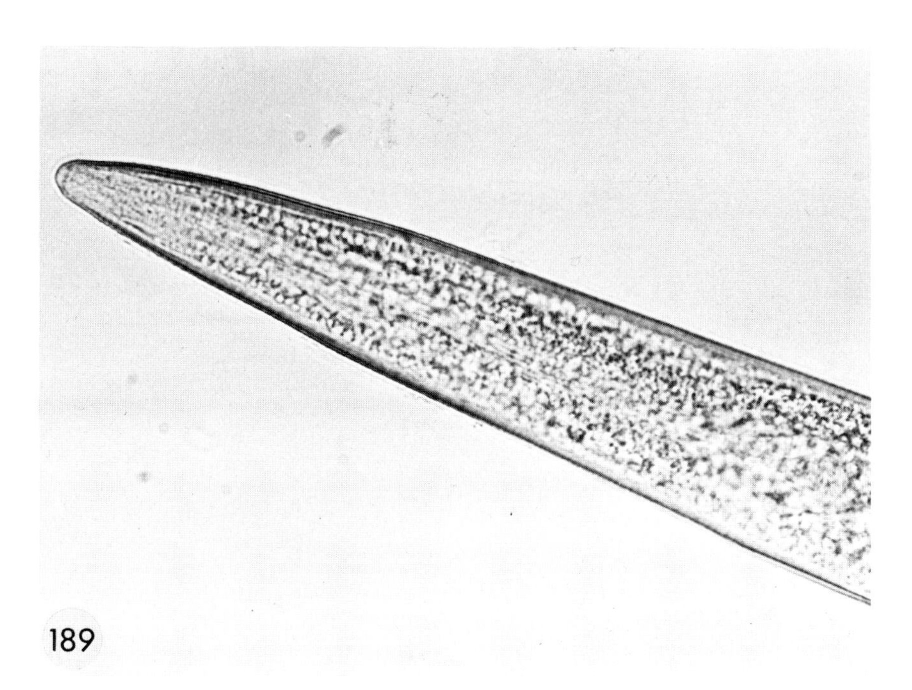

FIGURE 189. Anterior end of a *Parasitorhabditis* sp. (Rhabditidae) removed from the body cavity of a scolytid beetle, *Ips* sp. Note absence of a median bulb. (Specimen from O. Triggiani.)

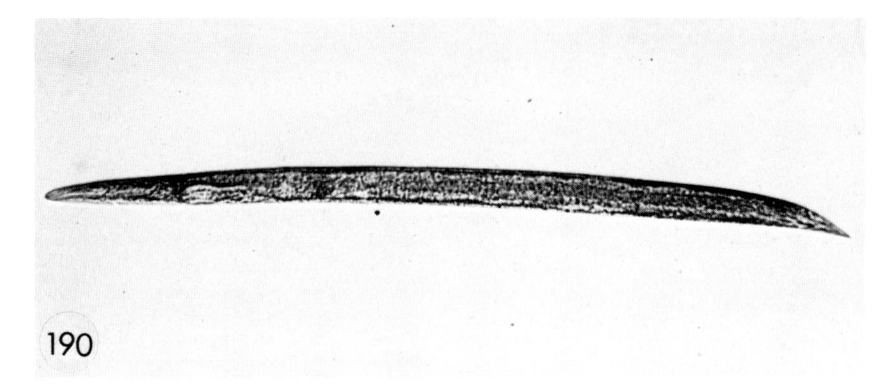

FIGURE 190. A juvenile *Parasitorhabditis* sp. (Rhabditidae) removed from the body cavity of a scolytid beetle, *Ips* sp. (Specimen from O. Triggiani.)

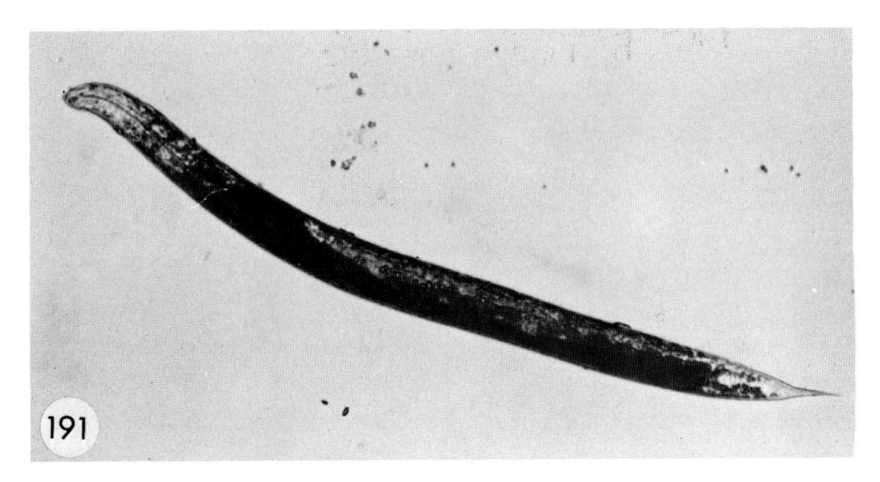

FIGURE 191. A juvenile of *Eudiplogaster aphodii* (Diplogasteridae) removed from the body cavity of *Aphodius fimetarius*.

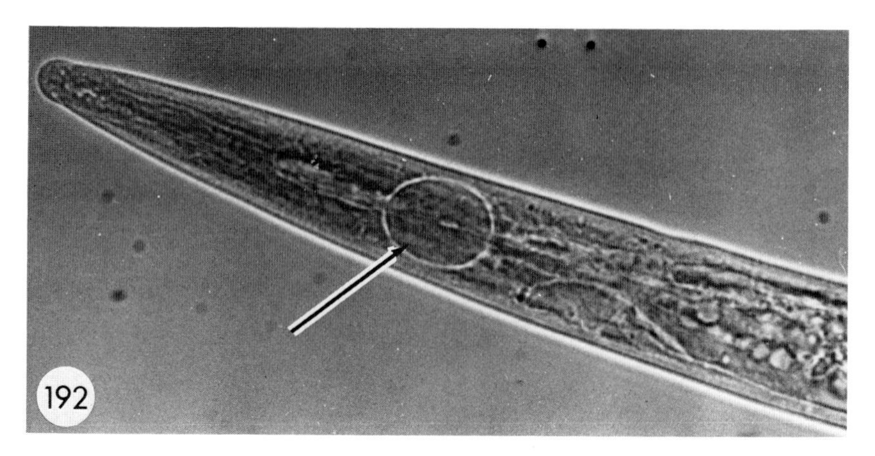

FIGURE 192. Anterior end of a *Parasitaphelenchus* sp. (Parasitaphelenchidae) removed from the body cavity of a scolytid beetle, *Ips* sp. Note presence of a median bulb (arrow). (Specimen from O. Triggiani.)

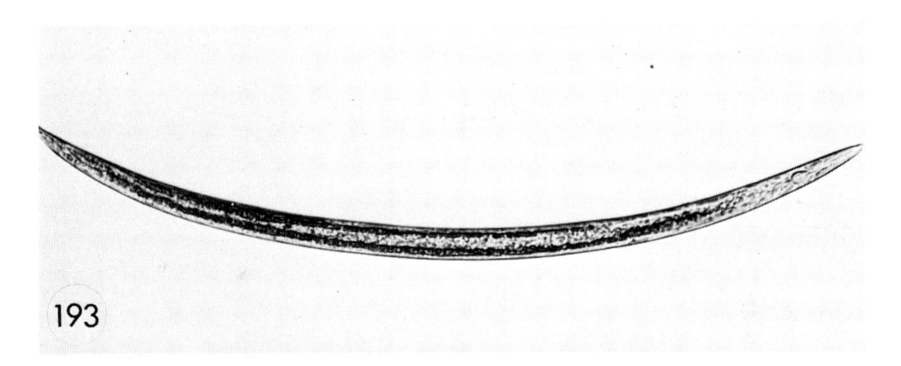

FIGURE 193. A juvenile of *Parasitaphelenchus* sp. (Parasitaphelenchidae) removed from the body cavity of a scolytid beetle, *Ips* sp. (Specimen from O. Triggiani.)

FIGURE 194. Uterine sacs (arrows) of *Sphaerularia bombi* (Sphaerulariidae) removed from the abdomen of infected *Bombus terrestris* bumblebee queens.

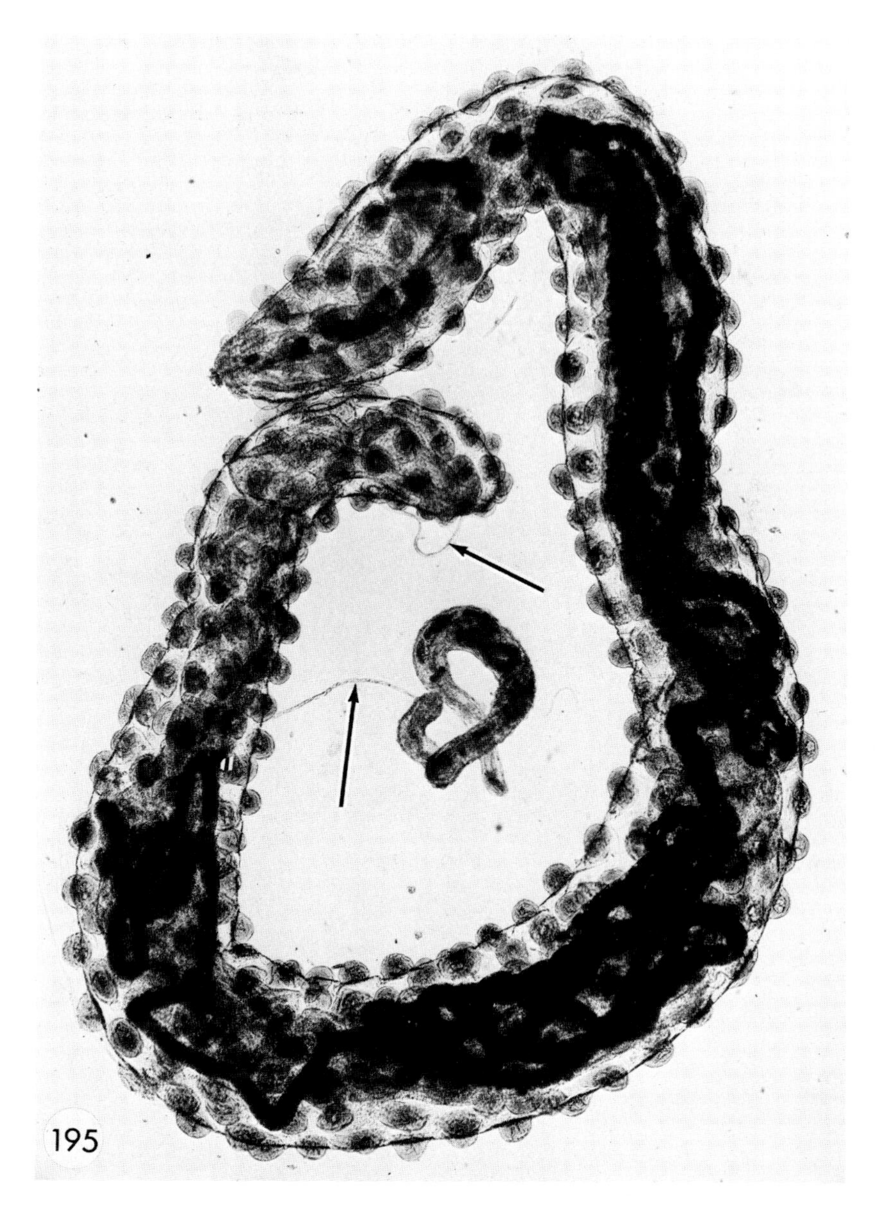

FIGURE 195. Two isolated specimens of *Sphaerularia bombi* (Sphaerulariidae) removed from the abdomen of infected *Bombus terrestris* bumblebee queens. Note attached bodies of original female nematode (arrows). The smaller (inner) specimen came from a heavily infected queen and the uterine sac did not expand as in the larger specimen.

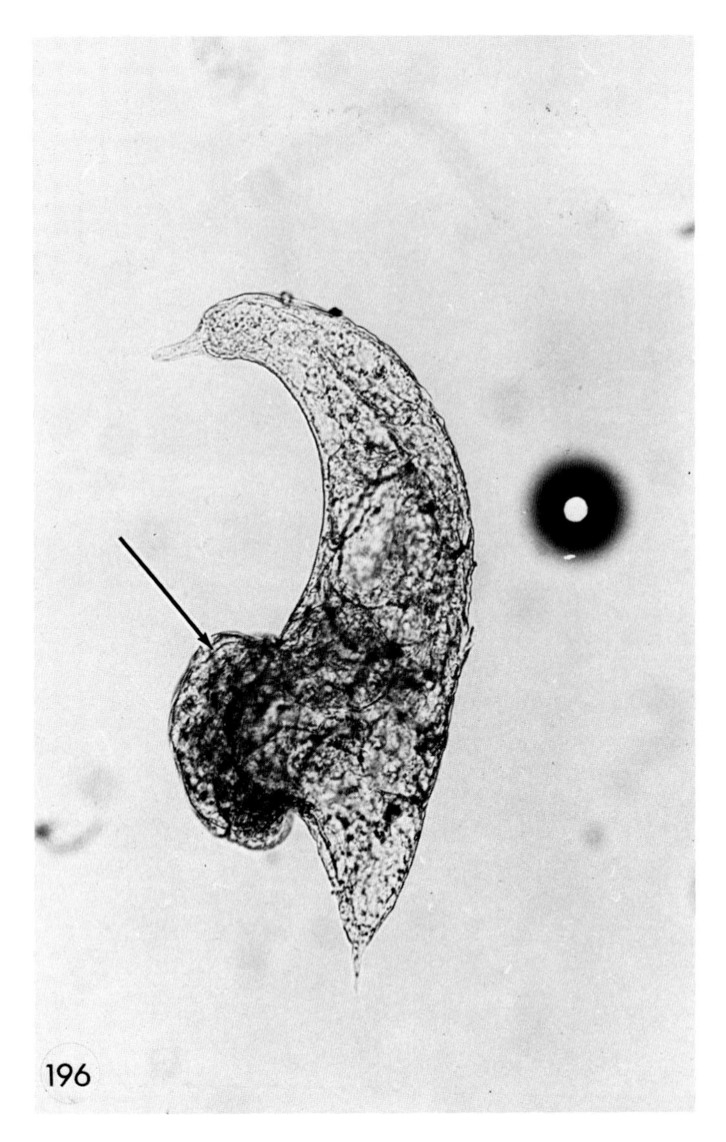

FIGURE 196. A mature female of *Tripius sciarae* (Sphaerulariidae) removed from the body cavity of a sciarid fly. Note the enlarged cells (arrow) of the partially extruded uterus emerging through the vulval opening.

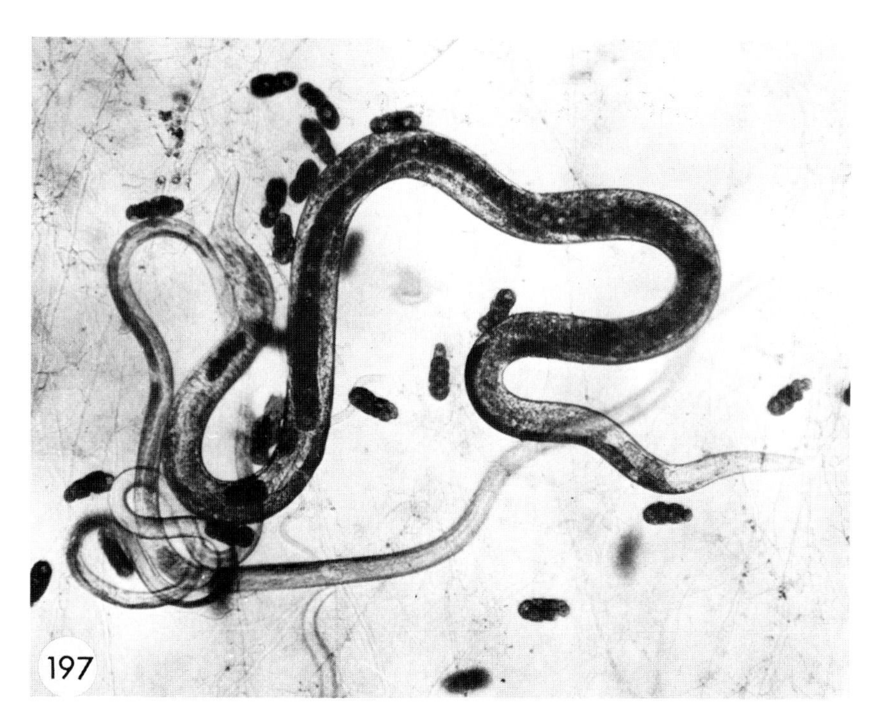

FIGURE 197. Mycetophagous free-living population of *Deladenus* sp. (Neotylenchidae) developing on their symbiotic fungus (*Amylostereum* sp.) on an agar plate. Note numerous cylindrical eggs.

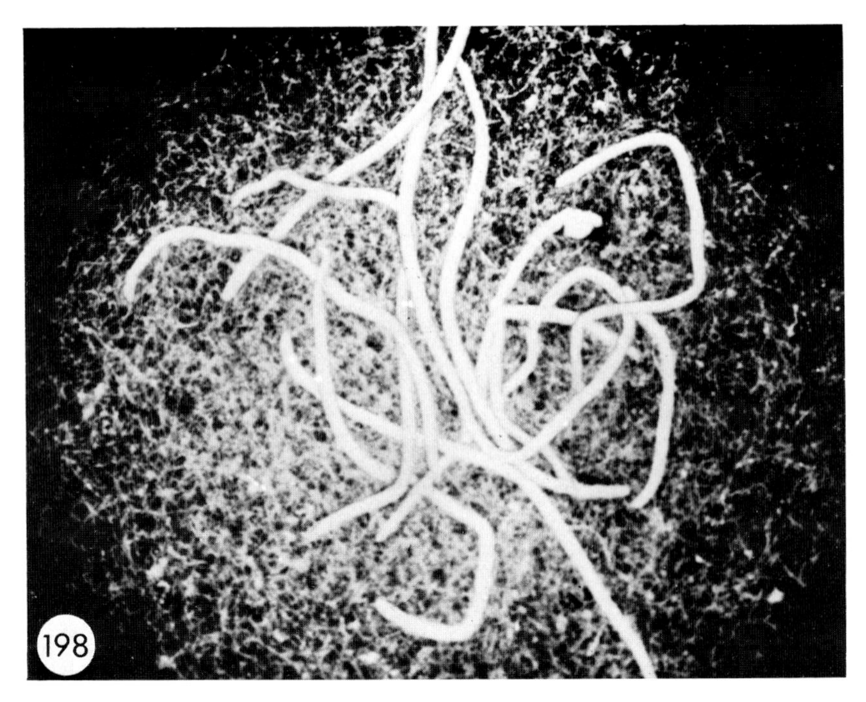

FIGURE 198. Parasitic adult females of *Deladenus siricidicola* removed from the body cavity of an adult *Sirex* wasp. (Photo courtesy of R. Bedding.)

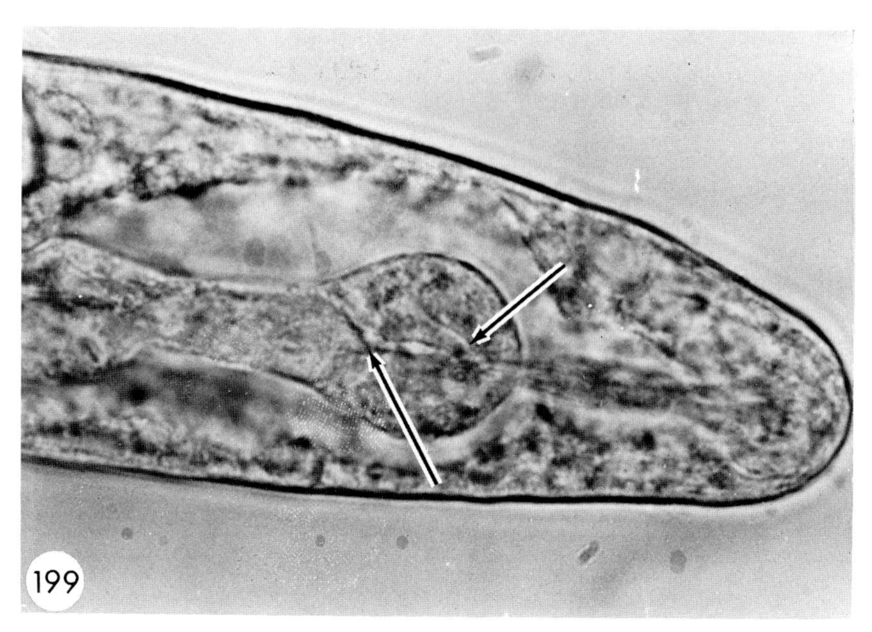

FIGURE 199. Head of *Praecocilenchus rhaphidophorus* (Entaphelenchidae) showing the enlarged median bulb containing the dorsal and ventral gland openings (arrows) in an arrangement typical of members of the Aphelenchida.

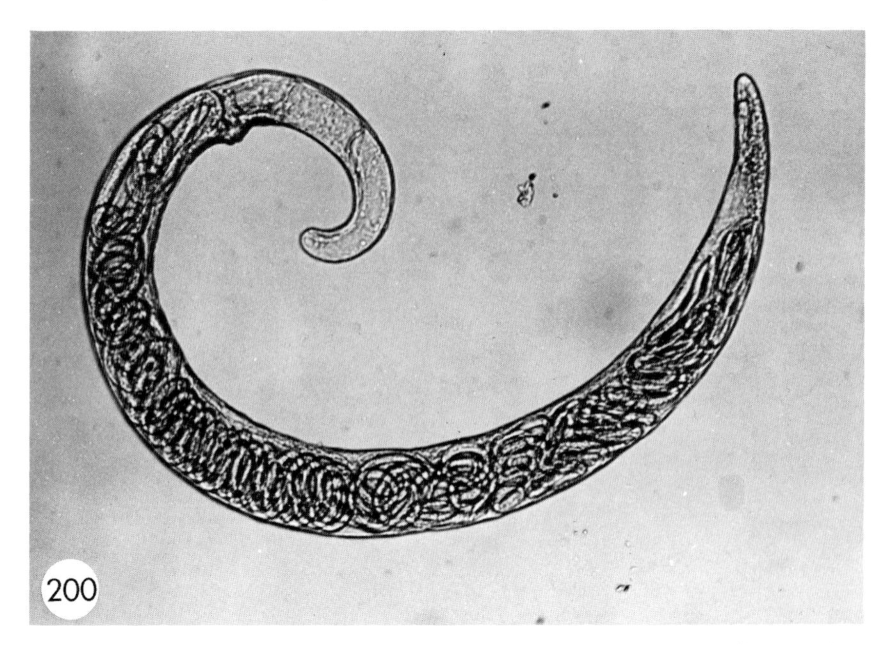

FIGURE 200. A mature female of *Praecocilenchus rhaphidophorus* (Entaphelenchidae) removed from the body cavity of the palm weevil, *Rhynchophorus ferrugineus.* The juvenile nematodes mature to the adult stage in the uterus of the female nematode.

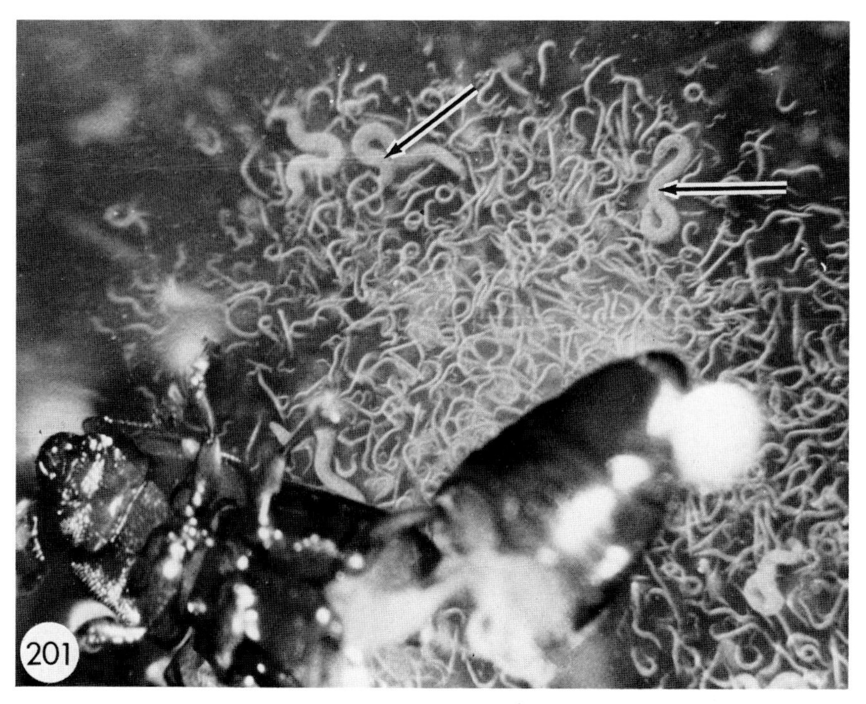

FIGURE 201. An infected *Carpophilus* beetle partially dissected to show the numerous juveniles and larger swollen adults (arrows) of *Howardula* sp. (Allantonematidae) in the body cavity. (Photo by J. Lindegren.)

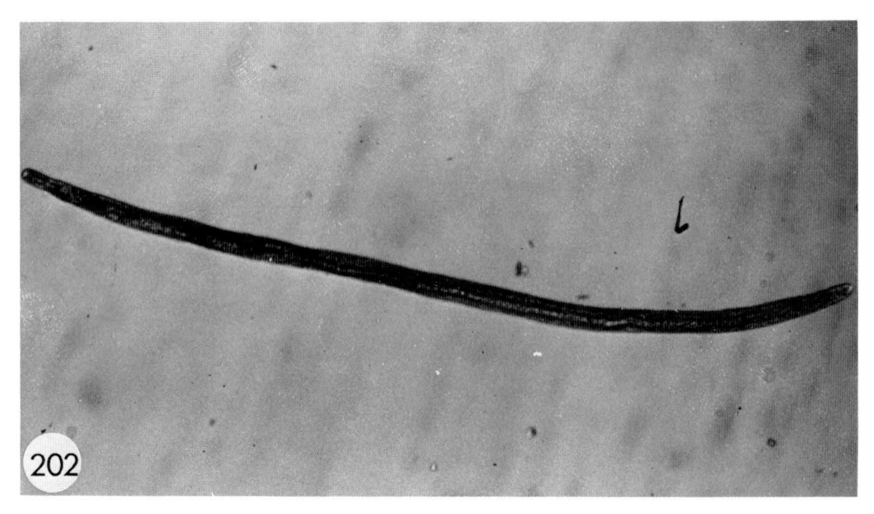

FIGURE 202. A mature female of *Contortylenchus brevicomi* (Allantonematidae) removed from the body cavity of the bark beetle, *Dendroctonus brevicomis*. Note elongate form different from many other Allantonematidae.

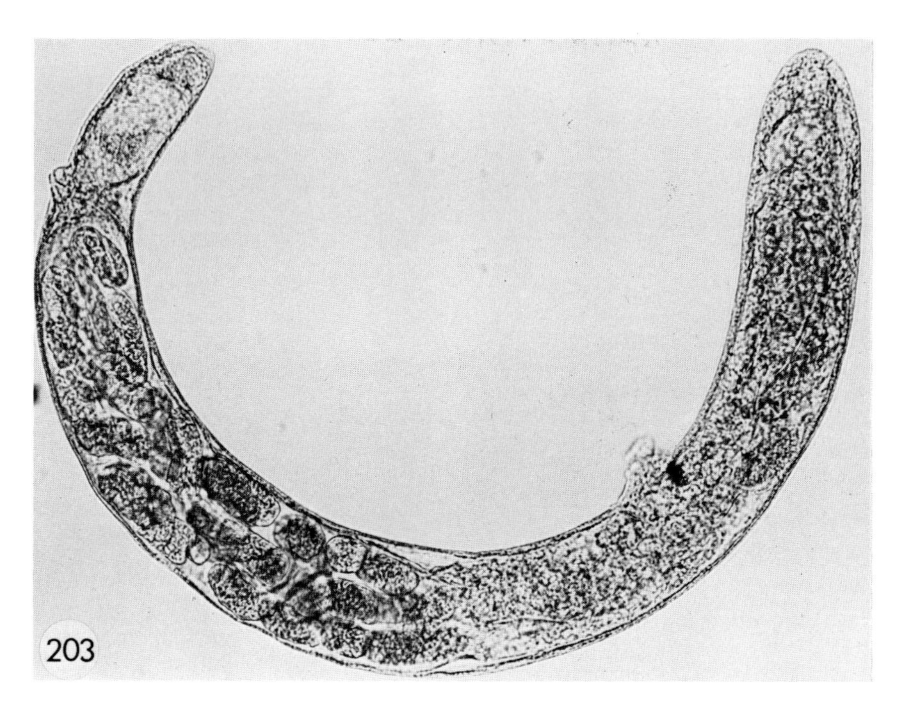

FIGURE 203. An amphimictic female of *Psyllotylenchus viviparus* (Allantonematidae) removed from the body cavity of an adult flea.

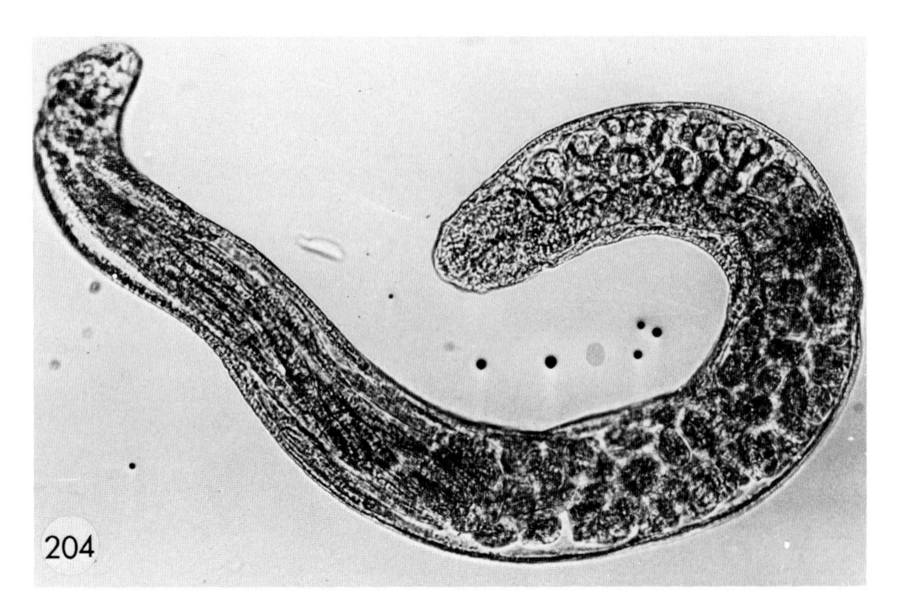

FIGURE 204. An amphimictic female of *Heterotylenchus autumnalis* (Allantonematidae) removed from the body cavity of the fly *Musca autumnalis.*

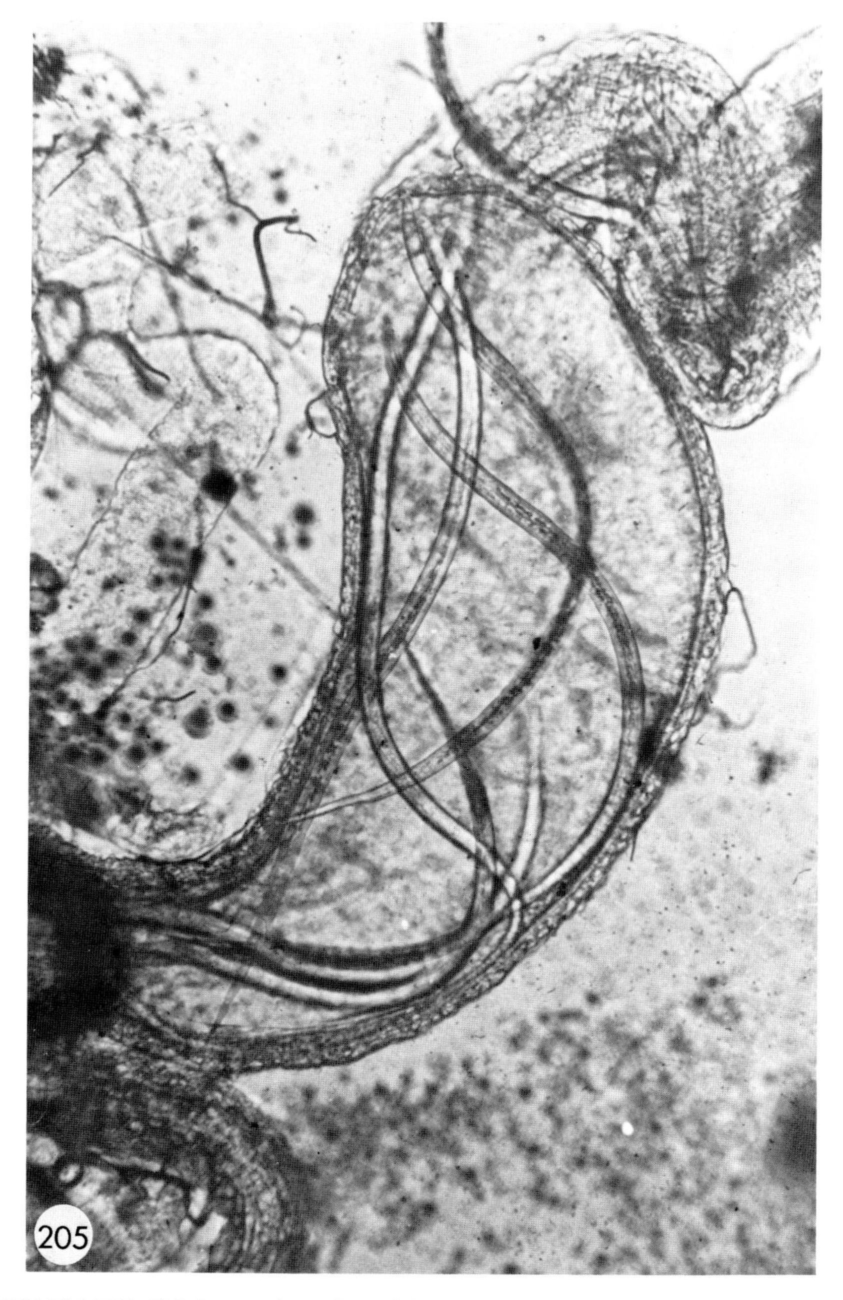

FIGURE 205. Third-stage juveniles of *Parasitorhabditis* sp. (Rhabditidae) inside the hindgut of an adult bark beetle (Scolytidae).

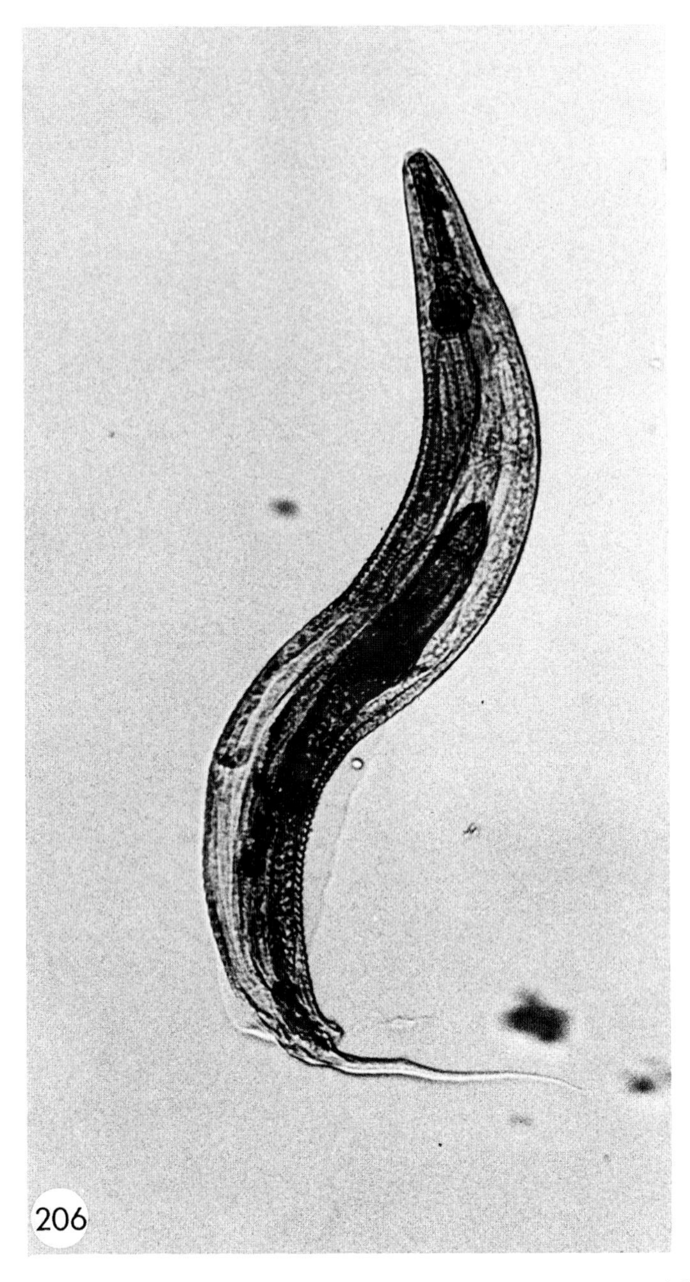

FIGURE 206. A male of *Thelastoma pterygoton* (Oxyurida) removed from the hindgut of a larva of *Oryctes monoceros* (Scarabaeidae).

IMMATURE STAGES OF INSECT ENDOPARASITES

INTRODUCTION

While dissecting insects, one may encounter the immature stages of insect parasites. These entomophagous (or entomogenous) insects belong to several orders and demonstrate a wide variation in habits and biology. They are sometimes called parasitoids rather than parasites because they develop at the expense of a single host, which is killed as a result of the attack. Insect parasites may feed and develop inside the insect host (endoparasite) or they may devour the host from the outside (ectoparasite). Only those of the former category will be discussed in this section. Both groups are extremely important regulators of insect populations. Many adult forms serve a predatory role by inflicting wounds and feeding on the hemolymph of insect pests or devouring the entire prey.

TAXONOMIC STATUS

Representatives of the orders Hymenoptera and Diptera are by far the most commonly encountered endoparasitic insects. Aside from these, one family of Coleoptera, the Ripiphoridae, are endoparasitic in cockroaches and hymenopterous larvae, and members of the order Strepsiptera occur in Orthoptera, Hemiptera, Homoptera, and Hymenoptera.

Among the parasites of agriculturally important insect pests, the Hymenoptera are probably the most commonly encountered. Unfortunately, there is no general agreement among experts regarding the systematic arrangement of the higher taxa, and the system followed here is taken from the catalog of Krombein *et al.* (1979). Using this classification, below are listed the major superfamilies and families of endoparasitic Hymenoptera:

1. Ichneumonoidea—Commonly occur in many plant pests, especially Lepidoptera. Includes:
   Braconidae   Ichneumonidae
   Aphidiidae   Stephanidae
   Hybrizontidae
2. Chalcidoidea—Commonly encountered in Lepidoptera, Diptera, Coleoptera, and Homoptera. Includes:
   Torymidae   Eupelmidae
   Pteromalidae   Encyrtidae
   Eurytomidae   Eulophidae
   Chalcididae   Mymaridae
   Leucospididae   Trichogrammatidae
   Eucharitidae
3. Cynipoidea—Attack members of the Siricidae and Aphidiidae (Hymenoptera), Diptera, and Homoptera. Includes:
   Ibaliidae   Eucoilidae
   Liopteridae   Alloxystidae
   Figitidae   Cynipidae
4. Evanioidea—Occur in egg capsules of Blattidae and in Hymenoptera and wood-boring Coleoptera. Includes:
   Evaniidae
   Aulacidae
   Gasteruptiidae
5. Pelecinoidea—Attack scarabaeid (Coleoptera) larvae. Includes the single family Pelecinidae.
6. Proctotrupoidea—Occur in insect eggs (Scelionidae), Diptera, Homoptera, Coleoptera, and Neuroptera. Includes:
   Vanhorniidae   Diapriidae
   Roproniidae   Scelionidae
   Heloridae   Platygasteridae
   Proctotrupidae

7. Ceraphronoidea—Most are hyperparasites of Homoptera and Diptera. Includes:
   - Ceraphronidae
   - Megaspilidae
8. Trigonaloidea—Occur in Hymenoptera and Diptera. Includes the single family Trigonalidae.

Representatives of the above groups attack a wide range of economically important plant pests and exert a profound influence on insect populations. There are many examples of classical biological control utilizing these parasites against introduced crop pests.

As with the Hymenoptera, there are many different classifications of the higher taxa of Diptera. The reference followed here is "A Catalogue of the Diptera of America North of Mexico" prepared under the direction of Stone *et al.* (1965).

The endoparasitic habit has not been recorded for members of the suborder Nematocera, but occurs in the Brachycera and is most common with the Tachinidae in the suborder Cyclorrhapha. Below are listed the families of endoparasitic Diptera attacking insects:

Suborder Brachycera
- Superfamily Asiloidea
  - Family Nemestrinidae—Occur in locusts (Orthoptera) and larvae of Coleoptera
  - Family Bombyliidae—Occur in immature stages of Lepidoptera, Coleoptera, and Diptera.

Suborder Cyclorrhapha
- Superfamily Phoroidea
  - Family Phoridae—Found in ants and bees (Hymenoptera), coccinellids (Coleoptera), Lepidoptera, Diptera, and crickets (Orthoptera)
- Superfamily Syrphoidea
  - Family Pipunculidae—Occur in Homoptera
  - Family Conopidae—Occur in bumblebees and wasps (Hymenoptera)
- Superfamily Tephritoidea
  - Family Pyrgotidae—Occur in scarabaeid beetles (Coleoptera)

Superfamily Oestroidea
Family Sarcophagidae—Occur in grasshoppers (Orthoptera)
Family Tachinidae—Common in Lepidoptera and Coleoptera
Unplaced family
Family Cryptochetidae—Occur in Homoptera

The majority of the parasitic Diptera are beneficial since they are primary parasites of plant-feeding insects. Members of the Tachinidae are most frequently encountered.

LIFE CYCLE

Hymenoptera

Most of the endoparasites pass the egg through their ovipositor into the body cavity of their host. Depending on the parasite, the egg may be deposited into the host egg, larva, pupa, or adult. Complete development in host eggs occurs in three families, the Trichogrammatidae, Mymaridae, and Scelionidae, and in the genus *Ooencyrtus* of the Encyrtidae and *Tetrastichus* of the Eulophidae. In all other cases of oviposition in the host egg, the parasite continues or passes most of its development in the larva or subsequent stages (egg–larva parasites).

The egg is normally placed free in the hemocoel and is therefore subject to host defense reactions, such as encapsulation. Some chalcidoidids have avoided this reaction by ovipositing in the brain or ganglion of the host. Other members of the above Encyrtidae (e.g., *Encyrtus*) attach the egg to the body wall of the host by a stalk fixed in the oviposition puncture (Fig. 217). Rarely, the egg may be stuck in a puncture made in the host's body wall or stuck on the host's cuticle. After hatching, the parasites burrow through the body wall and enter the hemocoel.

In other instances, the egg may be deposited away from the host but encountered upon ingestion. Members of the family Trigonalidae deposit their eggs on leaves, where they are eaten by the hosts. After hatching in the host's intestine, the young parasite larvae bore through the gut and enter the hemocoel.

The egg may hatch apart from the host. In these cases, the first instar parasite larva is usually different from the remainder (larval hypermetamorphosis), being modified to search out and enter the host. Such a primary larva is called a triungulin or planidium, depending on morphological characteristics. Triungulin larvae possess true segmented legs for locomotion, whereas planidia utilize body movement and elongate setae to reach their hosts. The body wall of these active forms is usually heavily sclerotized. After reaching their host and undergoing the first molt, these larvae transform into more typical grublike larvae characteristic of later instars. Not only representatives of the Hymenoptera possess these active larvae, but also members of the Diptera, Strepsiptera, and parasitic Coleoptera.

The great majority of endoparasitic Hymenoptera are monoembryonic (each egg develops into a single individual). However, the phenomenon of polyembryony occurs in the families Encyrtidae (*Ageniaspis, Litomastix,* and *Copidosoma*), Braconidae (*Macrocentrus*), Platygasteridae (*Platygaster*), and Dryinidae (*Aphelopus*). In polyembryony, the original egg divides and forms many more embryos (up to 1500 in *Litomastix*), with the result that a single oviposition can result in a multitude of emerging parasites.

Host nutrients may be ingested or taken up through the integument of developing larvae. In some forms, cells of the embryonic membrane break off and float free in the host's hemocoel. These trophic cells enlarge and serve as nourishment to the developing larvae.

In general, there are five larval instars among the endoparasitic Hymenoptera; however, there can be as few as three or as many as nine.

In many hymenopterous larvae, respiration occurs by diffusion through the integument since the tracheal system is not yet functional. Some forms have developed elaborate mechanisms of obtaining air from the outside or through the host's respiratory system. Some encyrtids form an egg-stalk which bears air tubes and extends through the host's integument.

After feeding has terminated, the larva enters a prepupal state. During this stage, the hindgut (proctodaeum) of the intestine joins the midgut (mesenteron), which up to now has been separate. When this function is completed the waste material stored up during the larval period is voided. This excretion is called the meconium and can often be found adjacent to the pupae and last larval skin.

Most hymenopterous endoparasites pupate within the host remains. This may occur inside the host cadaver or in the host cocoon or the host mines. The hymenopterous pupa is exarate and in most cases occurs in its own cocoon spun by the final instar larva.

Diptera

In contrast to the Hymenoptera, only a minority of the endoparasitic Diptera deposit their eggs directly through the host's integument into the hemocoel. The majority oviposit on the host's integument, on the host's food source, or in the vicinity of the host. The egg period is relatively short (hatching usually occurs within 48 hr) since most embryonic development occurs before the egg is deposited.

The first-stage larva is often modified for entering the host. Two larval types which differ from the normal maggotlike form are the microtype larva, which originates from microtype eggs deposited on plant surfaces, and the planidium larva. The latter occurs among the Bombyliidae, Nemestrinidae, Tachinidae, and Sarcophagidae, which deposit their eggs in the vicinity of the host. The active planidia frequently bear spines, scales, plates, or setae. They are extremely mobile and capable of passing an extended period without nourishment. After entering the body cavity of their host, the planidia molt and assume a more normal fly larva appearance.

With the exception of *Cryptochetum,* all dipterous larvae possess mouth parts in the form of a hook or a pair of hooks which are used to rasp away at the host tissues. Some forms absorb host nutrients directly through their cuticle, rather than ingest material through the mouth. Most of the endoparasitic Diptera possess three larval instars; however, the number can vary between three and eight.

Contrary to most Hymenoptera, the majority of dipterous larvae possess a well defined tracheal system and must make contact with outside air during their development. This is usually done by perforating the host's integument or tracheal system and forming some type of tube or funnel, which maintains a connection with the parasite during its development in the host.

The great majority of these endoparasites belong to the Cyclorrhapha, which possess a coarctate pupa with the third instar larval skin hardening and forming the pupal shell (puparium).

Most flies pupate within the host cadaver or host cocoon. Some pupate in the soil and others in completely exposed areas.

Strepsiptera

Members of this order, which is presently in a state of taxonomic flux, have fascinated biologists with their sexual dimorphism and unique parasitic habits. Most of the females of this order are saclike in shape, lack legs, and remain in the body of their host. The males emerge, possess wings, and fly around in search of hosts bearing females. Mating occurs when the males alight on hosts parasitized by the females and copulate with the partially extruded females (Fig. 207).

The first-instar larvae are planidia, capable of rapid movement over the plant tissue as they search for hosts. After reaching the hemocoel of a host, larval development is normally internal, with only the males leaving the host. The seventh-stage larva exserts its cephalothorax through the host's intersegmental membranes, thus giving evidence of parasitism.

The ten plus families of this order attack insects belonging to the Orthoptera, Hemiptera, Homoptera, and Hymenoptera. Members of this group have been commonly called "stylops" and parasitized hosts were considered to be "stylopized."

Coleoptera

The single family Ripiphoridae contains representatives that are endoparasitic in insects. The first-instar larvae of this family are active triungulins which search for a host. It is usually only the first stage which passes a period within the host's body. The parasite usually exits at the beginning of the second stage and completes its development as an external feeder. Most of the larval stages have either rudimentary or reduced legs and can be recognized by this feature.

The host groups for these parasites include Orthoptera (Blattidae) and Hymenoptera (Vespidae, Andrenidae, Scoliidae, and Tiphiidae).

CHARACTERISTICS OF INFECTED INSECTS

It is usually not possible to determine the presence of parasites in their host until the host is dissected or the parasites have completed most or all of their development. At this time, the host may die and turn a characteristic shape or color. Frequently the host darkens as the developing parasites begin to show through the integument.

Some parasites, such as the Dryinidae (Bethyloidea), which attack leafhoppers (Cicadellidae), plant hoppers (Membracidae), and fulgorids (Fulgoridae), have larvae which protrude out of the host's body and can be easily recognized (Figs. 218, 219).

Caterpillars attacked by the polyembryonic *Litomastix* and similar forms demonstrate a characteristic distorted appearance from the numerous parasites developing inside (Fig. 229).

Melanotic spots on larvae may indicate an oviposition puncture or a respiratory funnel, and the eggs of some dipterous parasites may remain on the host's cuticle long after the larvae have penetrated, indicating possible parasitism.

Encyrtid parasitism of some scales can be determined by observing the dark stalk of the egg protruding from or stuck into the host's integument (Fig. 216).

Some parasites, e.g., Dryinidae, Strepsiptera, and others, may alter the host's pigmentation, size of appendages, and primary and secondary sexual characteristics. Parasites which are carried into the adult stage of the host may cause host intersexes.

METHODS OF EXAMINATION

Dissections of insects to search for internal parasites should be performed in Ringer's solution or tap water (some of the delicate forms will burst in distilled water).

An examination of the internal organs is best done in living specimens recently removed from their hosts. Fixation can be done using one of the larval fixatives, such as the Weaver and Thomas fixative (see Techniques).

For a detailed examination of the head structures, the specimen should be cleared with chloralphenol. The specimens can be placed in this solution right after fixation but should not be kept more than two to three days. After clearing, the larval parasites can be mounted in Faure mounting medium. Adding a few crystals of iodine to this solution will compensate for overclearing. The iodine will be deposited on the more chitinized areas and produce a better resolution.

Both direct and reflected light should be used when examining the immature stages of endoparasites.

IDENTIFICATION

Aside from first-instar triungulin larvae which have segmented legs, most endoparasitic insect larvae are devoid of legs or functional eyes. Each parasite has one immature stage (egg, larva, pupa) that is unique from other species and can be used for identification once it has been noted. Eggs with prolonged stalks that extend through the host's integument belong to members of the family Encyrtidae. Large, oblong eggs laid on the surface of the host probably belong to members of the Tachinidae (Fig. 238).

First-instar larvae of endoparasites may appear quite similar to later instars or they may differ in varying degrees. Striking differences between first and later instar larvae occur in the Mymaridae (Figs. 212, 213).

First-instar or primary larvae of the Platygasteridae, Scelionidae, and Dryinidae are often referred to as protopod larvae since they show little evidence of segmentation and possess rudimentary cephalic and thoracic segments.

Larvae which possess abdominal extensions have been called polypod larvae and occur in the Cynipoidea and Proctotrupidae. Active leg-bearing first-instar triungulin larvae, characteristic among others of the Ripiphoridae and Strepsiptera, are called oligopod larvae. Under the heading of apodous larvae are placed the maggotlike (muscoidiform) and grublike (hymenopteriform) larvae of many Diptera and Hymenoptera, respectively.

The above four types of first-instar larvae have been further broken down into subgroups and include triungulin, planidium, sacciform, teleaform, mymariform, cyclopiform, eucoiliform, mandibulate, microtype, muscoidiform, encyrtiform, hymenopteriform, chrysidiform, agriotypiform, vesiculate, and caudate larvae. Some larvae may possess characters of two or more of the above types. Further discussion and figures of these types are available in the works of Clausen (1940) and Hagen (1969).

First-instar larvae that possess large head capsules and falcate mandibles usually lose these characters after the first molt. The tracheal system also tends to develop in the second-instar larva. The spiracles become functional in the last instar of hymenopterous larvae.

For detailed differentiation, the head structures and the tracheal system (branches and spiracles) are used. Tubercles, cuticular or segmental extensions, setae, fleshy processes, and granulations are also useful in recognizing specific genera and species.

The head of representatives of the Chalcidoidea is generally very reduced and sometimes only the mandibles are recognizable. In contrast, the Ichneumonidae possess a more distinct head with recognizable appendages, although the antennae are only one-segmented and rest in a circular antennal socket.

The mandibles may be unarmed or possess from one to three teeth.

The tracheal system has also been used as a tool for classification. Although the mature larvae of most Ichneumonoidea, Chalcidoidea, and Cynipoidea have nine pairs of spiracles, the position and structure can be variable. Most of the ichneumonid larvae possess an accessory longitudinal tracheal commissure that runs from the first spiracle to the third spiracle. This character is used to separate them from the Braconidae and Chalcidoidea. Using the spiracular plate and slits on the puparia of many tachinids and sarcophagids, a species identification may be possible.

Attempts should be made to identify the host insect, since many species of endoparasites are specific to one host or host group. For example, the only dipterous parasites of bumblebees are members of the family Conopidae. Adult holometabolous insects are never attacked by members of the Ichneumonidae, Chalcidoidea, Proctotrupoidea, or Cynipoidea.

ISOLATION AND TESTING FOR INFECTIVITY

When an exact identification of a parasite is required, it will be necessary to obtain the adult stage. This is best done by maintaining the host under laboratory conditions until the parasitic larva has pupated and the adult has emerged. This usually requires some degree of maintenance of the plant species upon which the host is feeding.

During the rearing process, maintaining the correct balance between temperature and humidity is very important. If conditions become too dry, the parasite will desiccate, and if they are too moist, it will be attacked by fungi.

Usually, if the parasite has already formed a cocoon or pupated inside the host remains, the internal humidity will generally be maintained unless the specimen is subjected to extremes of temperature or humidity. In the case of scales and other hosts that maintain a close association with the plant surface, the humidity will have to be raised to compensate for the reduction of water when the plant stems are cut and transferred to the laboratory.

If several hosts occur in a sample from which parasites have emerged, infectivity tests may be desired in order to determine the original host of the emerged parasites. These can be conducted by confining parasites together with cultures of separate host species and observing whether oviposition and development occur. It should be remembered that many parasites will oviposit in abnormal hosts if their normal hosts are not available. There may or may not be development in these abnormal hosts, so development in a confined host under insectary conditions does not mean that such development occurs in nature. In some instances, it may be prudent to expose the parasite to all the species that were in the original source and then observe whether a host preference is demonstrated.

STORAGE

Adult parasites can be maintained for longer periods under cool temperatures (60°F). With Hymenoptera, care should be taken not to lower the temperature below 55°F, since this may deactivate the sperm in the female's spermatheca and result in all-male progeny.

Maintenance at cool temperatures should be broken from time to time by returning the specimen to room temperature and giving it water and food (honey serves very well for most adult parasites).

Advantage should be taken of any natural diapause in the parasite or its host, since this is a period during which the parasite larva or pupa can be kept for longer periods.

Microtype eggs that are deposited in the host's environment and are ingested will often keep for considerable periods under cool (60°F) conditions.

LITERATURE

There are a considerable number of publications dealing with the biology and larval development of parasitic insects. Two of the best general works in this field include Clausen's *Entomophagous Insects* (1940) and Hagen's "Developmental Stages of Parasites" (1969). A good discussion of larval characters and many important references are given by Finlayson and Hagen (1979). Structures found in the larval head of parasitic Hymenoptera are discussed by Vance and Smith (1933). A general account of parasitic insects is presented by Askew (1971).

KEY TO THE HIGHER CATEGORIES OF COMMON ENDOPARASITIC INSECT LARVAE

While the first-instar larvae often show more characteristics than the subsequent immature stages, they are minute and it is the latter stages that are more routinely encountered. The larval characters given in the following key generally apply to the more mature larvae (last two stages).

1. Last larval stage and pupa with their head and/or prothorax exposed, protruding from the inside of the host's body—Strepsiptera (Figs. 207–210).
1. Larva and pupae without their head and/or prothorax protruding from the host—**2**

2. Body usually maggotlike; usually with paired mouth hooks; often a pair of conspicuous spiracles located on the posterior abdominal segment (Diptera)—**11**
2. Body usually grublike; mouth parts composed of a pair of laterally opposing mandibles—**3**

3. Legs (may be reduced) present—Ripiphoridae.
3. Legs absent (Hymenoptera)—**4**

4. Parasites developing completely in the host egg—**5**
4. Parasites developing and emerging from the host juvenile (larva) and/or pupa; may oviposit in the host egg—**6**

5. First-stage larvae saclike, lacking segmentation and a tracheal system; subsequent stages similar—Trichogrammatidae (Fig. 211).
5. First-stage larvae often with a distinct tail and head region; usually with cuticular setae; subsequent stages saclike and nondescript—Mymaridae (Figs. 212–215) and Scelionidae.

6. Tail of larvae modified as a tracheal extension which is attached to the egg corion of the host—Some Encyrtidae (Figs. 216, 217).
6. Tail of larvae not modified as a tracheal extension which is attached to the egg corion of the host—**7**

7. Mature larva protrudes from the host's body within a sac formed from its own exuviae; commonly found in Homoptera—Dryinidae (Figs. 218, 219).
7. Developing larva remains within confines of host's body—**8**

8. Head capsule reduced, often indistinct with only the mandibles visible; body segments may also be indistinct—Most Chalcidoidea (Figs. 220–229) and Proctotrupoidea.

8. Head capsule normally well formed, distinct; body segments usually distinct (Ichneumonoidea)—**9**

9. Anal opening located ventrally at the base of the tail; last segment of larval tail generally prolonged (caudate)—Aphidiidae (Fig. 230).

9. Anal opening located dorsally, larvae caudate or grublike—**10**

10. Larvae mostly caudate or, if grublike, then often with an anal vesicle; accessory longitudinal tracheal commissure absent—Braconidae (Figs. 231–233).

10. Larvae mostly grublike or spindle-shaped, anal vesicle absent; accessory longitudinal tracheal commissure present—Ichneumonidae (Fig. 234).

11. Segments indistinct; mouth parts reduced—Pipunculidae.

11. Segments distinct; mouth parts normal—**12**

12. Body pear-shaped—**13**

12. Body maggot-shaped, widest in the middle—**14**

13. Integument covered with minute setae or tubercles—Conopidae.

13. Integument smooth—Pyrgotidae.

14. Body tapering anteriorly; integument covered with minute spines—Sarcophagidae.

14. Body variable; integument usually with rings of setae on each segment—Tachinidae (Figs. 235–238).

FIGURE 207. The strepsipteran *Eurystylops sierrensis* (arrow) partially pulled out of the abdomen of its bee host, *Dufourea trochantera.* (Specimen courtesy of R. M. Bohart.)

FIGURE 208. An adult female of the wasp *Polistes fuscatus* with the heads (arrows) of the strepsipteran *Xenos peckii* protruding between the abdominal tergites. (Specimen courtesy of R. M. Bohart.)

FIGURE 209. Enlargement of Fig. 208 showing the female heads (arrows) of the strepsipteran *Xenos peckii* protruding between the abdominal tergites of the wasp *Polistes fuscatus*. (Specimen courtesy of R. M. Bohart.)

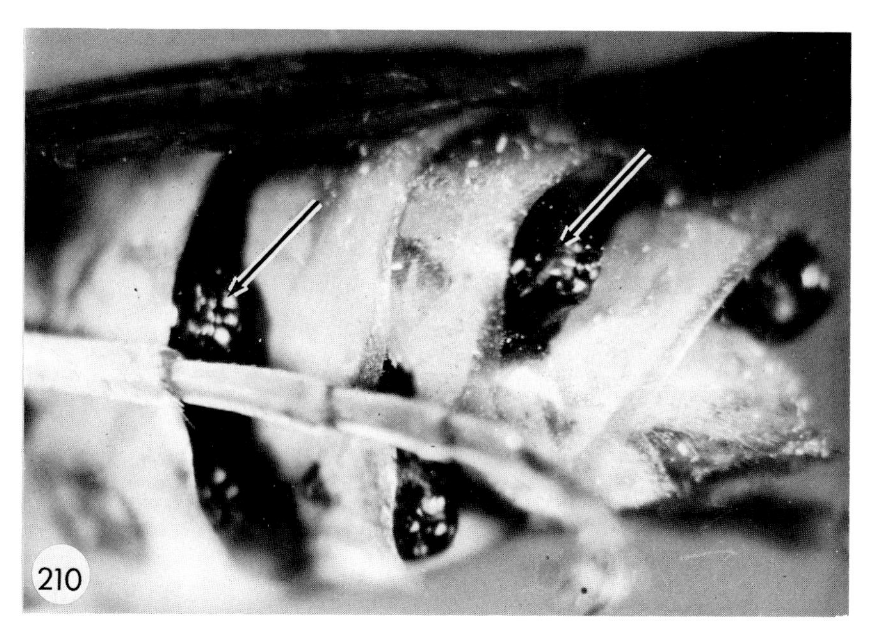

FIGURE 210. Protruding male heads (arrows) of the strepsipteran *Xenos peckii* in the abdomen of the wasp *Polistes fuscatus*. Note how much more massive they are than the females in Fig. 209. (Specimen courtesy of R. M. Bohart.)

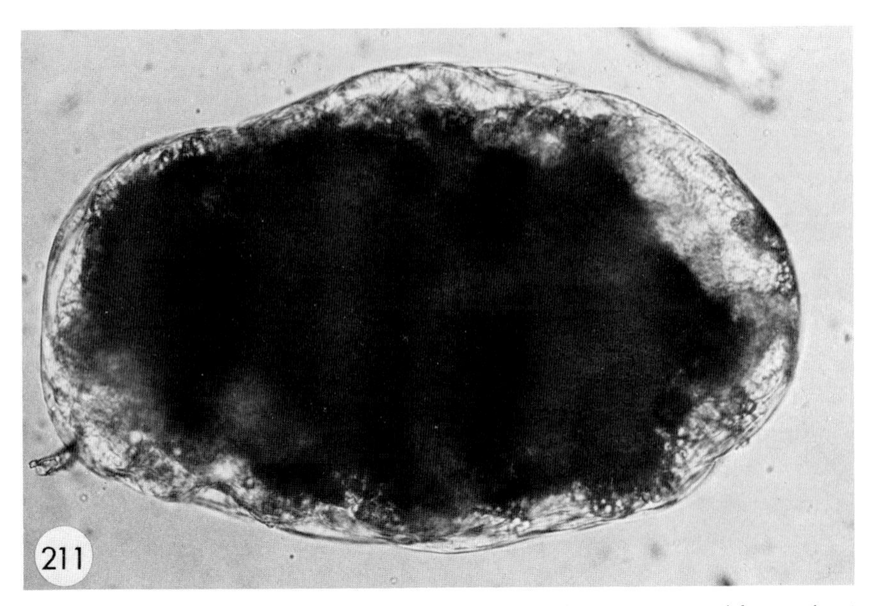

FIGURE 211. Final larval stage of *Trichogramma minutum* removed from a host insect egg. (Specimen courtesy of L. Caltagirone.) ×640.

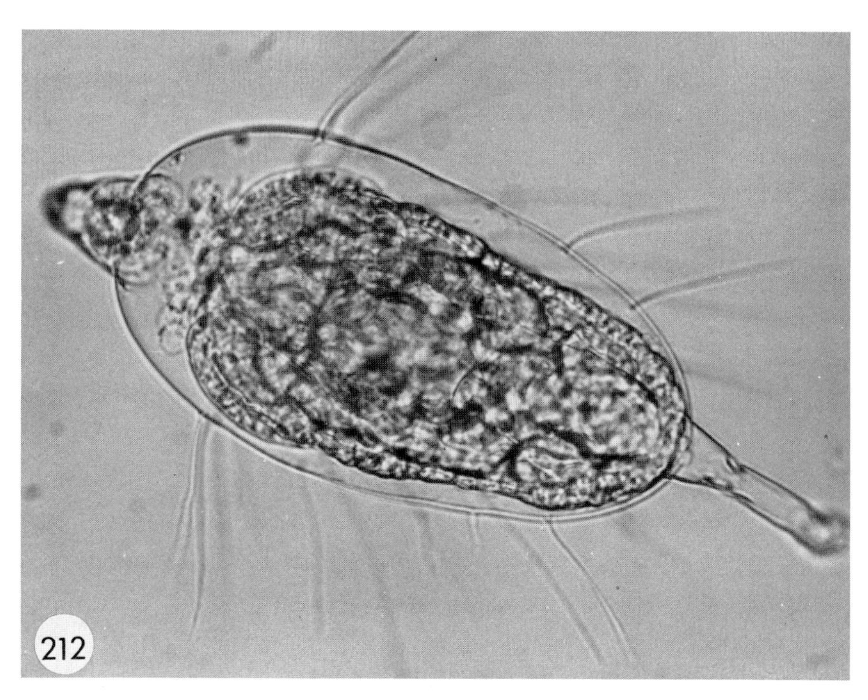

FIGURE 212. Dorsal view of first-stage larva of the mymarid *Anaphes* sp. removed from an egg of *Lygus* sp. Note elongate setae and head and tail region. (Specimen courtesy of Ken Lakin.) ×330.

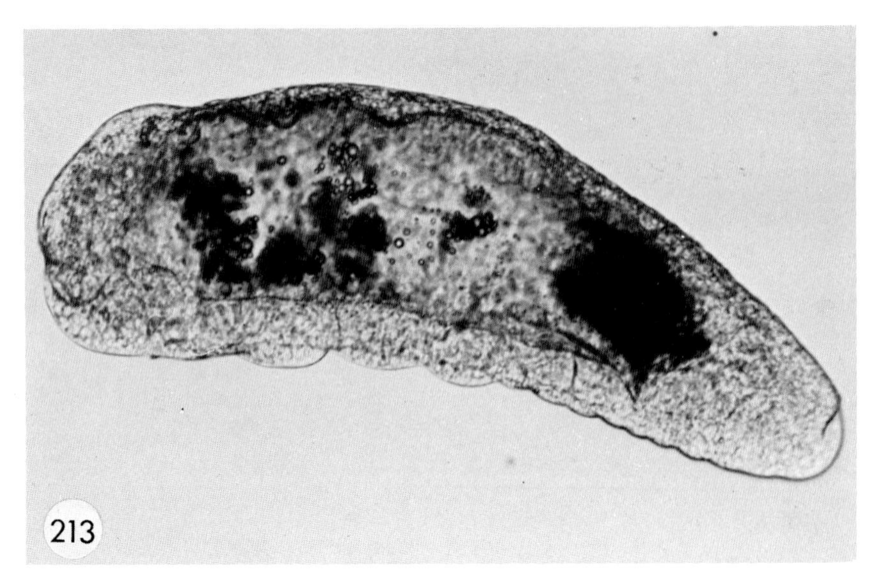

FIGURE 213. Third-stage larva of the mymarid *Anaphes* sp. removed from an egg of *Lygus* sp. Note more typical grublike appearance and lack of first-stage characters. (Specimen courtesy of Ken Lakin.) ×100.

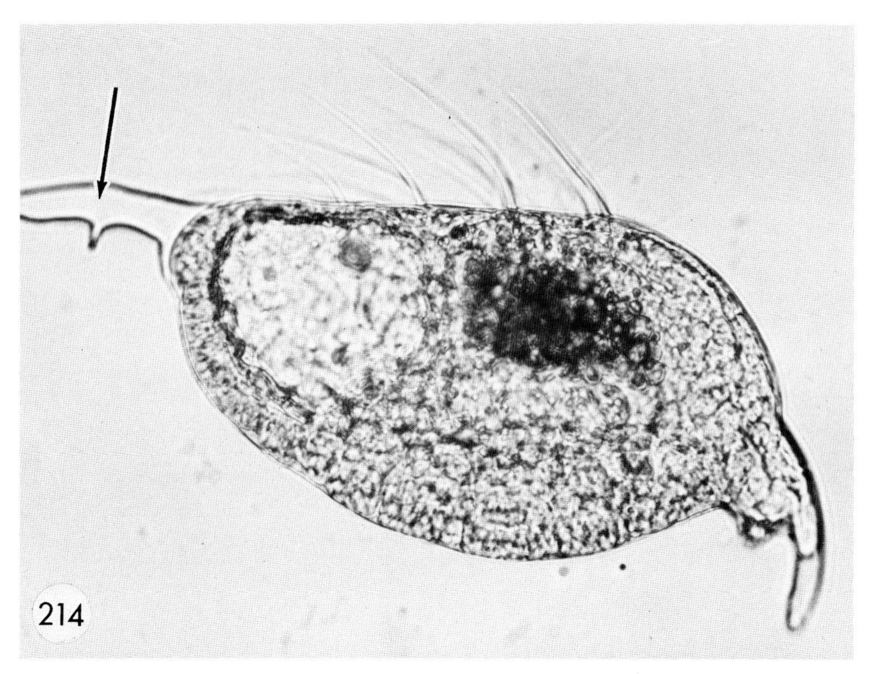

FIGURE 214. Lateral view of first-stage larva of the mymarid *Polynema goreum* removed from an egg of *Nabis* sp. Note elongate setae and head and tail (arrow) region. (Specimen courtesy of Ken Lakin.) ×400.

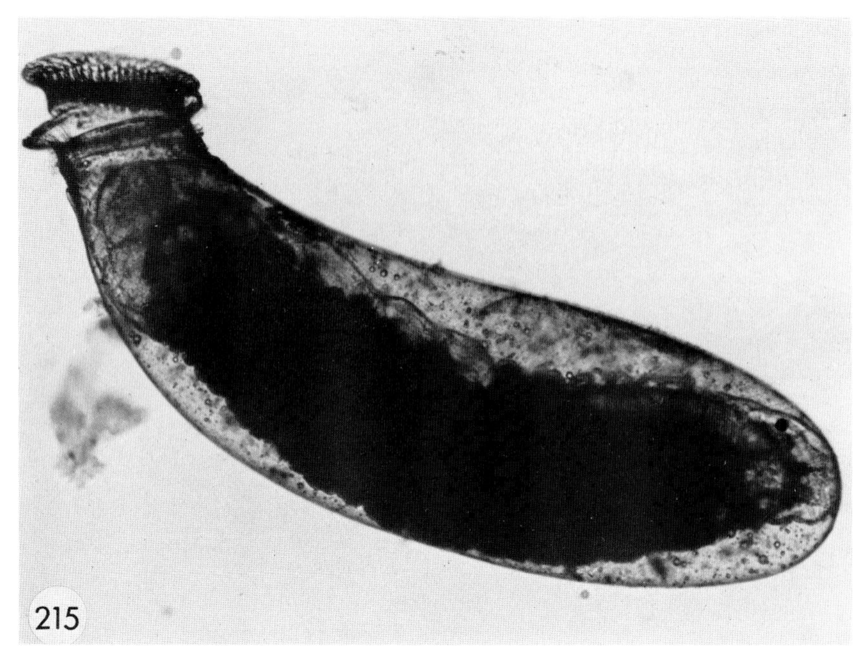

FIGURE 215. Egg of *Nabis* sp. containing the mature larva of the mymarid *Polynema goreum*. Note typical grublike appearance and absence of first-stage characters. (Specimen courtesy of Ken Lakin.) ×100.

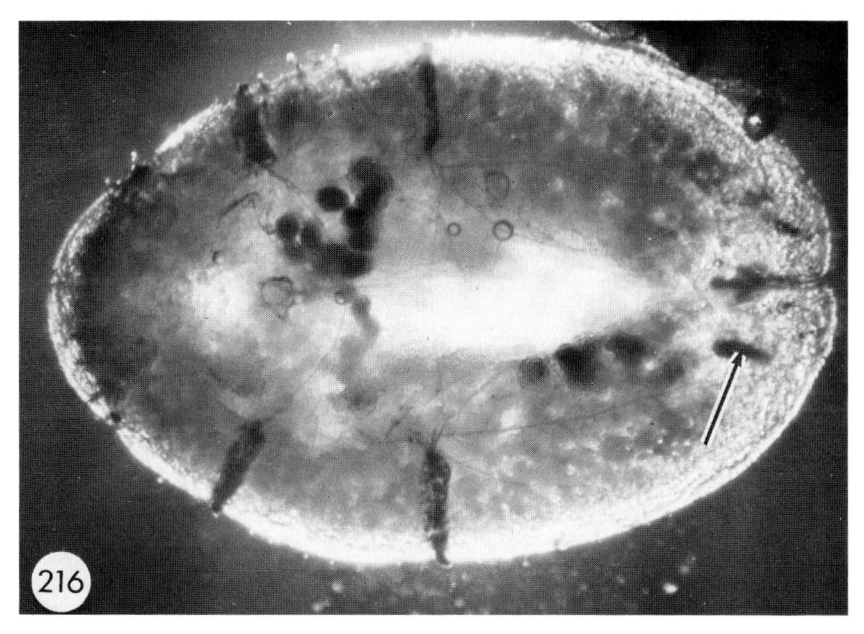

FIGURE 216. A scale, *Pulvinariella mesembryanthemi*, containing an egg stalk (arrow) of the encyrtid parasite *Encyrtus saliens*. (Specimen courtesy of Jane Wright.)

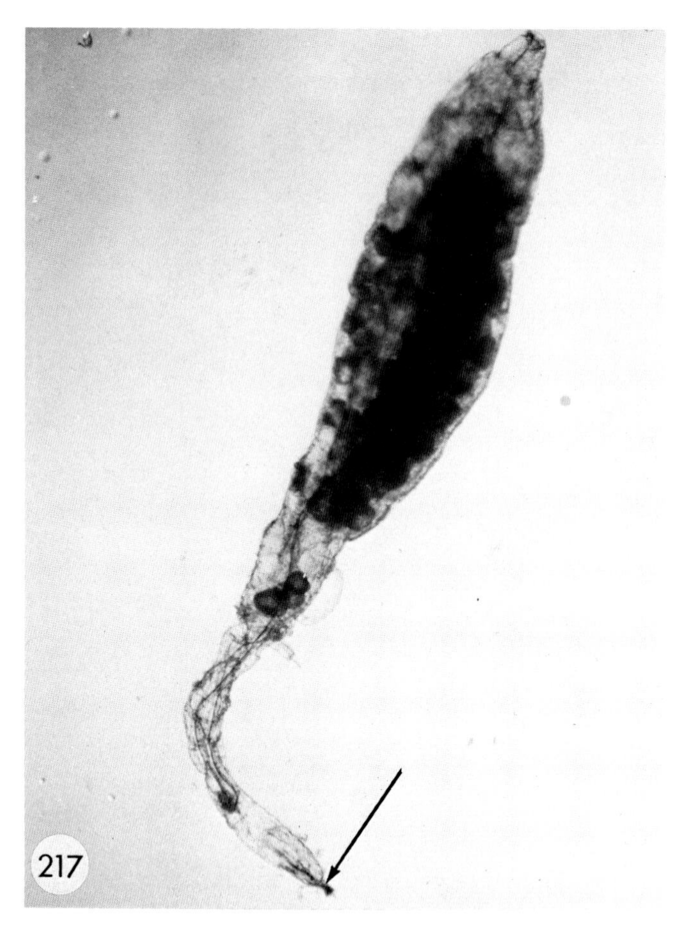

FIGURE 217. Fourth-stage larva of *Encyrtus saliens* showing the tail modified into a tracheal extension attached to the terminal stalk (arrow). (Specimen courtesy of Jane Wright.) ×50.

FIGURE 218. "Sac" (arrow) of a dryinid larva protruding from the abdomen of a leafhopper. (Photo by Jack K. Clark.)

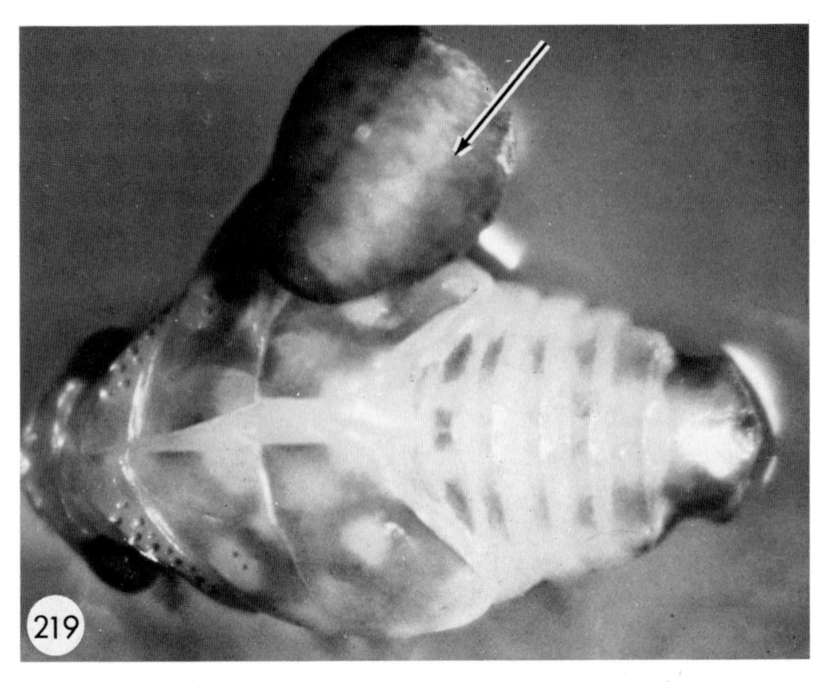

FIGURE 219. "Sac" (arrow) of a dryinid larva protruding from the mesothorax of a spittlebug nymph. (Specimen from J. Huber.)

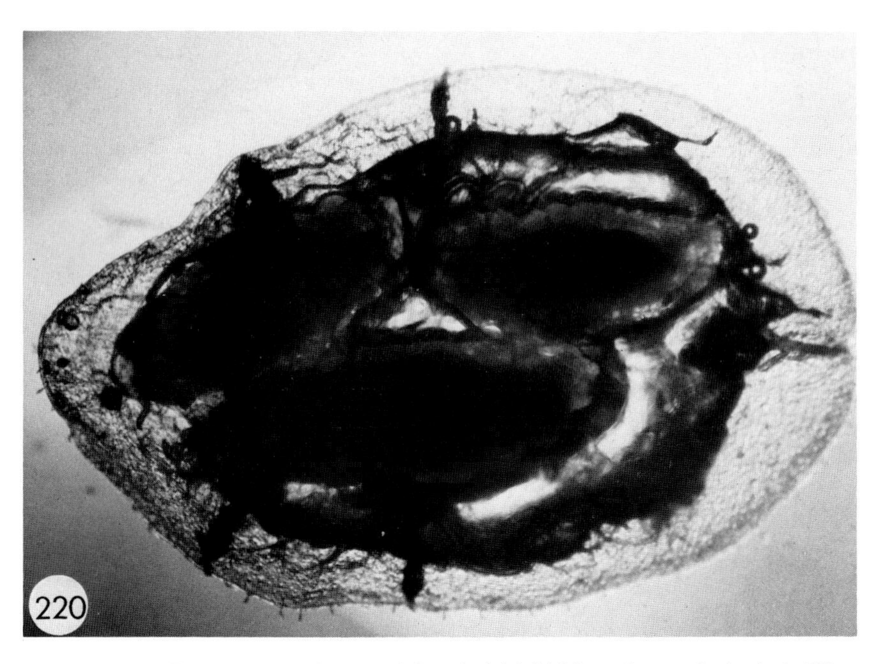

FIGURE 220. Three mature larvae of the chalcidoid *Metaphycus funicularis* filling the body cavity of the scale insect *Pulvinariella mesembryanthemi*. (Specimen from L. Etzel.)

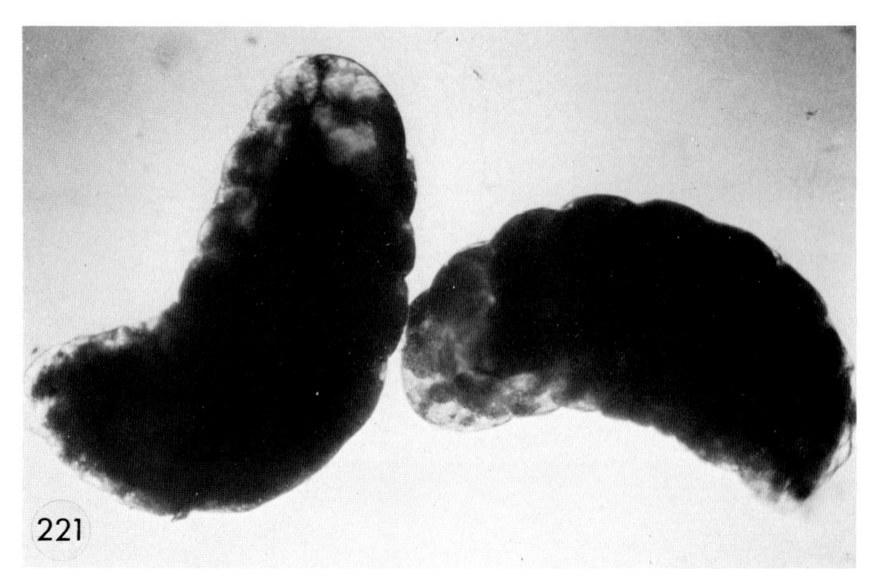

FIGURE 221. Two larvae of *Metaphycus funicularis* removed from the body cavity of the scale insect *Pulvinariella mesembryanthemi.* (Specimen from L. Etzel.) $\times$40.

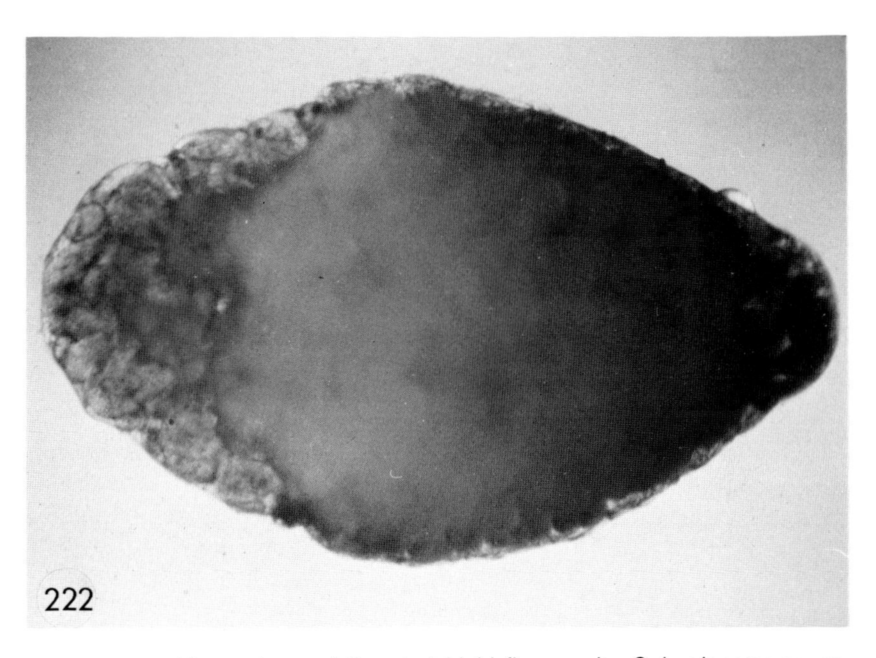

FIGURE 222. Mature larva of the chalcidoid fly parasite *Sphegigaster* sp. removed from a puparium of the housefly. (Specimen from Rincon Vitova Insectaries Inc.) ×50.

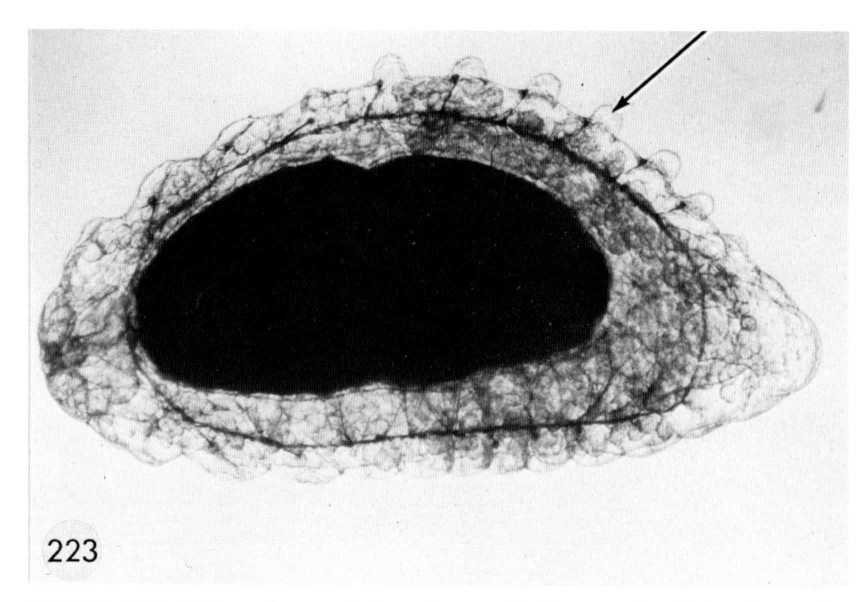

FIGURE 223. Mature larva of the pteromalid fly parasite *Splangia endius* removed from a puparium of the housefly. Note epidermal tubercles (arrow). (Specimen from Rincon Vitova Insectaries Inc.) ×50.

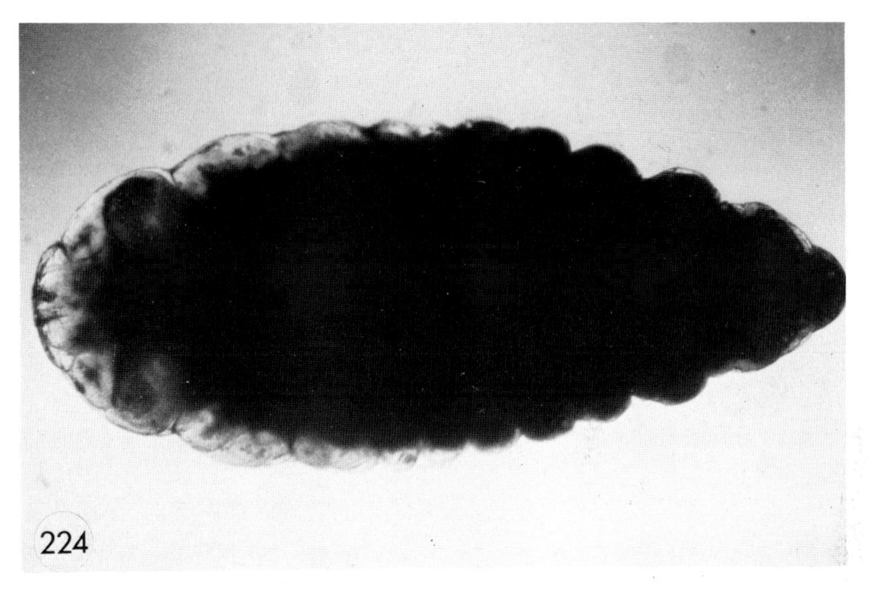

FIGURE 224. Mature larva of the encyrtid parasite *Comperia merceti* removed from the egg sac of a cockroach. (Specimen from A. Slater and Margaret Hurlbert.) ×50.

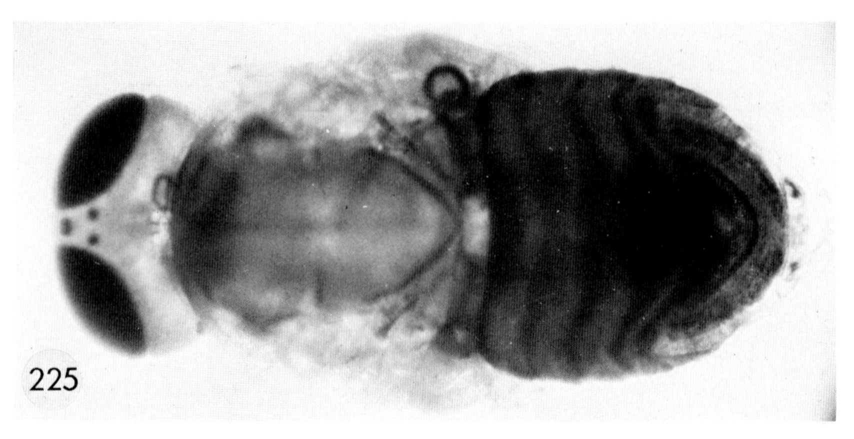

FIGURE 225. Pupa of the encyrtid parasite *Comperia merceti* removed from the egg sac of a cockroach. (Specimen from A. Slater and Margaret Hurlbert.) ×50.

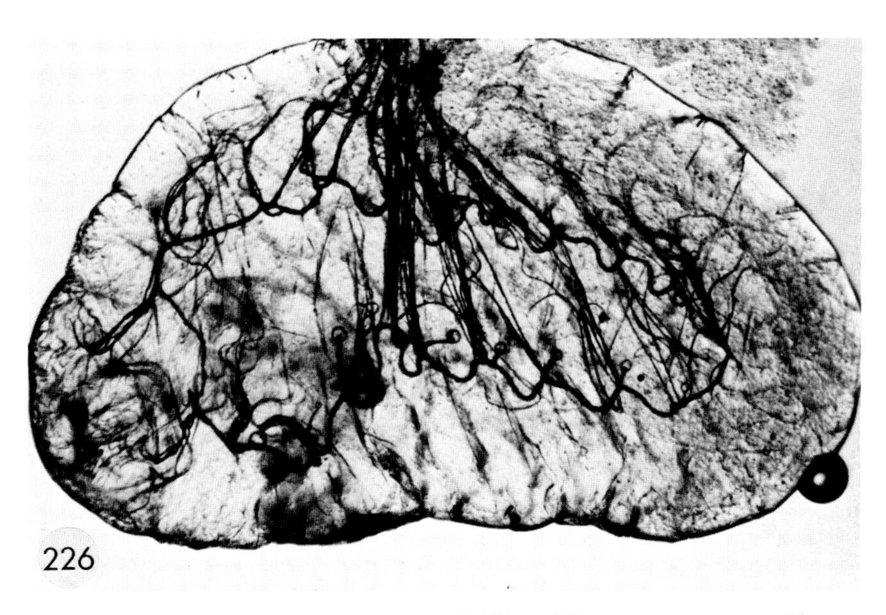

FIGURE 226. Mature larva of the pteromalid *Muscidifurax raptor* removed from a housefly pupa. (Specimen from Beneficial Biosystems.) ×120.

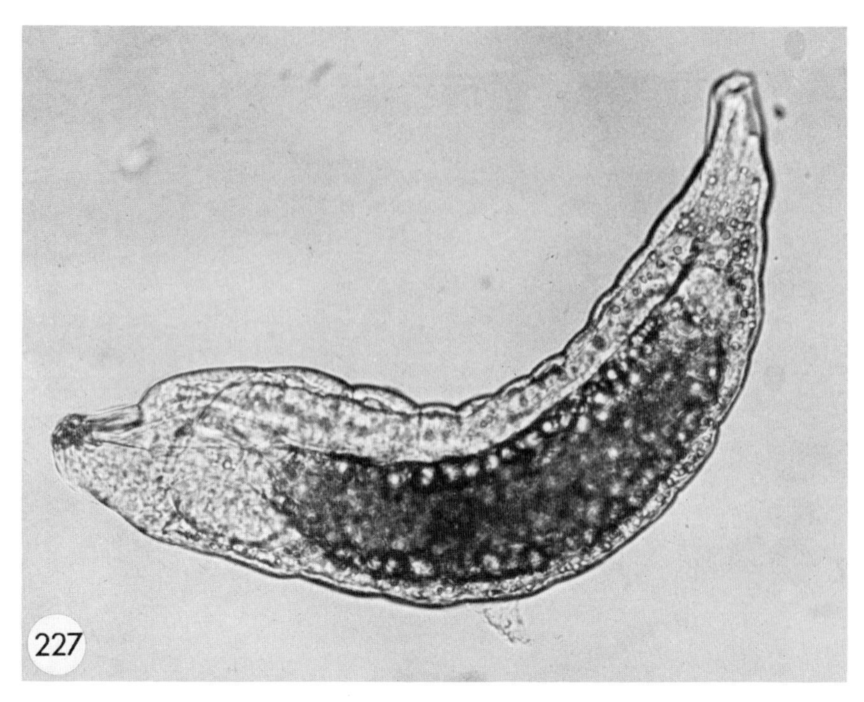

FIGURE 227. Mature larva of the eulophid *Encarsia formosa* removed from a whitefly. (Specimen from K. Hoelmer.) ×240.

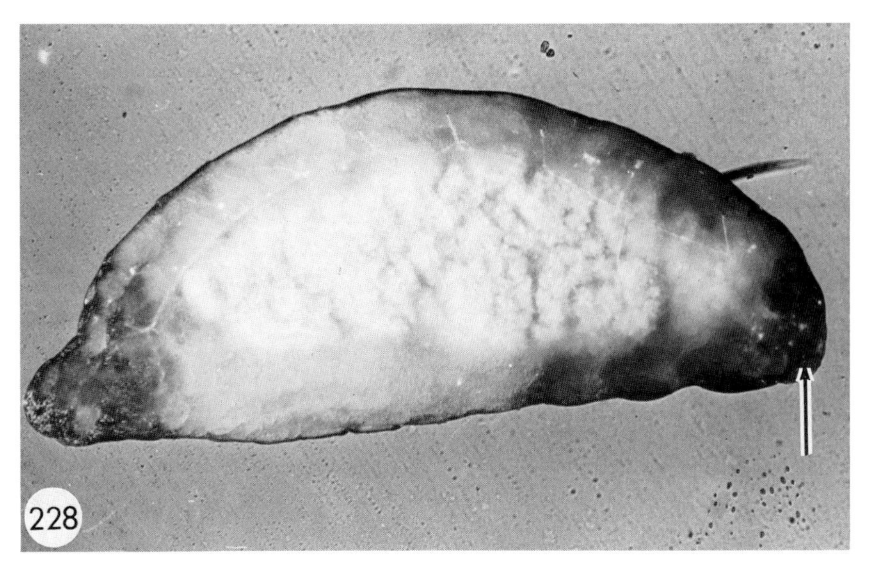

FIGURE 228. Mature larva of the encyrtid *Microterys ishii* removed from a scale insect. Note reduced head (arrow). (Specimen from L. Etzel.)

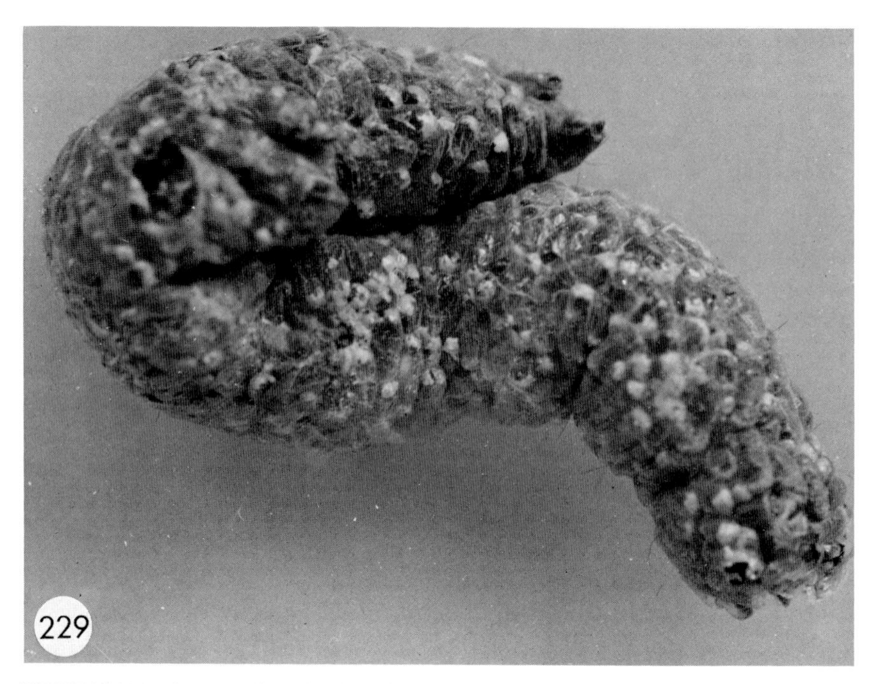

FIGURE 229. A caterpillar that has been parasitized by the polyembryonic encyrtid *Litomastix* sp. Note puffy appearance of the host. (Specimen from L. Caltagirone.)

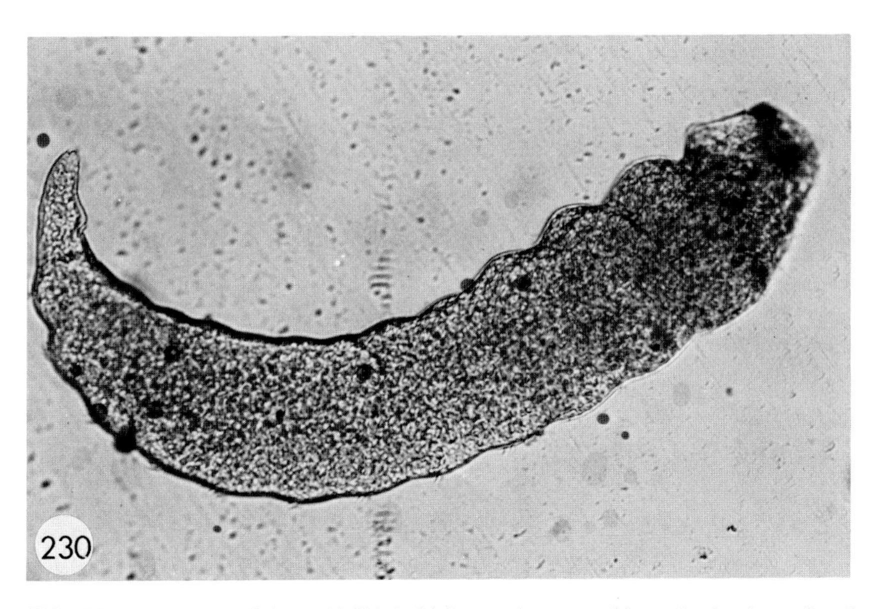

FIGURE 230. Larva of the aphidiid *Aphidius ervi* removed from the body cavity of an aphid. (Specimen from L. Etzel.) ×100.

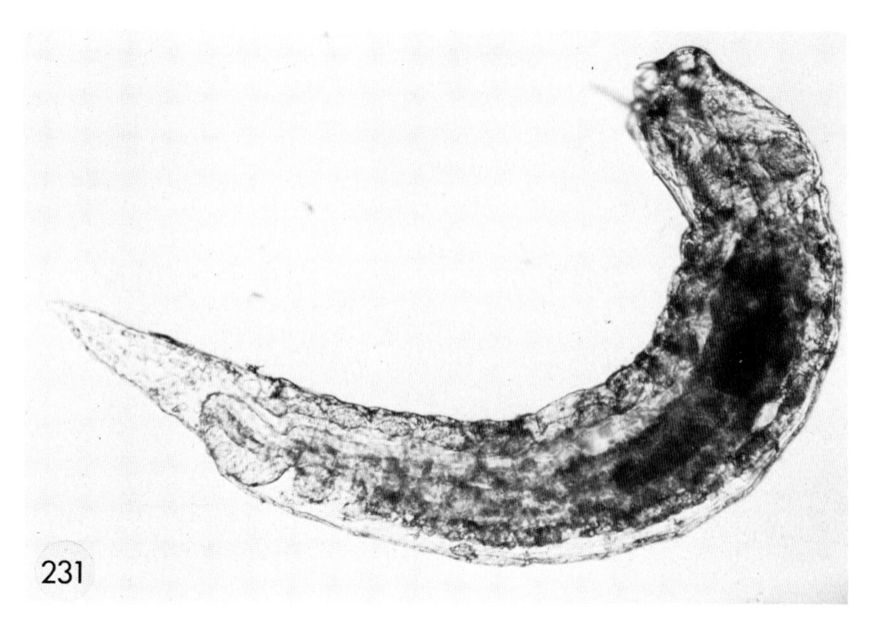

FIGURE 231. Caudate larva of the braconid *Microctonus aethiopoides* removed from the body cavity of an adult alfalfa weevil. (Specimen from L. Etzel.) ×50.

FIGURE 232. Larva of the braconid parasite *Phanerotoma flavitestacea* removed from the body cavity of the navel orangeworm. Note vesiculate appendage at the posterior end (arrow). (Specimen from L. Caltagirone.) ×100.

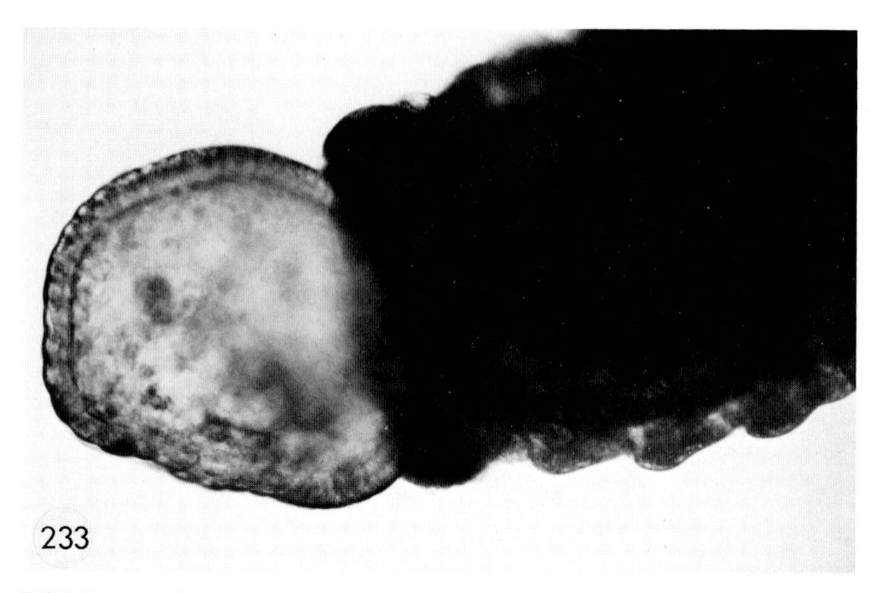

FIGURE 233. Posterior vesicle characteristic of many braconid larvae, on *Phanerotoma flavitestacea.* (Specimen courtesy of L. Caltagirone.)

FIGURE 234. Mature larva of the ichneumonid *Bathyplectes curculionis* removed from an alfalfa weevil larva. Note distinct head capsule (arrow). (Specimen courtesy of L. Etzel.) ×40.

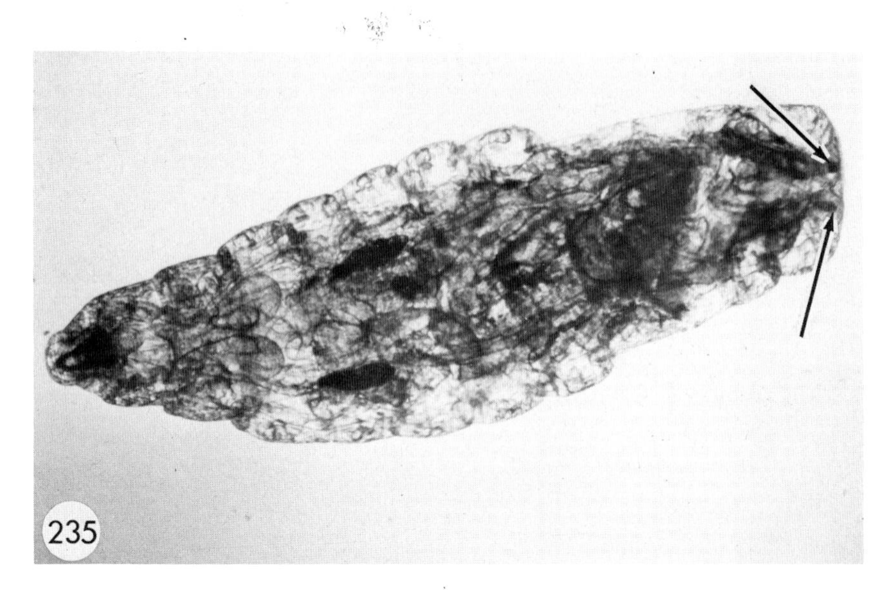

FIGURE 235. Larva of the tachinid parasite *Sturmia* sp. showing the maggotlike shape characteristic of most dipterous parasites. Note posterior spiracles (arrows). (Specimen courtesy of Baldomero Villegas.) ×60.

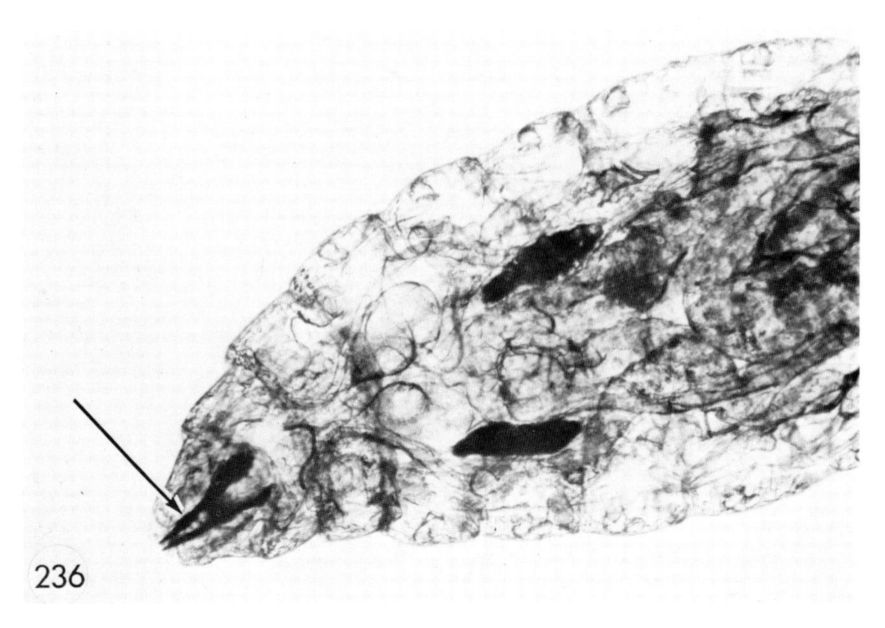

FIGURE 236. Head of the tachinid parasite *Sturmia* sp. showing the parallel mouth hooks (arrow) characteristic of most dipterous parasites. (Specimen courtesy of Baldomero Villegas.)

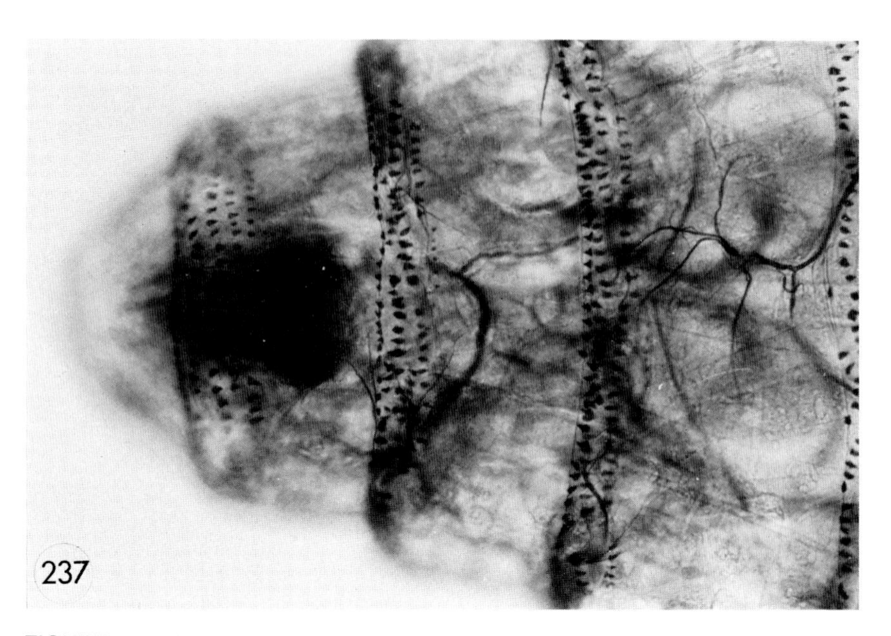

FIGURE 237. Integument of the tachinid parasite *Sturmia* sp. showing rings of setae on each segment. (Specimen courtesy of Baldomero Villegas.)

FIGURE 238. Tachinid eggs deposited on the integument of a lepidopterous larva. (Photo by Jack K. Clark.)

AVAILABILITY OF INSECT PATHOGENS AND ENDOPARASITES

This chapter was added for the practical purpose of indicating the biological control agents of insects that are commercially available and where they can be obtained. There is always the risk of a section such as this one becoming outdated, since some of the firms are relatively small and will not continue in this line indefinitely, and new suppliers are always emerging. However, we felt it would be useful to indicate sources for many of the agents listed in the *Guide*.

We emphasize that inclusion here does not imply our endorsement of the suppliers or their products, nor should criticism of those suppliers or products that we neglected to include be inferred.

By far the largest number of suppliers provide insect parasites, such as *Trichogramma* wasps for use against a range of pest insect eggs and chalcidoid wasps for control of filth flies.

The groups of available agents are listed in the following tables in the order in which they appear in the *Guide*. Addresses of the various suppliers are given at the end of this section.

Table 3. Availability of Insect Viruses

Virus	Target insect	Supplier
Nuclear polyhedrosis	Gypsy moth	Reuter Laboratories Inc.
Nuclear polyhedrosis, Elcar®	*Heliothis* sp.	Zoecon Corporation Peaceful Valley Farm Supply

Table 4. Availability of Entomogenous Bacteria

Bacterium	Target insect	Supplier
Bacillus thuringiensis thuriengiensis	Lepidopterous insects	
Bactosteine®		Biochem Products
Dipel®		Abbott Laboratories
Thuricide®		Zoecon Corporation
Certan®		Zoecon Corporation
Bacillus thuriengiensis israelensis (H-14) serotype	Mosquitoes	
Bactimos®		Biochem Products
Teknar®		Zoecon Corporation
Vectobac®		Abbott Laboratories
Bacillus popillae (milky spore)	Japanese beetle	Reuter Laboratories Inc. Kings Natural Pest Control Mellinger's Nursery Natural Pest Controls Peaceful Valley Farm Supply

Table 5. Availability of Entomogenous Fungi

Fungus	Target insect	Supplier
Verticillium lecanii		
Vertalec®	Aphids	Koppert Ltd.
Mycotal®	Whiteflies	Koppert Ltd.
Beauveria bassiana	Colorado potato beetle	USSR
Metarrhizium anisopliae		
Metaquino®	Spittlebugs	Codecap

Table 6. Availability of Insect-Parasitic Protozoa

Protozoan	Target insect	Supplier
Nosema locustae	Grasshoppers (also some crickets)	Peaceful Valley Farm Supply Mellinger's Nursery Beneficial Insectary Inc. Laporte Insectaries Colorado Insectory Gurney Seed and Nursery Co.

Table 7. Availability of Entomogenous Nematodes

Nematode	Target insect	Supplier
Neoaplectana carpocapse N. glaseri, Heterorhabditis spp.	A range of insects, including termites, root beetles, and carpenterworm	Biosis (formerly California Nematode Laboratories) Colorado Insectory Peaceful Valley Farm Supply TNF

Table 8. Availability of Endoparastic Insects

Species	Used to control	Supplier
Trichogramma wasps	Eggs of various insects	Rincon Vitova Insectaries Inc.
		Arizona Biological Control Inc.
		California Green Lacewings
		Foothill Agricultural Research Inc.
		Bio-Resources
		Gothard Inc.
		Bio Control Co.
		Gurney Seed and Nursery Co.
		Beneficial Insects Ltd.
		Biogenesis Inc.
		Bio-Insect Control
		Integrated Pest Management
		Kings Natural Pest Control
		Natural Pest Controls
		Unique Nursery
		Fossil Flower
		Harmony Farm Service and Supply
		Kunafin Trichogramma Insectaries
		Necessary Trading Co.
		Organic Pest Control Naturally
		Peaceful Valley Farm Supply
Encarsia formosa	Whitefly	Whitefly Control Co.
		Anticimex AB
		Bio-Resources
		Bio-Insect Control
		Organic Pest Control Naturally
		Peaceful Valley Farm Supply
		Pest Management Group
		Professional Ecological Services
		Rincon Vitova Insectaries
		Natural Pest Controls
		Better Yield Insects
		Integrated Pest Management
		Applied Bionomics
		Bunting and Sons
		English Woodlands
		Humber Growers

Table 8. (*Continued*)

Species	Used to control	Supplier
Encarsia formosa (*continued*)		Natural Pest Control
		CTIFL
		Koppert BV
		Resh Greenhouses and Hydroponic Garden Center
		Unique Nursery
Aphidius matricariae	Aphids	English Woodlands
Dacnusa sibirica	Leaf-mining flies	Natural Pest Control
		Koppert BV
Apanteles flavicoxis	Gypsy moth	Gypsy Moth Control Co.
		Peaceful Valley Farm Supply
Pachyrepoideus vindemiae	Houseflies	Natural Pest Controls
Nasonia vitripennis	Houseflies	Foothill Agricultural Research Inc.
		Beneficial Biosystems
		Arizona Biological Control Inc.
Sphegigaster sp.	Houseflies	Beneficial Biosystems
		Spaulding Laboratories
		Arizona Biological Control Inc.
		Rincon Vitova Insectaries Inc.
		Peaceful Valley Farm Supply
Tachinaephagus zealandicus	Houseflies	Rincon Vitova Insectaries Inc.
		Beneficial Insectary Inc.
		Spaulding Laboratories
		Bio-Resources
Metaphycus helvolus	Black scale	Bio-Insect Control
		Foothill Agricultural Research Inc.
		Arizona Biological Control Inc.
		Peaceful Valley Farm Supply
		Integrated Pest Management
Aphytis melinus	Red scale	Foothill Agricultural Research Inc.
		Arizona Biological Control Inc.
		Peaceful Valley Farm Supply
		Integrated Pest Management
		Natural Pest Controls
Microchelones blackburni	Pink bollworm	Natural Pest Controls

(*Continued*)

Table 8. (*Continued*)

Species	Used to control	Supplier
Muscidifurax raptor	Houseflies	Beneficial Insectary Inc.
		Spaulding Laboratories
		Rincon Vitova Insectaries Inc.
		Arizona Biological Control Inc.
		Bio-Insect Control
		Beneficial Biosystems
		Integrated Pest Management
		Natural Pest Controls
		Bio-Resources
		Peaceful Valley Farm Supply
Splangia endius	*Musca domestica*	Natural Pest Controls
		Integrated Pest Management
		Foothill Agricultural Research Inc.
		Beneficial Biosystems
		Rincon Vitova Insectaries Inc.
		Arizona Biological Control Inc.
		Beneficial Insectary Inc.
		Spaulding Laboratories
		Bio-Resources
		Peaceful Valley Farm Supply
		Resh Greenhouses and Hydroponic Garden Center
		Unique Nursery
Comperiella bifasciata	Red scale	Associates Insectary
		Peaceful Valley Farm Supply
Pentalitomastic plethoricus	Navel orangeworm	California Green Lacewings
		Peaceful Valley Farm Supply
Edovum puttleri	Colorado potato beetle	Peaceful Valley Farm Supply

SUPPLIER ADDRESSES

Abbott Laboratories, 900 West Route 70, Suite 6, Marlton, NJ 08053
Anticimex AB, c/o Trödgardeshallen, S-25229 Helsingborg, Sweden
Applied Bionomics, 8801 East Saanickton Road, Sidney, British Columbia, Canada V8L1H3
Arizona Biological Control Inc., Route 19, P.O. Box 363, Tucson, AZ 85704
Associates Insectary, P.O. Box 969, Santa Paula, CA 93060
Beneficial Biosystems, 1603-F 63rd Street, Emeryville, CA 94608
Beneficial Insectary Inc., 245 Oak Run Road, Oakrun, CA 96069
Beneficial Insects Ltd., P.O. Box 154, Banta, CA 95304
Better Yield Insects, 13310 Riverside Drive, Tecumseh, Ontario, Canada
Biochem Products, P.O. Box 264, Montchanin, DE 19710
Bio Control Co., P.O. Box 247, Cedar Ridge, CA 95924
Biogenesis Inc., P.O. Box 36, Mathis, TX 78368
Bio-Insect Control, 1710 South Broadway, Plainview, TX 79072
Bio-Resources, 1210 Birch Street, Santa Paula, CA 93060
Biosis, 3788 Fabian Way, Palo Alto, CA 94303
BR Supply Company, P.O. Box 845, Exeter, CA 93221
Bunting and Sons, The Nurseries, Great Horkesley, Colchester, Essex, England
California Green Lacewings, P.O. Box 2495, Merced, CA 95340
Codecap, Rua Vidalla Negreiros, 321-Vep 50000, Recife-Pe, Brazil
Colorado Insectory, P.O. Box 3266, Durango, CO 81301
CTIFL, Centre de Balandran, 30127 Bellegarde, France
English Woodlands, The Old Barn, Rohelane, Godalming, Surrey GU8 5NT, England
Foothill Agricultural Research Inc., 510 West Chase Drive, Corona, CA 91720
Fossil Flower, 463 Woodbine Avenue, Toronto, Ontario, Canada M4E2HS
Gothard Inc., P.O. Box 370, Canutillo, TX 79835
Gurney Seed and Nursery Co., Yonkton, SD 57079
Gypsy Moth Control Co., R.D. 1, Box 715, Landisburg, PA 17040
Harmony Farm Service and Supply, P.O. Box 451, Graton, CA 95444
Humber Growers, Common Lane, Welton, Brough, North Humberside, England
Integrated Pest Management, 305 Agostino Road, San Gabriel, CA 91176
Kings Natural Pest Control, 224 Yost Avenue, Spring City, PA 19475
Koppert BV, Vielingweg 8a, 2651 BE Berkel-Rodenrijs, The Netherlands
Koppert Ltd., P.O. Box 43, Turnbirdge Wells, Kent, England
Kunafin Trichogramma Insectaries, Route 1, P.O. Box 39, Quemado, TX 78877
Laporte Insectaries, 2220 North Highway 287, Fort Collins, CO 80524
Mellinger's Nursery, 2310 West South Range Road, North Lima, OH 44452
Natural Pest Control, Watermead, Yapton Road, Barnham, Bognor Regis, Sussex, England
Natural Pest Controls, 9397 Pemier Way, Sacramento, CA 95826
Necessary Trading Co., 328 Main Street, New Castle, VA 24127
Organic Pest Control Naturally, 19920 Forest Park Drive NE, Seattle, WA 98155
Peaceful Valley Farm Supply, 11173 Peaceful Valley Road, Nevada City, CA 95959

Pest Management Group, 810 Hollywood Road, Kelouna, British Columbia, Canada VIX359

Professional Ecological Services, 555 Hillside, Victoria, British Columbia, Canada V8T1Y8

Resh Greenhouses and Hydroponic Garden Center, 12626 Bridgeport Road, Richmond, British Columbia, Canada V6V1J5

Reuter Laboratories Inc., P.O. Box 347, Haymarket, VA 22O69

Rincon Vitova Insectaries Inc., P.O. Box 95, Oak View, CA 93022

Spaulding Laboratories, Route 2, P.O. Box 737, Arroyo Grande, CA 93420

TNF, 3335 Birch Street, Palo Alto, CA 94303

Unique Nursery, 5504 Sperry Drive, Citrus Heights, CA 95610

Whitefly Control Co., P.O. Box 986, Milpitas, CA 95035

Zoecon Corporation, 975 California Avenue, Palo Alto, CA 94034

TECHNIQUES

This section describes the various techniques and procedures that have been mentioned in the *Guide*. All bacteriological and mycological media discussed can be obtained in dehydrated form from commercial producers such as Difco Laboratories, Detroit, Michigan 48232; and BBL, Division of Bioquest, P.O. Box 175, Cockeysville, Maryland 21030. In addition, many are normally stocked by local scientific supply houses.

This portion is divided into four categories: (1) Culture Media, Differential Media, and Ringer's Solution; (2) Bacteriological Tests; (3) Stains and Staining Procedures; and (4) Miscellaneous Techniques. The contents of each section are arranged alphabetically.

CULTURE MEDIA, DIFFERENTIAL MEDIA, AND RINGER'S SOLUTION

AC Medium

AC medium is a good primary isolation medium possessing unique growth-promoting properties for both aerobic and anaerobic microorganisms. It is recommended as a general culture medium for anaerobes, microaerophiles, and aerobes. The medium contains an agar flux which retards air diffusion and, when sealed in snap-cap tubes immediately after autoclaving, it maintains an oxygen gradient ranging from anaerobic at the bottom to aerobic at the top. Within 24 hr after inoculation, anaerobic organisms will be growing near the bottom, microaerophiles near the center, and aerobic organisms near the top. Facultative organisms will grow throughout the tube. The composition is as follows:

Beef extract	3 g
Yeast extract	3 g
Malt extract	3 g
Proteose peptone No. 3 (Difco)	20 g
Dextrose	5 g
Agar	1 g
Ascorbic acid	0.2 g
Distilled water	1 liter

Mix all ingredients and heat to the boiling point while stirring. Dispense into tubes and autoclave at 15 lb for 15 min. Seal tubes immediately after autoclaving.

Azide Dextrose Broth

Azide dextrose broth is recommended for the isolation and identification of *Streptococcus* spp. Sodium azide is used as an inhibitor for Gram-negative organisms. The medium is reported to be selective for streptococci (Difco, 1953), but we found that staphylococci and micrococci, as well as some Gram-positive bacilli, also grow in this medium. *Streptococcus* may be separated from *Staphylococcus* and *Micrococcus* by subculturing on nutrient agar and testing for catalase. *Streptococcus* is catalase negative, while *Staphylococcus* and *Micrococcus* are catalase positive.

Broth

Beef extract	4.5 g
Tryptone	15 g
Dextrose	7.5 g
Sodium chloride	7.5 g
Sodium azide	0.2 g
Distilled water	1 liter

Dissolve the dry ingredients in the water, place 10 ml in each tube, and autoclave at 15 lb for 15 min.

Culture. The broth is inoculated from a pure culture of the test bacterium, or directly from diseased insect material, and incubated at 30°–37°C for 24 hr. The resultant growth can be examined microscopically for the presence of cocci. *Streptococcus* spp. usually form long chains in this broth.

Brain–Heart Infusion Agar (BHIA)

Brain–heart infusion agar is a solid medium for the cultivation of fastidious pathogenic bacteria.

Infusion from calf brains	200 g
Infusion from beef heart	250 g
Proteose peptone	10 g
Dextrose	2 g
Sodium chloride	5 g
Disodium phosphate	2.5 g
Agar	15 g
Distilled water	1 liter

Mix the dry ingredients with the water and autoclave for 15 min at 15 lb. Pour plates aseptically, when medium is cool enough to handle.

Brain–Heart Infusion Broth (BHIB)

Brain–heart infusion broth is a liquid medium recommended for the cultivation of fastidious pathogenic bacteria.

Formula. The formula for this broth is the same as that for brain–heart infusion agar, with the agar deleted. Dissolve the dry ingredients in water, dispense in tubes or culture flasks as needed, and autoclave at 15 lb for 15 min.

Dextrose Agar (DA)

Dextrose agar is recommended as a general all-purpose medium which supports the growth of a wide range of bacteria and fungi.

Beef extract	3 g
Tryptose	10 g
Dextrose	10 g
Sodium chloride	5 g
Agar	15 g
Distilled water	1 liter

Dissolve the dry ingredients in the water, autoclave for 15 min at 15 lb, cool to about 50°C, and dispense aseptically into petri dishes.

Dextrose Broth (DB)

The formula for dextrose broth is similar to that for dextrose agar, without the agar and with 5 g of dextrose rather than 10. Dissolve the dry ingredients in water, dispense into tubes or culture flasks, and autoclave at 15 lb for 15 min.

Nutrient Agar (NA)

Nutrient agar is recommended as a general solid medium for the cultivation of many bacteria.

Beef extract	3 g
Peptone	5 g
Agar	15 g
Distilled water	1 liter

Mix the dry ingredients with the water and autoclave for 15 min at 15 lb. Pour the plates aseptically when the medium is cool enough to handle.

Nutrient Broth (NB)

Nutrient broth is a liquid medium recommended for the cultivation of bacteria with general food requirements.

Beef extract	3 g
Peptone	5 g
Distilled water	1 liter

Dissolve the dry ingredients in the water, dispense into tubes or culture flasks with appropriate closures to protect from contamination, and autoclave at 15 lb for 15 min.

Nutrient Gelatin

Nutrient gelatin is used for the determination of gelatin liquefaction in bacterial identifications.

Beef extract	3 g
Peptone	5 g
Gelatin	120 g
Distilled water	1 liter

Dissolve the dry ingredients in the water while stirring and dispense into tubes. Cover with morton caps or cotton stoppers and autoclave at 15 lb for 15 min. Cool to solidify and store in air-tight containers. To test for gelatin liquefaction, the medium is inoculated by a deep stab and incubated at 20°C or room temperature.

Ringer's Solution, One-Quarter-Strength

One-quarter-strength Ringer's solution is often preferred in preparing bacterial suspensions, since it is generally more osmotically compatible with bacterial cells than sterile distilled water. Quarter-strength Ringer's may be sterilized by autoclaving, or preferably by millipore filtration, since autoclaving may cause a precipitate to form.

Sodium chloride	1.12 g
Potassium chloride	0.05 g
Calcium chloride	0.06 g
Sodium bicarbonate	0.02 g
Distilled water	500 ml

Sabouraud Dextrose Agar with Yeast Extract (SDA+Y)

SDA+Y is an enriched medium especially suited for the cultivation of fungi, and many of the more fastidious entomopathogenic fungi grow very well on it.

Neopeptone	10 g
Dextrose	40 g
Yeast extract	2 g
Agar	15 g
Distilled water	1 liter

Mix the dry ingredients with water and autoclave at 15 lb for 15 min. When cool, but still liquid (about 50°C), pour plates aseptically.

Sabouraud Dextrose Broth with Yeast Extract (SDB+Y)

SDB+Y is similar to SDA+Y but does not contain agar, and the dextrose is reduced to 20 g. Dissolve the dry ingredients in the water, dispense into tubes or culture flasks, and autoclave at 15 lb for 15 min.

Sabouraud Maltose Agar with Yeast Extract (SMA+Y)

This medium is similar to SDA+Y, with maltose substituted for dextrose. It is particularly useful in promoting spore production, where an abundance of spores is necessary for pathogenicity tests, or when difficulty is encountered in obtaining reproductive structures for indentification.

Neopeptone	10 g
Maltose	40 g
Yeast extract	2 g
Agar	15 g
Distilled water	1 liter

Mix the dry ingredients with the water, autoclave at 15 lb for 15 min, and dispense aseptically into petri dishes.

Streptococcus faecalis Differential Medium

Streptococcus faecalis is a potential bacterial pathogen often encountered in insects suffering wounds or environmental stress. *Streptococcus* spp. may first be isolated from suspected material by culturing in azide dextrose broth. A specific identification of *S. faecalis* may then be made on the following medium, devised by Meade (1963).

Medium

1. Dissolve in 1 liter of distilled water:
   10 g peptone
   1 g yeast extract
   2 g sorbitol
   5 g tyrosine
   12 g agar
2. Autoclave for 10 min at 10 lb (no further autoclaving is necessary).
3. Adjust pH to 6.2.
4. Add 0.01% 2,3,5-triphenyl tetrazolium chloride (TTC) and 1.0% thallous acetate.
5. Pour a shallow layer of this medium into sterile petri dishes and allow to set.
6. Add an additional 4 g of tyrosine to the remaining medium and keep warm, but near the setting point. Swirl before pouring to keep the tyrosine in suspension.

7. Pour a layer of medium with additional tyrosine over the original layer in each petri dish, and allow to set.
8. The medium may be stored in airtight plastic bags or containers under refrigeration until needed.

Test

1. Inoculate a plate of the above medium with a suspension of the suspected bacterium and incubate at 37°C for 4 hr.
2. Transfer culture to 45°C and incubate for 3 days.

Results. Uniformly dark maroon colonies encircled by clear zones are regarded as *Streptococcus faecalis*.

Tergitol-7+TTC Agar (T-7+TTC)

Tergitol-7 agar with the addition of 2,3,5-triphenyltetrazolium chloride (TTC) is a good medium for the isolation of Gram-negative coliform bacteria. Gram-positive bacteria are inhibited. It is diagnostic for *Escherichia coli* and *Xenorhabdus nematophilus*. *E. coli* produces yellow colonies surrounded by yellow zones. *X. nematophilus* forms deep blue colonies against a blue background. *Enterobacter* forms red colonies surrounded by yellow zones, while other Enterobacteriaceae and *Pseudomonas* form red colonies against a blue background.

Peptone No. 3 (Difco)	5 g
Yeast extract	3 g
Lactose	10 g
Agar	15 g
Tergitol-7	0.1 ml
Bromthymol blue	0.025 g
Distilled water	1 liter

Mix the above ingredients, autoclave at 15 lb for 15 min, and allow to cool to about 50°C. Dissolve 40 mg TTC in a few milliliters of water, and add this after millipore filtration to the cooled medium. Aseptically dispense the solution into petri dishes and allow to solidify.

BACTERIOLOGICAL TESTS

Carbohydrate Fermentation Studies

Purple Broth Base. Recommended for the preparation of carbohydrate broths used in fermentation studies of pure bacterial cultures. The concentration of carbohydrate generally employed for testing the fermentation reactions of bacteria is 0.5% or 1.0%. The use of 1.0% rather than 0.5% helps to insure against reversion of the reaction due to depletion of the carbohydrate by some bacteria.

Beef extract	1 g
Proteose peptone	10 g
Sodium chloride	5 g
Bromcresol purple	0.015 g
Distilled water	1 liter
Carbohydrate	10 g

Dissolve the dry ingredients in the water, tube the medium, and autoclave it at 15 lb for 15 min. Simple sugars such as glucose may be added before autoclaving. More complex carbohydrates, which are susceptible to hydrolysis during autoclaving, should be sterilized with a minimum amount of heat or added by millipore filtration after sterilization of the base broth. In the latter case, part of the water is used to dissolve the carbohydrate, and the remaining water is used to make the base broth. The base broth is then autoclaved in bulk, and, when it is cool, the dissolved carbohydrate is added aseptically by millipore filtration. The medium is then aseptically dispensed into previously sterilized tubes. To detect the production of gas, a very small tube (Durham tube) is inverted in the culture tube with the broth. Upon autoclaving, the small tube will fill with broth and sink. If gas is produced during fermentation, some of it will collect in the small tube.

Carbohydrates mentioned in this guide which can be used with this broth are glucose, arabinose, glycerol, and inositol. Tubed medium is inoculated from a pure culture of the test bacterium and incubated at 30°C for 16–24 hr. A positive reaction is indicated by the development of a yellow color, while uninoculated medium and negative results are indicated by retention of the original purple color of the medium.

Catalase Test

First, culture the organism on nutrient agar or other suitable medium, then flood some of the colonies with 3% hydrogen peroxide. The immediate production of gas bubbles indicates a positive reaction.

Cytochrome Oxidase Test

Reagents

1. 1% aqueous *N,N*-dimethyl-*p*-phenylenediamine (solution is stable for only 1–2 months under refrigeration, but may be kept 3–4 years if frozen).
2. Ethanolic napthol (1% alpha-naphthol in 95% ethyl alcohol).

Procedure

1. Culture test organism on a suitable solid medium such as nutrient agar.
2. Mix reagents: three drops solution 1 to two drops solution 2.
3. Flood some of the colonies with this mixture.
4. Colonies of cytochrome-oxidase-positive organisms turn blue.

Hydrogen Sulfide Production: Triple Sugar Iron Agar (TSI Agar)

This medium is recommended for determining several characteristics of Gram-negative enteric bacteria, including the production of hydrogen sulfide. (For details see Difco, 1953, p. 166.)

Beef extract	3 g
Yeast extract	3 g
Peptone	15 g
Proteose peptone	5 g
Lactose	10 g
Saccharose	10 g
Dextrose	1 g
Ferrous sulfate	0.2 g
Sodium chloride	5 g
Sodium thiousulfate	0.3 g
Agar	12 g
Phenol red	0.024 g
Distilled water	1 liter

Dissolve the dry ingredients in the water, heated to the boiling point. Tube the medium and autoclave at 15 lb for 15 min. The medium may be stored under refrigeration in airtight containers.

Culture. The tubed medium is inoculated by the stab method from a pure culture of the test organism and incubated at 30°–37°C for 16–24 hr.

Results. The production of hydrogen sulfide is indicated by the formation of a distinct black color.

Indole Production

1. Culture the bacterium under investigation in 5 ml 1.0% tryptone for 24 hr.
2. Add 0.2–0.3 ml Kovac's reagent to the above culture.
3. A dark red color in the surface layer constitutes a positive test for indole production. The original yellow color of the solution constitutes a negative test.

Kovac's Reagent. Kovac's reagent is made by dissolving 5 g *p*-dimethylaminobenzaldehyde in 75 ml amyl alcohol and adding 25 ml concentrated hydrochloric acid.

Methyl Red–Voges Proskauer (MR-VP) Test for Gram-Negative Enteric Bacteria

This is a test for the production of acetylmethylcarbinol or acetoin.

Culture Medium

Peptone	7 g
Dextrose	5 g
Dipotassium phosphate	5 g
Distilled water	1 liter

Dissolve dry ingredients in water, tube at 5 ml per tube, autoclave at 15 lb for 15 min, and store in airtight containers under refrigeration.

Culture. Inoculate the tubed medium from a pure culture of the test organism and incubate at 30°C for 5–7 days.

Test Reagents

MR Test. Dissolve 0.1 g methyl red in 300 ml 95% ethyl alcohol and dilute to 500 ml with distilled water.

VP Test. (1) 40% aqueous sodium hydroxide. (2) Solid creatin.

Test

MR. Add five drops of the methyl red solution to 5 ml of a 5- to 7-day culture. A positive reaction is indicated by a distinct red color, showing the presence of acid. A negative reaction is indicated by a yellow color.

VP. To 5 ml of a 7-day culture, add 25 mg solid creatin, then 5 ml 40% NaOH. Stopper and shake for 1 min. A positive reaction is shown by development of a red color a few minutes after agitation.

Nitrate Reduction Test

This is a test for the ability of bacteria to reduce nitrates (NO_3) to nitrites (NO_2).

Nitrate Culture Broth

Beef extract	3 g
Peptone	5 g
Potassium nitrate	1 g
Distilled water	1 liter

Dissolve the dry ingredients in the water, tube, and autoclave at 15 lb for 15 min. Store in an airtight container under refrigeration.

Culture. Inoculate tubed broth from a culture of the test organism and incubate at 30°–37°C for 12–24 hr.

Test Reagent Solutions

Sulfanilic Acid. Dissolve 8 g sulfanilic acid in 1000 ml 5 N acetic acid.

Alpha-naphthylamine. Dissolve 5 g alpha-naphthylamine in 1000 ml 5 N acetic acid.

Test. The medium is tested for the presence of nitrites by adding a few drops each of the above reagent solutions. A distinct pink, red, or rust color indicates the presence of nitrite reduced from the original nitrate. If an organism grows rapidly and reduces nitrate actively, it is suggested that the test for nitrite be performed at an early incubation period, since the reduction may be carried beyond the nitrite stage. The test must always be controlled by comparison with a tube of uninoculated medium.

Oxidase Test

1. Culture the bacterium in question on nutrient agar or another suitable solid culture medium.

2. Flood some of the colonies with a 1% aqueous solution of *N,N,N',N*-tetramethyl-*p*-phenylenediamine dihydrochloride. (CAUTION: this chemical is extremely toxic; avoid contact with skin or clothing.) This solution should be stored under refrigeration and discarded after one month, or when it turns blue; however, it may be kept much longer (3–4 years) if stored frozen.
3. Colonies of oxidase-positive bacteria turn blue.

Phenylalanine Deaminase Test

This test is recommended for the separation of the Proteus and Providence groups from other members of the Enterobacteriaceae.

Phenylalanine Agar

Yeast extract	3 g
Dipotassium phosphate	1 g
Sodium chloride	5 g
dl-Phenylalanine	2 g
Agar	12 g
Distilled water	1 liter

Dissolve the dry ingredients in the water by heating to the boiling point. Distribute in tubes and autoclave at 15 lb for 15 min. Allow autoclaved medium to solidify in a slanted position.

Test Reagent. Dissolve 2.0 g ammonium sulfate and 1 ml 10% sulfuric acid in 5 ml half-saturated ferric ammonium sulfate.

Culture. Inoculate slant from a pure culture of the test bacterium and incubate at 30°–37°C for 18–24 hr.

Test. Add five drops of the test reagent to the slant culture and rotate the tubes to wet and loosen the growth. A characteristic green color develops in Proteus and Providence cultures.

Voges Proskauer (VP) Test for Bacillus spp., Other Gram-Positive Bacteria, and Nonenteric Gram-Negative Bacteria

[For Gram-negative enteric bacteria see Methyl Red–Voges Proskauer (MR-VP) test.] This is a test for the production of acetylmethylcarbinol or acetoin.

Culture Medium

Proteose peptone	7 g
Sodium chloride	5 g
Glucose	5 g
Distilled water	1 liter

Dissolve dry ingredients in water, tube at 6 ml per tube, and autoclave at 15 lb for 15 min. Store in airtight container under refrigeration.

Culture. Inoculate tubed medium from a pure culture of the test organism and incubate at 30°C for 5 days.

Test Reagents. 1. 5% alpha-naphthol in 100% ethyl alcohol
2. 40% aqueous potassium hydroxide

Test. To 6 ml of a 5-day-old culture, add 2.4 ml alpha-naphthol solution and 0.8 ml 40% KOH. Stopper and shake the tube for one minute. Slope the tube and examine at 15 min and 1 hr.

Results. Positive reaction is indicated by the development of a strong red color.

STAINS AND STAINING PROCEDURES

Acid-Fast Staining (Ziehl–Neelsen Method)

Staining Solutions

1. Ziehl's carbol fuchsin

Solution A

Basic fuchsin (90% dye content)	0.3 g
Ethyl alcohol (95%)	10 ml

Solution B

Phenol	5 g
Distilled water	95 ml

Mix solutions A and B

2. Acid alcohol: 95% ethyl alcohol with 3% by volume of concentrated HCl.
3. Counterstain

Methylene blue (80% dye content)	0.3 g

Ethyl alcohol (95%) .... 30 ml
Distilled water .... 100 ml

Staining Procedure

1. Stain dried smear for 3–5 min with Ziehl's carbol fuchsin; apply enough heat for gentle steaming.
2. Rinse in tap water.
3. Decolorize in acid alcohol until only a suggestion of pink remains.
4. Wash in tap water.
5. Counterstain with the methylene blue solution for 1 min.
6. Wash in tap water.
7. Air dry and examine under oil.

Results. Acid-fast organisms red; others blue.

Analine Blue–Lactophenol Analine Blue

This is a mounting medium and stain for fungi. Prepare lactophenol and add 0.5% analine blue. (See Lactophenol.)

Cotton Blue (Methyl Blue)–Lactophenol Cotton Blue

This is a mounting medium and stain for fungi. Prepare lactophenol and add 0.5% cotton blue (methyl blue). (See Lactophenol.)

Cytoplasmic Polyhedrosis Virus Detection (Sikorowski et al., 1971)

Staining Solution

The staining solution should be freshly prepared before use.

Buffalo black NBR (Allied Chemical) .. 0.1 g
100% Methyl alcohol.............. 50.0 ml
Distilled water .... 20.0 ml
Glacial acetic acid .... 30.0 ml

Preparation of Smear

1. Larvae 4 days old or older are starved for 12–24 hr.
2. Place larvae in 100% ethyl alcohol for 5 min.
3. Remove larva from alcohol, place on paper towel, and allow to dry.

4. Dissect out the midgut from between the last thoracic legs and the first pair of prolegs.
5. Mash the midgut with a flat toothpick to prepare a thin smear.
6. Air dry smear for 1–2 hr at room temperature.

Staining Procedures

1. Cover smear with stain for 5 min at 40°C (or 20 min at room temperature). Do not allow stain to dry during this period.
2. Drain slide and allow to air dry.
3. Wash gently in tap water for 5 sec.
4. Air dry and examine without cover glass under oil immersion (×1000).

Results. Polyhedra are stained navy blue, while background material is stained light blue.

Feulgen–Schiff Reaction for Granulosis Virus Capsules in Tissue

Reagents

1. *Fixative.* 10% Formalin in phosphate buffer at pH 6.7–7.0. Fix material at 4.0°C for up to 72 hr.
2. *Schiff's Reagent.* Dissolve 1 g of basic fuchsin in 200 ml boiling distilled water. Shake for 5 min and cool to 50°C. Filter and add to the filtrate 20 ml 1 N HCl. Cool to 25°C and add 1 g of sodium or potassium metabisulfite ($Na_2S_2O_5$). Place this solution in the dark for 14–24 hr. Add 2 g activated charcoal and shake for 1 min. Filter. Keep filtrate in the dark at 4.0°C. Allow to reach room temperature before use.
3. *Acid for Hydrolysis.* 1 N hydrochloric acid.
4. *Metabisulfite Solution* (prepare fresh). Dilute 5 ml 10% aqueous potassium or sodium metabisulfate and 5 ml 1 N HCl in 90 ml distilled water.

Sections. Dehydrate fixed material through alcohol series to xylol and embed in paraffin or paraplast. Cut sections 5 μm thick and fix to slides.

Method

1. Bring sections to water.
2. Rinse briefly in cold 1 N HCl.
3. Place in 1 N HCl at 60°C (preheated) for hydrolysis for 8 min.
4. Rinse briefly in cold 1 N HCl and then distilled water.

5. Transfer to Schiff's reagent for 30–60 min at room temperature.
6. Drain and rinse in three changes of freshly prepared metabisulfite solution.
7. Rinse in water.
8. Dehydrate in alcohol series, clear in xylol, and mount in balsam or synthetic resin.

Results. Granulosis virus capsules and virogenic stroma appear reddish-purple (violet) in color.

Flagella Stain

Bacterial flagella, often used as taxonomic characters, are very fine organelles of locomotion. In order to be seen under the light microscope, they must be treated in some manner to increase their dimension. The manner described here is taken from Leifson (1960).

Culture. Flagellated bacterial cells are more readily found in young cultures, especially in a phosphate-enriched broth medium. Nutrient broth with 0.1% potassium phosphate added is a good general flagella broth. Culture test organisms for 16 hr or less at 20°C in 3.0 cc flagella broth.

Fixation. Add 6.0 cc 10% formalin to the above broth culture.

Wash

1. Dilute the fixed culture with distilled water and centrifuge for 30 min at 3000 rpm in a clinical centrifuge.
2. Discard supernatant, suspend pellet in distilled water, and centrifuge as above.
3. Repeat.
4. Suspend final pellet in distilled water so that it is barely turbid.

Slide Preparation

1. Clean slides overnight in hot (70°–80°C) sulfuric acid saturated with potassium dichromate.
2. Rinse slides thoroughly in tap water, then distilled water, and then air dry. Do not touch cleaned slides with anything but clean forceps. They must be kept absolutely grease free. Store in a clean, dry, airtight container.
3. Just before use, heat a slide in the colorless flame of a bunsen

burner (the side to be used against the flame) and draw a line with a wax pencil across the slide from side to side about one third of the distance from one end. The slide should be handled only by this end.

4. Place a drop of the final bacterial suspension on the distal end of the cooled slide; tilt the slide to cause the suspension to run down to the wax line. Two such smears, side by side, are readily made on each slide. When the smear has air dried, it is ready to be stained.

Stain

1. Prepare three stock solutions:
   a. 1.2% Basic fuchsin in 95% ethyl alcohol
   b. 3.0% Tannic acid in distilled water
   c. 1.5% Sodium chloride in distilled water
2. Prepare the stain by mixing equal parts of the three stock solutions. The stain solution may be stored for 1 week at room temperature, 1–2 months under refrigeration, and indefinitely if frozen.

Stain Application

1. Place the prepared slide on a staining rack and flood with the staining solution for 5–15 min. (Shorter time for new and/or warm stain, longer for old and/or cold stain.)
2. Wash all the stain off the slide at once with running tap water. Do not allow the stain to run off the slide before it is placed under running water.
3. Air dry and examine under oil for flagella.

Giemsa Stain

Giemsa is a commercially prepared blood stain which may be purchased from many biological supply houses. It has many uses in microbiology. For general use in diagnostic work, the stock solution may be diluted at the rate of one drop per cc of distilled water or buffer.

Giemsa with HCl Hydrolysis

This stain is used to differentiate rickettsias from granulosis virus capsules.

1. Air dry smear.
2. Fix in any suitable fixative, e.g., Carnoy's for 5 min, or methanol for 3–4 min.
3. Hydrolyze in 0.1% HCl for 2 min.
4. Stain in Giemsa diluted 3 drops stock to 2 cc HOH for 15–45 min.
5. Wash in several changes of HOH.
6. Air dry.
7. Examine under oil, or dehydrate and mount.

Results. Rickettsias are strongly colored (positive) in contrast to granulosis virus capsules.

*Giemsa Stain for Microsporidian Spores**

1. Air dry smear.
2. Fix in methyl alchol for 3–4 min.
3. Air dry or blot dry.
4. Dilute stock solution 1 drop to 1 cc distilled water.
5. Stain smear for 15 min.
6. Wash in distilled water.
7. Air dry, or blot dry.
8. Examine under oil, or dehydrate and mount.

Slow Giemsa Staining with Acid Hydrolysis for Virus Inclusion Bodies

Buffer. In 185.0 ml distilled water, mix 1 ml 0.5 M KH_2PO_4 and 1.5 ml 0.5 M Na_2HPO_4. Final pH should be 6.89.

Giemsa Stain. Mix 0.5 ml Giemsa with 50 ml buffer.

Method

1. Fix sections, smear, etc. in Bouin Dubosq Brasil (see Iron Hematoxylin Stain) for 10 min.
2. Wash in 70% ethyl alcohol for three changes, one change every hour.
3. Rinse in distilled water for 5 min.
4. Rinse in Giemsa buffer for 5 min.

*Also see Giemsa with HCl Hydrolysis.

5. Stain in buffered Giemsa overnight.
6. Rinse in distilled water.
7. Differentiate in 80% ethyl alcohol with 5% glacial acetic acid (for hydrolysis) until the general blue color of the slide disappears and a spotty blue color remains.
8. Drain and immerse in a 1:1 mixture of acetone–xylene.
9. Pass through two changes of xylene.
10. Mount in balsam or synthetic resin.

Results. Inclusion bodies are purple in color and virus rods (if they can be seen) are red.

Giemsa Stain for Virus Polyhedrosis Inclusions

1. Air dry smear.
2. Treat air-dried smear with 0.1% HCl for 2–5 min.
3. Stain in diluted Giemsa for 5–10 min.
4. Rinse in running water for 5–10 sec.
5. Air dry and examine under oil.

Results. Polyhedrosis inclusions stain blue and are, therefore, more visible than normal under bright field illumination. Fat globules will stain purple to red, while other crystals, such as ureates, will not stain.

Gram Stain

The method of gram staining offered here is based on the work of Bartholomew (1962) and is recommended for producing less variable results than other methods.

Dye Formulas

1. *Ammonium Oxalate Crystal Violet* (Hucker Modification)
   Solution A. Dissolve 4 g crystal violet (90% dye content) in 40 ml 95% ethyl alcohol.
   Solution B. Dissolve 1.6 g ammonium oxalate in 160 ml distilled water. Mix solutions A and B. It is recommended that the resulting solution be allowed to stand 48 hr before use.
2. *Gram's Iodine.* Place 2 g potassium iodine into a mortar, add 1 g

of iodine, and grind with a pestle for 5–10 sec. Add 1 ml of distilled water and grind. The iodine and potassium iodide should now be in solution. Add 10 ml water and mix. Pour into a reagent bottle and rinse the mortar and pestle with sufficient water to bring the final volume to 200 ml.

3. *Counterstain.* Add 20 ml of 2.5% safranin (86% dye content) in 95% ethyl alcohol to 180 ml distilled water.

Procedure

1. Air dry smear and heat fix by passing lightly through a bunsen flame a few times (smear side up).
2. Place slide on a staining rack and flood with ammonium oxalate crystal violet for 1 min.
3. Rinse in tap water running in a beaker for 5 sec.
4. Rinse slide with Gram's iodine, then flood slide with this solution for 1 min.
5. Rinse in running tap water for 5 sec.
6. Pass the wet slide through three changes of *n*-propyl alcohol in separate coplin jars, 1 min each.
7. Rinse in running tap water for 5 sec.
8. Rinse slide with safranin counterstain then flood with counterstain for 1 min.
9. Rinse in running tap water for 5 sec, then air dry.
10. Examine under oil immersion.

Results. Gram-positive organisms, blue-violet; Gram-negative organisms, red. CAUTION: Many organisms are Gram variable, particularly in older cultures. Bacteria used for Gram stain should be from young cultures, preferably 8–16 hr old and no older than 24 hr.

Guegen's Solution

This is a recommended mounting medium and stain for fungi.

1. Prepare lactophenol and heat. (See Lactophenol.)
2. Saturate hot lactophenol with Sudan III, let cool, and filter.
3. Add 0.5% cotton blue (methyl blue) or analine blue.

Iron Hematoxylin Stain for Granulosis Virus Capsules in Tissue (Hunger, 1961)

Iron Alum. 2.5% aqueous ferric ammonium sulfate.

Heidenhain's Iron Hematoxylin

Stock Solution. Freshly prepared hematoxylin is not good for staining and must be "ripened." A stock solution may be prepared by dissolving 10% of hematoxylin crystals in 95% ethyl alcohol. It will ripen after several months and may be good for more than a year, but it will not keep indefinitely.

Stain. 0.5–1.0% aqueous hematoxylin. This may be prepared from the above ripened stock, or hematoxylin crystals may be dissolved directly in water and then allowed to stand ("ripen") for 3–6 weeks.

Fixative (Bouin Dubosq Brasil)

80% Ethyl alcohol	150 cc
Formalin	60 cc
Glacial acetic acid	15 cc
Picric acid	1 g

Method 1

a. Fix infected larvae or tissues in Bouin Dubosq Brasil overnight.
b. Wash in 70% ethyl alcohol, dehydrate, clear, and embed according to the usual paraffin or paraplast methods.
c. Cut sections 3–5 μm in thickness and fix them to slides.
d. Dewax in xylol and pass through descending series of ethyl alcohol to distilled water.
e. Hydrolyze in 50% acetic acid at room temperature for 5 min.
f. Rinse well in distilled water.
g. Mordant in 2.5% iron alum for 2 hr.
h. Rinse thoroughly in distilled water.
i. Stain in Heidenhain's iron hematoxylin for 5 hr.
j. Differentiate carefully in 2.5% iron alum solution with frequent rinsing and microscopic examinations until the black nuclei and the characteristic deep black network of diseased cells become clearly outlined.
k. Wash in running tap water for 1 hr and rinse in distilled water for 10 min.
l. Counterstain in 0.5% aqueous erythrosin for 2 min.
m. Dehydrate in ascending series of ethyl alcohol (70%, 80%, 95%, 100%), passing quickly to 95%, clear in xylol, and mount in balsam or synthetic resin.

Results. The capsules are selectively stained in a beautiful intense bright red color, and even single ones phagocytosed by blood cells are stained. At first sight they are clearly contrasted from the pale gray violet cytoplasm. Moreover, the typical network mentioned above is stained deep black, similar to the nuclei.

Method 2

a–d. Same as Method 1.
e. Hydrolyze in 1 N HCl for 10–20 min at 60°C.
f. Rinse well in distilled water.
g. Mordant in 2.5% iron alum for 5–10 hr.
h. Rinse thoroughly in distilled water.
i. Stain in Heidenhain's iron hematoxylin for 15–20 hr.
j. Differentiate carefully in 2.5% iron alum with frequent rinsing and microscopic examinations until the excess coloration is removed so that the capsules appear blue-black to black and structural details of the black network become visible.
k–m. Same as Method 1.

Results. The capsules show a striking blue-black color. If the sections are thin enough, the individual capsules become sharply outlined; otherwise they appear as blue-black to black masses. The network of infected cells is stained deep black. The nuclei appear black if hydrolysis was for 10–15 min and red if hydrolysis was for 20 min or more. The cytoplasm is colored pale red violet (10-min hydrolysis) to red (20-min hydrolysis).

This method is also convenient for staining smears, with hydrolysis extended to about 30 min in order to obtain deep black capsules.

Lactophenol

This is a mounting medium for fungi, and a base for Cotton Blue, Analine Blue, and Guegen's stains.

Phenol crystals	100 g
Lactic acid (USP 85%)	80 ml
Glycerine	159 ml
Distilled water	100 ml

Macchiavello Stain

1. Air dry smear.
2. Fix for 5–10 min in methanol.

3. Stain for 60 min in 1% basic fuchsin.
4. Rinse in HOH.
5. Differentiate for 5 sec in 0.5% citric acid.
6. Rinse in HOH.
7. Counterstain for 20 sec in 1% methylene blue.
8. Rinse in HOH.
9. Air dry and examine under oil, or dehydrate to xylene and mount in permount.

Results. Rickettsia, crystals, and globules of spheroidocytes, red. Albuminoid crystals and bacteria, blue.

Microsporidia Spore Stain for Determining the Number of Nuclei in Spores

Stain. Two percent lacto-aceto-orcein. 2 g synthetic orcein, 50 cc 85% lactic acid, and 50 cc glacial acetic acid.

Method

1. Remove some of the infected tissue (fat body, etc.) from the diseased host and place it on a cover slip.
2. Add 1–2 drops of 2% lacto-aceto-orcein stain and macerate the tissue with needles.
3. Let tissue and stain stand for 2 min.
4. Flip cover slip over on a microscope slide.
5. Blot excess stain away with filter paper while pushing down on the cover slip.
6. Examine under microscope.

Neutral Red Stain

1. Air dry smear.
2. Stain for 10–15 min in 0.5–1.0% neutral red in 100% ETOH.
3. Rinse.
4. Air dry.
5. Examine under oil or dehydrate and mount

Results. NR bodies in rickettsial infections stain red.

Spore Stain—Bacterial Spores

This is the Conklin modification of the Wirtz method (Wirtz, 1908; Conklin, 1934).

Staining Solutions

1. 5% aqueous malachite green.
2. 5% aqueous mercurichrome or 0.25% aqueous safranin.

Staining Procedure

1. Fix air-dried smear by passing slide rapidly through a bunsen flame (smear side up) two or three times.
2. Flood slide with malachite green solution and steam for 10 min, keeping slide flooded by addition of fresh staining solution.
3. Wash for 30 sec in running water.
4. Counterstain for 1 min with mercurichrome or for 15 sec with safranin.
5. Wash briefly in running water.
6. Air dry and examine.

Results. Spores stain green and the rest of the cell stains red. Cellular inclusions other than spores will stain red or not at all.

Sudan III

This stain is used to differentiate virus polyhedra from fat droplets

1. Air dry smear.
2. Stain for 10–15 min in saturated aqueous Sudan III.
3. Rinse for 5–10 sec in running tap water.
4. Air dry and examine under oil.

Results. Fat droplets stain red while polyhedra remain unstained.

MISCELLANEOUS TECHNIQUES

Bacteriological Loop (Fig. 239)

Wire inoculating or transfer loops are used for such purposes as making streak plates, inoculating broth cultures, and transferring inocula to slides for microscopic examination. The loop is made from 24- to 26-gauge platinum or nichrome wire fixed to a standard bacteriological wire loop holder. The wire is formed into a loop about 3 mm in diameter with a shaft about 8 cm long.

Bacteriological Needle or Wire for Stab Cultures

A needle about 8 cm long may be used for stab cultures, or a suitable substitute may be made from loop wire by twisting a double length of

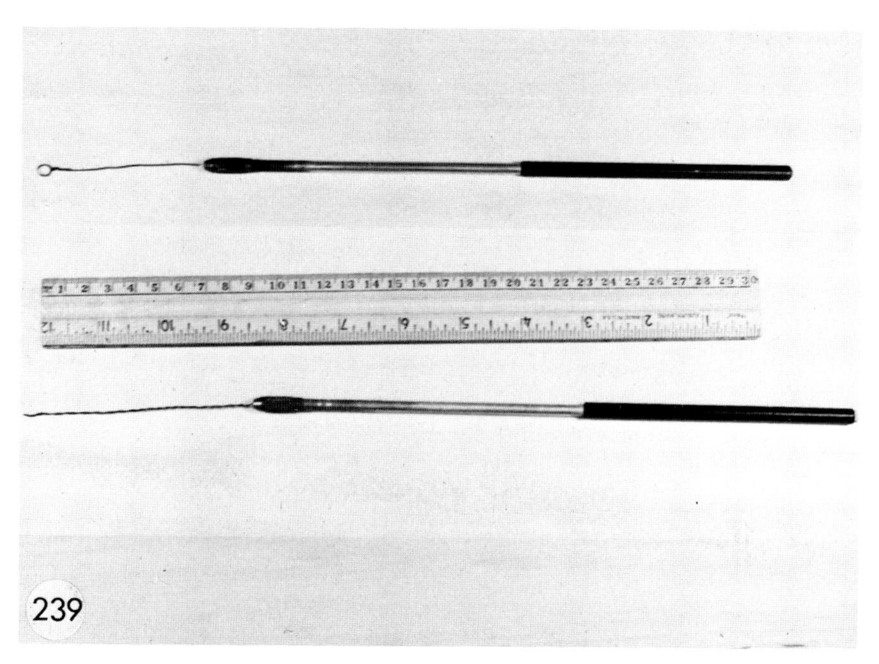

FIGURE 239. Bacteriological loop for streak plate cultures and twisted wire for stab cultures.

wire back on itself and leaving the last 15 mm single for insertion into the holder (Fig. 239).

Chloralphenol (Clearing Medium)

This is an excellent clearing agent for small arthropods. Specimens will not discolor; wings will not be destroyed (one of the advantages over KOH).

Phenol crystals	10 g
Water	3 ml
Chloralhydrate	10 g

Dissolve phenol in water and add chloralhydrate. Store in amber bottle.

Specimens can be placed in chloralphenol as soon as killed, directly from Weaver and Thomas fixative, or dry. Specimens kept in this medium for too long (3 or more days in the case of small chalcids) tend to burst at the intersegemental membranes.

Faure Mounting Medium

This is a medium of high refractive index, very good for mounting small arthropods.

Gum arabic (lump)	12 g
Chloralhydrate	20 g
Glacial acetic acid	5 ml
50% glucose syrup w/w	5 ml
Water	35 ml

Dissolve the gum in the water (may take several days), add other ingredients, and stir occasionally until chloralhydrate is dissolved. Filter through glass wool in a funnel or through filter paper (takes several days).

Specimens can be mounted immediately after being killed, or directly from Weaver and Thomas fixative or chloralphenol. Soft-bodied specimens, like larvae of endoparasites, should be mounted in Faure medium in which a few crystals of iodine have been dissolved. The iodine will compensate for the excessive clearing of the specimen by depositing itself in the more chitinized areas (mouth parts, buccal armatures, spiracles, sculptures on the integument).

Galleria Mellonella—Rearing Method

Rearing Medium

1. Mix together 100 cc water, 100 cc honey, 100 cc glycerine, and 5 cc Decca-Vi-Sol or equivalent vitamins.
2. Pour the liquid mixture into 1200 cc dry Pablum or Gerber's mixed cereal.
3. Mix until homogenous and place in a 1-gallon mason jar with a screen top. (A circle may be cut from a regular top and a piece of ordinary window screen cut to fit.)
4. Place 200–300 freshly collected eggs on the medium and incubate at 30°C or room temperature (30°C is preferable).

The larvae will reach the last instar in 4–5 weeks at 30°C and take another week or more at room temperature. The culture may be stored at 10°C for up to 3 months, after which it should be returned to rearing temperature and the remaining larvae allowed to pupate. Emerging adults are collected by anesthetizing with CO_2 or chilling in a refrigerator and are transferred to a clean 1-gallon mason jar with a screen top. New adults are added to the jar as they emerge. It is not necessary to provide the adults with food or water. A piece of accordion-pleated wax paper provides an oviposition site and unfolding the wax paper causes the eggs to fall off.

Koch's Postulates

The following postulates, devised by the German bacteriologist Robert Koch (1843–1910), provide a general outline which may be used when conducting infectivity tests. If these steps are carried out with positive results, it may be considered conclusive evidence that the microorganism in question is the cause of the disease.

1. The microorganism must be present in every case of the disease.
2. The microorganism must be isolated in pure culture.
3. The microorganism in pure culture, when inoculated into a susceptible animal, must give rise to the disease.
4. The same microorganism must be present in, and recoverable from, the experimentally diseased animal.

Microscopic Preparations—Wet Mount and Tissue Smear

Wet Mount. A wet mount is simply made by placing a bit of tissue or hemolymph in a drop of water or Ringer's solution on a microscope slide and covering it with a cover glass. The preparation can be kept longer by ringing the cover glass with Vaspar or nail polish.

Vaspar. Vaspar is prepared by melting together equal parts of Vaseline (petroleum jelly) and paraffin. It may be kept in a small crucible or other container and melted over a bunsen flame when needed. It is applied with a small watercolor brush in a thin line to seal the edges of a cover glass.

Tissue Smear. Tissue smears are prepared by dissecting out a small amount of tissue, which is squashed and spread on a microscope slide and allowed to air dry. The thinner the preparation the better it will be for microscopic examination. A wet mount may be used for this purpose by removing the cover slip and allowing it to air dry. Smears may be fixed by gentle heating or by chemical fixation as prescribed in the staining procedure.

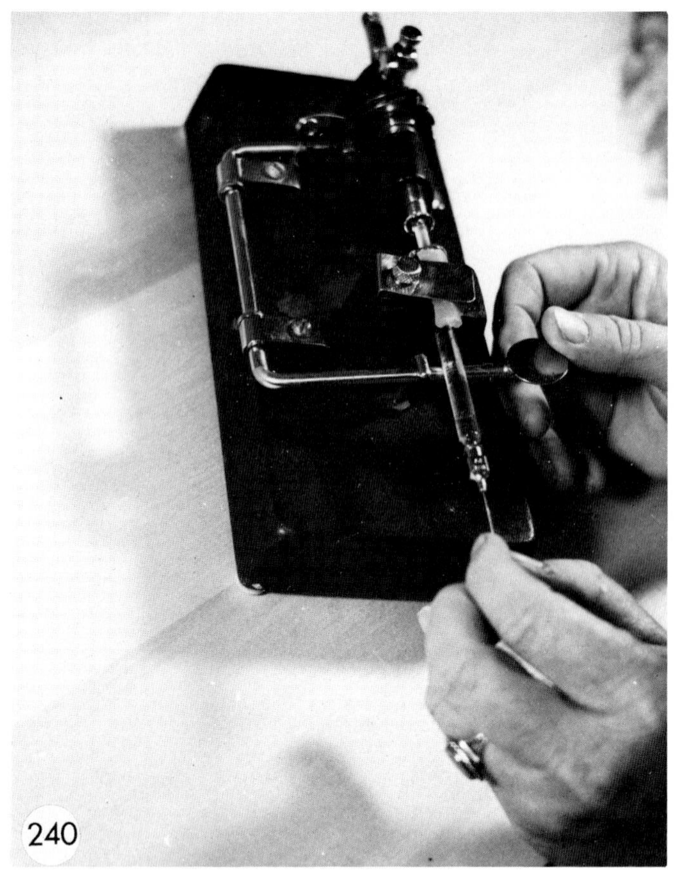

FIGURE 240. The Dutky–Fest microinjector, showing a hypodermic syringe fitted with a glass-tipped needle used for per os injections.

Per Os Injection

Viruses, rickettsias, bacteria, protozoa, and some fungi normally enter their insect host through the mouth and penetrate the gut wall into the hemocoel. This route of infection is termed per oral, or per os, and when testing such microorganisms for pathogenicity, a method for per os inoculation should be used. The simplest method is to contaminate the insect's food with a suspension of the suspected pathogen. A more precise method is to force-feed the inoculum to an insect by per os injection. One way this can be done is with a Dutky–Fest microinjector equipped with a 1-cc tuberculin syringe and a fine glass-tipped needle (Fig. 240).

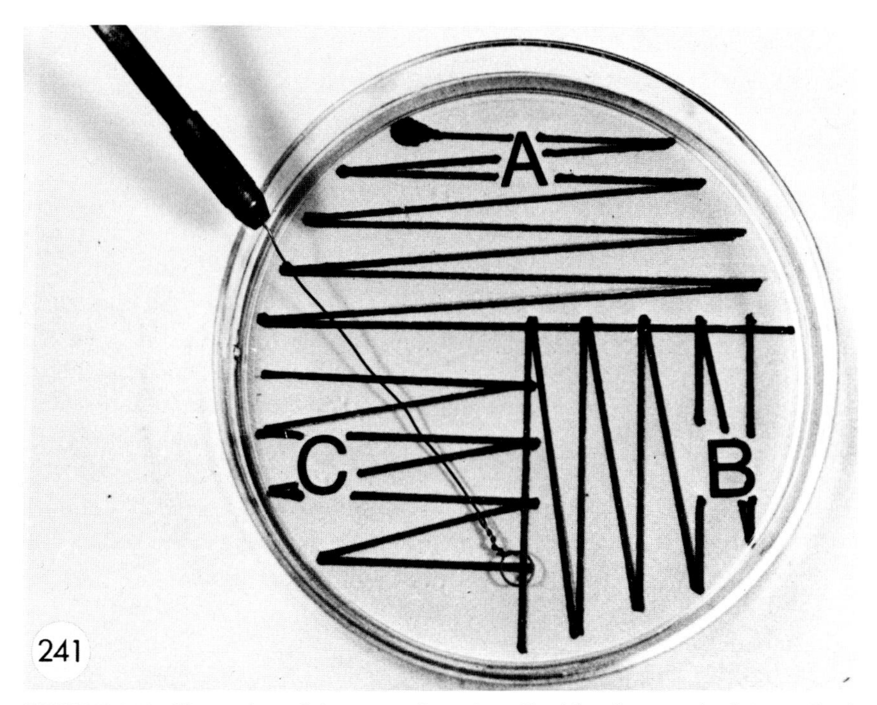

FIGURE 241. Illustration of the procedure described for the streak plate method.

Stab Cultures

Certain bacteriological media, such as nutrient gelatin and triple sugar iron agar, require inoculaion by the stab method. This is accomplished by taking a wire inoculating needle, flaming it red hot, cooling it in alcohol, burning off the alcohol, dipping the needle in the inoculum, and inserting it deeply into the test medium.

Streak Plate Method (Fig. 241)

The streak plate method is used to obtain separate colonies of a single bacterial species from a mixed culture or from a specimen containing a mixed flora, or to subculture pure isolates from one culture to another. Suspensions used for streak plates should be barely turbid.

Materials Needed

Petri plate with solid agar medium
Bacteriological inoculating loop (Fig. 239)

Suspension of inoculum
Bunsen burner, or alcohol burner

Procedures. Flame the loop along its entire length until red hot, then quickly pass the holder through the flame for a few inches. Cool the loop by dipping it into 70% alcohol and flaming off the excess.

1. Place a loopful of inoculum close to the edge of the agar in a petri dish and spread on area A as shown in Fig. 241. In doing this, raise the cover of the petri dish at one side just high enough to allow manipulation of the inoculating loop.
2. Flame the loop and, turning the plate 90 degrees, cover area B as illustrated, passing over the streak in A each time.
3. Turn the plate again and streak over area C without flaming the loop.
4. Incubate the plate in the inverted position. Isolated colonies should be found in area C if the inoculum was heavy, and in area B and possibly area A if the inoculum was lighter.

Virus Inclusion Bodies—Dissolution in Alkali

The dissolution of virus inclusion bodies in alkali may be demonstrated as follows: Prepare a wet mount of the material under investigation and place a drop of 1 N NaOH at the edge of the cover slip, allowing it to enter while observing under high dry (×400) magnification. As the alkali flows through, most virus inclusion bodies will swell and dissolve. The swelling is difficult to see in granulosis capsules, but is quite evident with polyhedrosis inclusions (Fig. 19).

Weaver and Thomas Fixative

A good medium to preserve small insects. It has the advantage over alcohol that the specimens will not shrink. Also suitable to preserve soft-bodied specimens (larvae).

Formaldehyde solution (approximately 37%)	20 ml
Glacial acetic acid	6.25 ml
Chloralhydrate	50 ml
Water	250 ml

REFERENCES

Afrikian, E. K. 1973. *Entomopathogenic Bacteria and Their Significance.* Armenian Academy of Sciences, Erevan. 418 pp. [in Russian].

Ainsworth, G. C. 1961. *Ainsworth and Bisby's Dictionary of the Fungi.* 5th Ed. Commonwealth Mycological Institute, Kew, Surrey, England. 547 pp.

Ainsworth, G. C., F. K. Sparrow, and A. S. Sussman. 1973. *The Fungi; An Advanced Treatise,* Vols. IVA (621 pp.) and IVB (504 pp.). Academic Press, New York.

Aruga, H., and Y. Tanada (Eds.). 1971. *The Cytoplasmic-Polyhedrosis Viruses of the Silkworm.* University of Tokyo Press, Tokyo. 234 pp.

Askew, R. R. 1971. *Parasitic Insects.* Elsevier, New York. 316 pp.

Assmuss, E. P. 1858. Verzeichniss einiger Insecten, in denen ich Gordiaceen antraf. *Wien. Entomol. Monatsschr.* **2:**171–181.

Bailey, L. 1973a. Control of invertebrates by viruses. In: *Viruses and Invertebrates* (Ed. A. J. Gibbs). North-Holland, Amsterdam. 533–553.

Bailey, L. 1973b. Viruses and Hymenoptera. In: *Viruses and Invertebrates* (Ed. A. J. Gibbs). North-Holland, Amsterdam. 442–454.

Bailey, L. 1976. Viruses attacking the honey bee. *Adv. Virus Res.* **20:**271–304.

Bailey, L., and R. D. Woods. 1977. Bee viruses. In: *The Atlas of Insect and Plant Viruses.* Vol. 8 of *Ultrastructure in Biological Systems* (Ed. K. Maramorosch). Academic Press, New York. 141–158.

Bailey, L., A. J. Gibbs, and R. D. Woods. 1964. Sacbrood virus of the larval honeybee (*Apis mellifera* Linnaeus). *Virology* **23:**425–429.

Bailey, L., J. F. E. Newman, and J. G. Porterfield. 1975. The multiplication of Nodamura virus in insect and mammalian cell cultures. *J. Gen. Virol.* **26:**15–20.

de Barjac, H., and A. Bonnefoi. 1973. Mise au point sur la classification des *Bacillus thuringiensis. Entomophaga* **18:**5–17.

Barnett, H. L., and B. B. Hunter. 1972. *Illustrated Genera of Imperfect Fungi.* 3rd Ed. Burgess Publishing, Minneapolis, Minnesota. 241 pp.

Bartholomew, J. W. 1962. Variables influencing results and precise definition of steps in gram staining as a means of standardizing the results obtained. *Stain Technol.* **37:**139–155.

Baudoin, J. 1969. Nouvelles espèces de Microsporidies chez des larves de Trichoptères. *Protistologica* **5:**441–446.

Bell, J. V. 1974. Mycoses. In: *Insect Diseases.* Vol. I (Ed. G. E. Cantwell). Marcel Dekker, New York. 185–236.

Bell, J. V., and R. J. Hamalle. 1974. Viability and pathogenicity of entomogenous fungi after prolonged storage on silica gel at −20°C. *Can. J. Microbiol.* **20:**639–642.

Bergoin, M., and S. Dales. 1971. Comparative observations on poxviruses of invertebrates and vertebrates. In: *Comparative Virology* (Eds. K. Maramorosch and E. Kurstak). Academic Press, New York. 171–205.

Bishop, D. H. L., and R. E. Shope. 1979. Bunyaviridae. *Comp. Virol.* **14:**1–156.

Bland, C. E., J. N. Couch, and S. Y. Newell. 1981. Identification of *Coelomomyces,* Saprolegniales and Lagenidiales. In: *Microbial Control of Pests and Plant Diseases, 1970–1980* (Ed. H. D. Burges). Academic Press, New York. 129–162.

Bovien, P. 1937. Some types of associations between nematodes and insects. *Vidensk. Medd. Dan. Naturhist. Foren. Khobenhavn* **101.** 114 pp.

Brooks, M. A. 1974. Genus VIII *Symbiotes* and Genus IX *Blattabacterium.* In: *Bergey's Manual of Determinative Bacteriology.* 8th Ed. (Eds. R. E. Buchanan and N. E. Gibbons). Williams & Wilkins, Baltimore. 900–901.

Brooks, W. M. 1974. Protozoan infections. In: *Insect Diseases.* Vol. I (Ed. G. E. Cantwell). Marcel Dekker, New York. 237–300.

Brown, A. H. S., and G. Smith. 1957. The genus *Paecilomyces* Bainier and its perfect stage *Byssochlamys* Westling. *Trans. Br. Mycol. Soc.* **40:**17–89.

Buchanan, R. E., and N. E. Gibbons (Eds.). 1974. *Bergey's Manual of Determinative Bacteriology.* 8th Ed. Williams and Wilkins, Baltimore. 1268 pp.

Bucher, G. E. 1959. Bacteria of grasshoppers of western Canada. III. Frequency of occurrence, pathogenicity. *J. Insect Pathol.* **1:**391–405.

Bucher, G. E. 1961. Control of the eastern tent caterpillar, *Malacosoma americanum* (Fabricius), by distribution of spores of two species of *Clostridium. J. Insect Pathol.* **3:**439–445.

Bucher, G. E. 1963. Nonsporulating bacterial pathogens. In: *Insect Pathology: An Advanced Treatise.* Vol. II (Ed. E. A. Steinhaus). Academic Press, New York. 117–147.

Bulla, L. A., Jr., and T. C. Cheng (Eds.). 1976. *Biology of the Microsporidia.* Vol. 1 of *Comparative Pathobiology.* Plenum Press, New York. 371 pp.

Burges, H. D. (Ed.). 1981. *Microbial Control of Pests and Plant Diseases, 1970–1980.* Academic Press, New York. 949 pp.

Burges, H. D., and N. W. Hussey. 1971. *Microbial Control of Insects and Mites.* Academic Press, New York and London. 861 pp.

Burkholder, W. E., and R. J. Dicke. 1964. Detection by ultraviolet light of stored product insects infected with *Mattesia dispora. J. Econ. Entomol.* **57:**878–879.

Canning, E. U. 1956. A new eugregraine of locusts, *Gregarina garnhami* n. sp., parasitic in *Schistocera gregaria* Forsk. *J. Protozool.* **3:**50–62.

Cantwell, G. E. 1974. Honey bee diseases, parasites and pests. In: *Insect Diseases.* Vol. 2 (Ed. G. E. Cantwell). Marcel Dekker, New York. 501–547.

Christie, J. R. 1950. Parasites of invertebrates. In: *Introduction to Nematology* (Eds. B. G. Chitwood and M. B. Chitwood). Baltimore, University Park Press, 1974. 246–266.

Clark, T. B. 1978. A filamentous virus of the honey bee. *J. Invert. Pathol.* **32:**332–340.

Clark, T. B., and T. Fukuda. 1971. *Plistophora chapmani* n. sp. in *Culex territans* Louisiana. *J. Invert. Pathol.* **18:**400–404.

Clark, T. B., W. R. Kellen, and P. T. M. Lum. 1965. A mosquito iridescent virus (MIV) from *Aedes taenorhynchus* (Wiedeman). *J. Invert. Pathol.* **7:**519–521.

Clark, T. B., W. R. Kellen, J. E. Lindegren, and R. D. Sanders. 1966. *Pythium* sp. (Phycomycetes: Pythiales) pathogenic to mosquito larvae. *J. Invert. Pathol.* **8:**351–354.

Clausen, C. P. 1940. *Entomophagous Insects.* McGraw-Hill, New York and London. 688 pp.

Codreanu, M. 1940. Sur quatre grégarines nouvelles du genre *Enterocystis,* parasites des Ephémères torrenticoles. *Arch. Zool. Exp. Gen.* **81:**113–122.

Codreanu, R. 1966. On the occurrence of spore or sporont appendages in the Microsporida and their taxonomic significance. *Proc. First Cong. Parasitol. Rome.* 602–603.

Codreanu, R., and S. Vavra. 1970. The structure and ultrastructure of the microsporidan *Telomyxa glugeiformis* Léger and Hesse 1910 parasite of *Ephemera danica* (Mull) nymphs. *J. Protozool.* **17:**374–384.

Conklin, M. E. 1934. Mercurochrome as a bacteriological stain. *J. Bacteriol.* **27:**30.

Corliss, T. O. 1960. *Tetrahymena chironomini* sp. nov., a ciliate from midge larvae, and the current status of facultative parasitism in the genus *Tetrahymena. Parasitology* **50:**11–153.

Couch, J. N. 1937. A new fungus intermediate between the rusts and *Septobasidium. Mycologia* **29:**665–673.

Couch, J. N. 1938. *The Genus Septobasidium.* University of North Carolina Press, Chapel Hill, North Carolina. 480 pp.

Couch, J. N., S. V. Romney, and B. Rao. 1974. A new fungus which attacks mosquitos and related diptera. *Mycologia* **66:**374–379.

Couch, J. N., R. V. Andreeva, M. Laird, and R. A. Nolan. 1979. *Tabanomyces milkoi* (Dudka and Koval) emended *genus novum,* a fungal pathogen of horseflies. *Proc. Natl. Acad. Sci. USA* **76**(5)**:**229–230.

Delgarno, L., and M. W. Davey. 1973. Virus replication. In: *Viruses and Invertebrates* (Ed. A. J. Gibbs). North-Holland, Amsterdam. 245–270.

David, W. A. L. 1975. The status of viruses pathogenic for insects and mites. *Annu. Rev. Entomol.* **20:**97–117.

Davidson, E. W. 1981. *Pathogenesis of Invertebrate Microbial Diseases.* Allanheld, Osmun, Montclair, New Jersey. 562 pp.

DeHoog, G. S. 1972. The genera *Beauveria, Isaria, Tritirachium* and *Acrodontium* gen. nov. In: *Studies in Mycology.* Vol. 1. Centraalbureau voor Schimmelcultures, Baarn. 1–41.

Diesing, K. M. 1851. *Systema helminthum 2.* Wilhelm Braumüller, Vienna, Austria. 588 pp.

Difco Laboratories. 1953. *Difco Manual of Dehydrated Media and Reagents for Microbiological and Clinical Laboratory Procedures.* 9th Ed. Difco Laboratories, Detroit, Michigan. 350 pp.

Dissanaike, A. S. 1955. A new Schizogregarine *Triboliocystis garnhami* n.g., n.sp., and a new Microsporidian *Nosema buchleyi* n. sp. from the fat body of the flour beetle *Tribolium castaneum. J. Protozool.* **2:**150–156.

Doane, C. C., and J. J. Redys. 1970. Characteristics of motile strains of *Streptococcus faecalis* pathogenic to larvae of the gypsy moth. *J. Invert. Pathol.* **15:**420–430.

Dobos, P., B. J. Heil, R. Hallett, D. T. C. Kells, T. Brecht, and D. Teninges. 1979. Biophysical and biochemical characterization of five animal viruses with bisegmented double stranded RNA genomes. *J. Virol.* **32:**593–605.

Doby, J. M., and F. Saguez. 1964. *Weiseria,* genre nouveau de Microsporidies et *Weiseria laurenti* n. sp., parsites de larves de *Prosimulium inflatum* Davies, 1957 (Diptères Paranématocères). *C. R. Acad. Sci. Paris* **259:**3614–3617.

Dutky, S. R. 1963. The milky diseases. In: *Insect Pathology: An Advanced Treatise.* Vol. II (Ed. E. A. Steinhaus). Academic Press, New York. 75–115.

Entwistle, P. F., and J. S. Robertson. 1968. Rickettsia pathogenic to two saturnid moths. *J. Invert. Pathol.* **10:**345–354.

Evlakhova, A. A. 1974. *Entomogenous Fungi: Classification, Biology and Practical Significance.* Nauka, Leningrad. 260 pp. [in Russian].

Falcon, L. A. 1971. Use of bacteria for microbial control. In: *Microbial Control of Insects and Mites* (Eds. J. D. Burges and N. W. Hussey). Academic Press, New York. 67–95.

Falcon, L. A., and R. T. Hess. 1977. Electron microscope study on the replication of *Autographa* nuclear polyhedrosis virus and *Spodoptera* nuclear polyhedrosis virus in *Spodoptera exigua. J. Invert. Pathol.* **29:**36–43.

Faulkner, P. 1981. Baculovirus. In: *Pathogenesis of Invertebrate Microbial Diseases* (Ed. E. W. Davidson). Allanheld, Osmun, Montclair, New Jersey. 3–37.

Faust, R. M. 1974. Bacterial diseases. In: *Insect Diseases.* Vol. I (Ed. G. E. Cantwell). Marcel Dekker, New York. 87–183.

Federici, B. A. 1975. *Cyclops vernalis* (Copepoda: Cyclopoida), an alternate host for the fungus *Coelomomyces punctatus.* In: *Proceedings of the 43rd Conference of the California Mosquito Control Association.* 172–174.

Finlayson, T., and K. S. Hagen. 1979. Final-instar larvae of parasitic Hymenoptera. Pest Management Paper No. 10. Simon Fraser University, Burnaby, British Columbia. 111 pp.

Fraenkel-Conrat, H., and R. R. Wagner (Eds.). 1981. *Comprehensive Virology.* Vol. 17. *Methods Used in the Study of Viruses.* Plenum Press, New York, 463 pp.

Gams, W. 1971. *Tolypocladium,* eine Hyphomycetengattung mit geschwollenen Phialiden. *Persoonia* **6**(2)**:**171–294.

Glaser, R. W. 1932. Studies on *Neoaplectana glaseri,* a nematode parasite of the Japanese beetle (*Popillia japonica*). New Jersey Department of Agriculture Circular No. 211. 34 pp.

Glinski, Z. 1968. Bacteriological diagnostics of the bee foul brood. *Med. Weter.* **24:**346–349.

Granados, R. R. 1973. Insect poxviruses: Pathology, morphology and development. *Misc. Publ. Entomol. Soc. Am.* **9:**73–94.

Granados, R. R. 1981. Entomopoxvirus infections in insects. In: *Pathogenesis of Invertebrate Microbial Diseases* (Ed. E. W. Davidson). Allanheld, Osmun, Montclair, New Jersey. 101–126.

Greenberg, B. 1971. *Flies and Disease: Ecology, Classification and Biotic Associations.* Vol. I. Princeton University Press, Princeton, New Jersey. 856 pp.

Hagen, K. S. 1969. Developmental stages of parasites. In: *Biological Control of Insect Pests and Weeds* (Eds. P. DeBach and E. I. Schlinger). Reinhold, New York. 168–246.

Hagmeier, A. 1912. Beiträge: zur Kenntnis der Mermithiden. *Zool. Jahrb. Abt. Syst.* **32:**521–612.

Hall, R. A. 1981. The fungus *Verticillium lecanii* as a microbial insecticide against aphids and scales. In: *Microbial Control of Pests and Plant Diseases, 1970–1980* (Ed. H. D. Burges). Academic Press, New York. 483–498.

Harrap, K. A. 1973. Virus infection in invertebrates. In: *Viruses and Invertebrates* (Ed. A. J. Gibbs). North-Holland, Amsterdam. 271–299.

Harrap, K. A., and C. C. Payne. 1979. The structural properties and identification of insect viruses. *Adv. Virus Res.* **25:**273–355.

Hassan, S., and C. Vago. 1972. The pathogenicity of *Fusarium oxysporum* to mosquito larvae. *J. Invert. Pathol.* **20:**268–271.

Hazard, E. T., and D. W. Anthony. 1974. A redescription of the genus *Parathelohania* Codreanu 1966 (Microsporida: Protozoa) with a reexamination of previously described species of *Thelohania* Henneguy 1892 and descriptions of two new species of *Parathelophania* from anopheline mosquitos. U.S. Department of Agriculture Technical Bulletin No. 1505. 26 pp.

Hazard, E. T., and S. W. Oldacre. 1975. Revision of Microsporida (Protoza) close to *Thelohania,* with descriptions of one new family, eight new genera and thirteen new species. *USDA Tech. Bull.* **1530:**1–104.

Hazard, E. T., and K. E. Savage. 1970. *Stempellia lunata* sp. n. in larvae of the mosquito *Culex pilosus* collected in Florida. *J. Invert. Pathol.* **15:**49–54.

Hazard, E. T., E. A. Ellis, and D. J. Joslyn. 1981. Identification of microsporida. In: *Microbial Control of Pests and Plant Diseases, 1970–1980* (Ed. H. D. Burges). Academic Press, New York. 163–182.

Heimpel, A. M. 1967. A taxonomic key proposed for the species of the "Crystalliferous Bacteria." *J. Invert. Pathol.* **9:**364–375.

Heimpel, A. M., and T. A. Angus. 1963. Diseases caused by certain spore-forming bacteria. In: *Insect Pathology: An Advanced Treatise.* Vol. II (Ed. E. A. Steinhaus). Academic Press, New York. 21–73.

Henry, J. E., and E. A. Oma. 1981. Pest control by *Nosema locustae,* a pathogen of grasshoppers and crickets. In: *Microbial Control of Pests and Plant Diseases, 1970–1980* (Ed. H. D. Burges). Academic Press, New York. 573–586.

Hesse, E. 1935. Sur quelques microsporidies parasites de *Megacyclops viridis. Arch. Zool. Exp. Gen,* **75:**651–661.

Hillman, B. T., J. Morris, W. R. Kellen, D. Hoffman, and P. E. Schlegel. 1982. An invertebrate calici-like virus: Evidence for partial virion disintegration in host excreta. *J. Gen. Virol.* **60:**115–123.

Holldobler, K. 1930. Uber eine merkwurdige Parasitenerkrankung von *Solenopsis fugas. Z. Parasitenkd.* **2:**67–72.

Honigberg, B. (chairman), *et al.* 1964. A revised classification of the phylum Protozoa. *J. Protozool.* **11:**7–20.

Hope, F. W. 1840. Journal of the proceedings. *Proc. Entomol. Soc. London* **2:**84–86

Howard, C. R. (Ed.) 1982. *New Developments in Practical Virology.* Alan R. Liss, New York. 343 pp.

Huger, A. 1961. Methods for staining capsular virus inclusion bodies typical of granuloses of insects. *J. Insect Pathol.* **3:**338–341.

Huger, A. 1966. A virus disease of the Indian rhinoceros beetle, *Oryctes rhinoceros* (Linnaeus), caused by a new type of insect virus, *Rhabdionvirus oryctes* gen. n., sp. n. *J. Invert. Pathol.* **8:**38–51.

Hunter, B. F. 1970. Ecology of waterfowl botulism toxin production. In: *Transactions of the 35th North American Wildlife Natural Resources Conference.* 9 pp.

Ignoffo, C. M. 1968. Viruses—Living insecticides. In: *Insect Viruses* (Ed. K. Maramorosch). *Curr. Top. Microbiol. Immunol.* **42:**129–167.

Ignoffo, C. M. 1974. Microbial control of insects: Viral pathogens. In: *Proceedings of the Summer Institute on Biological Control of Plant Insects and Diseases* (Eds. F. G. Maxwell and F. A. Harris). University Press, Jackson, Mississippi. 541–557.

Ishihara, R. 1967. Stimuli causing extrusion of polar filaments of *Glugea fumiferanae* spores. *Can. J. Microbiol.* **13:**1321–1332.

Jahn, T. L. 1949. *How to Know the Protozoa.* Wm. C. Brown, Dubuque, Iowa. 234 pp.

Jamnback, H. A. 1970. *Caudospora* and *Weiseria,* two genera of Microsporidia parasitic in blackflies. *J. Invert. Pathol.* **16:**3–13.

Jouvenaz, D. P., and E. I. Hazard. 1978. New family, genus and species of Microsporida (Protozoa: Microsporida) from the tropical fire ant, *Solenopsis geminata* (Fabricius) (Insecta: Formicidae). *J. Protozool.* **25:**24–29.

Kamburov, S. S., D. J. Nadel, and R. Kenneth. 1967. Observations on *Hesperomyces virescens* Thaxter (Laboulbeniales), a fungus associated with premature mortality of *Chiolocorus bipustulatus* L. in Israel. *Isr. J. Agric. Res.* **17:**131–134.

Karling, J. S. 1948. Chytridiosis of scale insects. *Am. J. Bot.* **35:**246–254.

Keilin, D. 1920. On two new gregarines. *Allantocystis dasyhelei* n. g., n. sp., and *Dendrohyrhynchus systeni* n. g., n. sp., parasitic in the alimentary canal of the dipterous larvae, *Dasyhelea obscura* Winn. and *Systenus* sp. *Parasitology* **1:**154–158.

Kellen, W. R., and J. E. Lindegren. 1974. Life cycle of *Helicosporidium parasiticum* in the naval orangeworm, *Paramyelois transitella. J. Invert. Pathol.* **23:**202–208.

Kellen, W. R., T. B. Clark, J. E. Lindegren, B. C. Hoe, M. H. Rogoff, and S. Singer. 1965. *Bacillus sphaericus* Neide as a pathogen of mosquitoes. *J. Invert. Pathol.* **7:**442–448.

Kellen, W. R., J. E. Lindegren, and D. F. Hoffman. 1972. Developmental stages and structure of a *Rickettsiella* in the naval orangeworm, *Paramyelois transitella* (Lepidoptera: Phycitidae). *J. Invert. Pathol.* **20:**193–199.

Kelly, D. C. 1981. Non-occluded viruses. In: *Pathogenesis of Invertebrate Microbial Diseases* (Ed. E. W. Davidson). Allanheld, Osmun, Montclair, New Jersey. 39–60.

Kelly, D. C., and J. S. Robertson. 1973. Icosohedral cytoplasmic deoxyriboviruses. *J. Gen. Virol* **21:**17–41.

King, D. S., and R. A. Humber. 1981. Identification of the entomophthorales. In: *Microbial Control of Pests and Plant Diseases, 1970–1980* (Ed. H. D. Burges). Academic Press, New York. 107–127.

Kobayashi, M. 1971. Replication cycle of cytoplasmic polyhedrosis virus as observed with the electron microscope. In: *The Cytoplasmic Polyhedrosis Virus of the Silkworm* (Eds. H. Aruga and Y. Tanada). University of Tokyo Press, Tokyo. 103–128.

Koval, E. Z. 1974. *Guide Book to Entomophilic Fungi of the USSR*. Naukova Dumka, Kiev. 260 pp. [in Russian].

Krieg, A. 1963. Rickettsiae and rickettsioses. In: *Insect Pathology: An Advanced Treatise*. Vol. I (Ed. E. A. Steinhaus). Academic Press, New York. 577–617.

Krieg, A. 1971. Possible use of Rickettsiae for microbial control of insects. In: *Microbial Control of Insects and Mites* (Eds. H. D. Burges and N. W. Hussey). Academic Press, New York. 173–179.

Krombein, K. V., P. D. Hurd, Jr., D. R. Smith, and B. D. Burks. 1979. Catalog of Hymenoptera in America North of Mexico. USDA, Agricultural Research Service. U.S. Government Printing Office, Washington, D.C.

Kudo, R. 1924. A biologic and taxonomic study of the Microsporidia. III. *Biol. Monogr.* **9:**1–268.

Kudo, R. 1942. On the microsporidian, *Duboscqia legeri,* parasitic in *Reticulitermes flavipes. J. Morphol.* **7:**307–333.

Kudo, R. 1966. *Protozoology*. 5th Ed. Charles C. Thomas, Springfield, Illinois. 1174 pp.

Kurstak, E. 1972. Small DNA densonucleosis virus (DNV). *Adv. Virus Res.* **17:**207–241.

Kurstak, E., and S. Garzon. 1977. Entomopoxviruses (poxviruses of invertebrates) In: *The Atlas of Insect and Plant Viruses.* Vol. 8 of *Ultrastructure in Biological Systems* (Ed. K. Maramorosch). Academic Press, New York. 29–66.

Kurstak, E., P. Tijsen, and S. Garzon. 1977. Densonucleosis viruses (Parvoviridae). In: *The Atlas of Insect and Plant Viruses.* Vol. 8 of *Ultrastructure in Biological Systems* (Ed. K. Maramorosch). Academic Press, New York. 67–92.

La Rivers, I. 1949. Entomic nematode literature from 1926 to 1946, exclusive of medical and veterinary titles. *Wasmann Collect.* **7:**177–206.

Lee, P. E. 1977. Iridoviruses (Iridoviridae). In: *The Atlas of Insect and Plant Viruses.* Vol. 8 of *Ultrastructure in Biological Systems* (Ed. K. Maramorosch). Academic Press, New York. 93–104.

Leifson, E. 1960. *Atlas of Bacterial Flagellation.* Academic Press, New York. 171 pp.

Léger, L. 1926. Sur *Trichoduboscqia epori* Léger, Microsporidie parasite des larves d'Ephémerides. *Trav. Lab. Hydrobiol. Piscicult. Univ. Grenoble* **18:**9–14.

Léger, L., and E. Hesse. 1905. Sur un nouveau protiste parasite des Otiorrhynques. *C. R. Soc. Biol.* **58:**92–94.

Lipa, J. J. 1963. Infections caused by protozoa other than sporozoa. In: *Insect Pathology: An Advanced Treatise.* Vol. II (Ed. E. A. Steinhaus). Academic Press, New York. 348–351.

Longworth, J. F. 1973. Viruses and lepidoptera. In: *Viruses and Invertebrates* (Ed. A. J. Gibbs). North-Holland, Amsterdam. 428–441.

Longworth, J. F. 1978. Small isometric viruses of invertebrates. *Adv. Virus Res.* **23:**103–157.

Ludwig, F. W. 1947. Studies on the protozoan fauna of the larvae of the crane fly *Tipula abdominalis.* II. The life history of *Ithania wenrichi* n. gen., n. sp., a coccidian found in the caeca and mid-gut, and a diagnosis of Ithaniinae n. subfamily. *Trans. Am. Microsc. Soc.* **66:**22–33.

Lysenko, O. 1963. The taxonomy of entomogenous bacteria. In: *Insect Pathology: An*

Advanced Treatise. Vol. II (Ed. E. A. Steinhaus). Academic Press, New York. 1–20.

McEwen, F. L. 1963. Cordyceps infections. In: *Insect Pathology: An Advanced Treatise* (Ed. E. A. Steinhaus). Academic Press, New York. 273–290.

Machado, A. 1913. Zytologie und Entwicklungszyklus der *Chagasella alydi,* einer neuen Kokzidienart aus einer Wanze vom "Genus Alydus." *Mem. Inst. Oswaldo Cruz* **5**:32–44.

McLaughlin, R. E. 1965. *Mattesia grandis* n. sp., a sporozoan pathogen of the boll weevil, *Anthonomis grandis* Boheman. *J. Protozool.* **12**:405–413.

McLaughlin, R. E. 1971. Use of protozoans for microbial control of insects. In: *Microbial Control of Insects and Mites* (Eds. H. D. Burges and N. W. Hussey). Academic Press, New York. 151–172.

McLaughlin, R. E. 1973. Protozoa as microbial control agents. *Misc. Publ. Entomol. Soc. Am.* **9**:95–98.

MacLeod, D. M. 1963. Entomophthorales infections. In: *Insect Pathology: An Advanced Treatise*. Vol. II (Ed. E. A. Steinhaus). Academic Press, New York. 189–231.

Maddox, J. V., W. M. Brooks, and J. R. Fuxa. 1981. *Vairimorpha necatrix,* a pathogen of agricultural pests: Potential for pest control. In: *Microbial Control of Pests and Plant Diseases, 1970–1980* (Ed. H. D. Burges). Academic Press, New York. 587–594.

Madelin, M. F. 1963. Diseases caused by hyphomycetous fungi. In: *Insect Pathology: An Advanced Treatise* (Ed. E. A. Steinhaus). Academic Press, New York. 233–271.

Madelin, M. F. 1966. Fungal parasites of insects. *Annu. Rev. Entomol.* **11**:423–448.

Mains, E. B. 1950. Entomogenous species of *Akanthomyces, Hymenostilbe* and *Insecticola* in North America. *Mycologia* **42**:566–589.

Mains, E. B. 1951. Entomogenous species of *Hirsutella, Tilachlidium and Synnematium. Mycologia* **43**:691–718.

Mains, E. B. 1959. North American species of *Aschersonia* parasitic on Aleyrodidae. *J. Insect. Pathol.* **1**:43–47.

Maramorosch, K. 1968a. Plant pathogenic viruses in insects. In: *Insect Viruses* (Ed. K. Maramorosch). *Curr. Top. Microbiol. Immunol.* **42**:94–107.

Maramorosch, K. (Ed.). 1968b. *Insect Viruses. Curr. Top. Microbiol. Immunol.* **42.** 192 pp.

Marshall, I. D. 1973. Viruses and Diptera. In: *Viruses and Invertebrates* (Ed. A. J. Gibbs). North-Holland, Amsterdam. 406–427.

Martignoni, M. E. 1981. A catalogue of viral diseases of insects, mites and ticks. In: *Microbial Control of Pests and Plant Diseases, 1970–1980* (Ed. H. D. Burges). Academic Press, New York. 899–912.

Martignoni, M. E., and P. J. Iwai. 1975. A catalog of viral diseases of insects and mites. USDA Forest Service General Technical Report No. 40. 23 pp.

Martignoni, M. E., P. J. Iwai, and L. J. Wickerham. 1969. A candidiasis in larvae of the Douglas-fir tussock moth, *Hemerocampa pseudotsugata. J. Invert. Pathol.* **14**:108–110.

Matthews, R. E. F. 1982. Classification and nomenclature of viruses. *Intervirology* **17**(3):1–199.

Maurand, J., A. Fize, B. Fenwick, and R. Michel. 1971. Etude au microscope électronique de *Nosema infirmum* Kudo 1921, microsporidie parasite d'un copépode cyclopoide: Création du genre nouveau *Tuzetia,* à propos de cette èspece. *Protistologica* **7:**221–225.

Meade, S. C. 1963. A medium for the isolation of *Streptococcus faecalis,* sensu strictu. *Nature* 197:1323–1324.

Miller, J. H. 1940. The genus *Myriangium* in North America. *Mycologia* **32:**587–600

Miller, M. W., and N. van Uden. 1970. In: *The Yeasts* (Ed. J. Lodder). North-Holland, Amsterdam. 408–429.

Mims, C. A., M. F. Day, and I. D. Marshall. 1966. Cytopathic effect of Semliki Forest virus in the mosquito, *Aedes aegypti. Am. J. Trop. Med. Hyg.* **15:**775–784.

Mittenburger, H. G. (Ed.). 1980. *Safety Aspects of the Baculoviruses as Biological Insecticides.* Symposium Proceedings, Bundesministerium für Forschung und Technologie, Bonn. 301 pp.

Morrill, A. W., and E. A. Black. 1912. Natural control of white flies in Florida. USDA Bureau of Entomology Bulletin No. 102. 78 pp.

Morris, T. J., R. T. Hess, and D. E. Pinnock. 1979. Physiochemical characterization of a small RNA virus associated with baculovirus infections *Trichoplusia ni. Intervirology* **11:**238–247.

Moulder, J. W. 1974. Order I Rickettsiales. In: *Bergey's Manual of Determinative Bacteriology.* 8th Ed. (Eds. R. E. Buchanan and N. E. Gibbons). Williams & Wilkins, Baltimore. 882–928.

Murphy, F. A. 1980. Togavirus morphology and morphogenesis. In: *The Togaviruses* (Ed. R. W. Schlesinger). Academic Press, New York. 241–316.

Nenninger, U. 1948. Die Peritrichen der Umgebung von Erlangen mit besonderer Berucksichtigung ihrer Wirtsspezifität. *Zool. Jahrb. Syst.* **77:**169–266.

Niklas, O. F. 1957. Zur temperaturabhängigkeit der vertikalbewegungen Rickettsiose-kranker Maikafer-engerlinge (*Melolontha* spec.). *Anz. Schaedlingskd.* **30:**113–116.

Noland, L. E., and H. E. Finley. 1931. Studies on the taxonomy of the genus *Vorticella. Trans. Am. Microsc. Soc.* **50:**81.

Ormières, R., and V. Sprague. 1973. A new family, new genus, and new species allied to the Microsporida. *J. Invert. Pathol.* **21:**224–240.

Payne, C. C. 1981. Cytoplasmic polyhedrosis viruses. In: *Pathogenesis of Invertebrate Microbial Diseases* (Ed. E. W. Davidson). Allanheld, Osmun, Montclair, New Jersey. 61–100.

Payne, C. C., and K. A. Harrap. 1977. Cytoplasmic polyhedrosis viruses. In: *The Atlas of Insect and Plant Viruses.* Vol. 8 of *Ultrastructure in Biological Systems* (Ed. K. Maramorosch). Academic Press, New York. 105–130.

Payne, C. C., and D. C. Kelly. 1981. Identification of insect and mite viruses. In: *Microbial Control of Pests and Plant Diseases, 1970–1980* (Ed. H. D. Burges). Academic Press, New York. 61–92.

Petch, T. 1921. Fungi parasitic on scale insects. Presidential Address. *Trans. Br. Mycol. Soc.* **7:**18–24.

Pilley, B. M. 1976. A new genus, *Vairimorpha* (Protozoa: Microsporidia), for *Nosema necatrix* Kramer 1965: Pathogenicity and life cycle in *Spodoptera exempta* (Lepidoptera: Noctuidae). *J. Invert. Pathol.* **28:**177–183.

Poinar, G. O., Jr. 1975. *Entomogenous Nematodes.* E. J. Brill, Leiden. 317 pp.

Poinar, G. O., Jr. 1979. *Nematodes for Biological Control of Insects.* CRC Press, Boca Raton, Florida. 277 pp.

Poinar, G. O., Jr. 1983. *The Natural History of Nematodes.* Prentice-Hall, Englewood Cliffs, New Jersey. 323 pp.

Poinar, G. O., Jr., and G. Thomas. 1967. The nature of *Achromobacter nematophilus* as an insect pathogen. *J. Invert. Pathol.* **9:**510–514.

Poinar, G. O., Jr., R. Hess, and L. C. Caltagirone. 1976. Virus-like particles in the calyx of *Phanerotoma flavitestracea* (Hymenoptera: Braconidae) and their transfer into host tissue. *Acta. Zool. (Stockholm)* **57:**161–165.

Prasertphon, S., and Y. Tanada. 1968. The formation and circulation in *Galleria* of hyphal bodies of entomophthoraceous fungi. *J. Invert. Pathol.* **11:**260–280.

Reinganum, C., G. T. O'Loughlin, and T. W. Hagan. 1970. A non-occluded virus of the field crickets *Teleogryllus oceanicus* and *T. commodus* (Orthoptera: Gryllidae). *J. Invert. Pathol.* **16:**214–220.

Rioux, J. A., and F. Achard. 1956. Entomophytose mortelle à *Saprolegnia diclina* Humphrey 1892 dans un élevage d'*Aedes berlandi* Seguy 1921. *Vie Milieu* **7:**326–337.

Ristic, M., and J. P. Krier. 1974. Family III Anaplasmataceae. In: *Bergey's Manual of Determinative Bacteriology.* 8th Ed. (Eds. R. E. Buchanan and N. E. Gibbons). Williams & Wilkins, Baltimore. 906–914.

Roberts, D. W., and W. G. Yendol. 1971. The use of fungi for microbial control of insects. In: *Microbial Control of Insects and Mites* (Eds. H. O. Burges and N. W. Hussey). Academic Press, New York. 125–149.

Rudolphi, C. A. 1809. *Entozoorum Sive Vermium Intestinalium Historia Naturalis.* Vol. 2. Sumtibus Tabernae Librariae et Artium, Amsterdam. 386 pp.

Rühm, W. 1956. Die Nematoden der Ipiden. *Parasitol. Schrift.* **6:**1–437.

Samson, R. A. 1981. Identification: Entomopathogenic deuteromycetes. In: *Microbial Control of Pests and Plant Diseases, 1970–1980* (Ed. H. D. Burges). Academic Press, New York. 93–106.

Samson, R. A., and H. C. Evans. 1973. Notes on entomogenous fungi from Ghana. I. The genera *Gibellula* and *Pseudogibelulla. Acta. Bot. Neerl.* **22:**522–528.

Sanders, R. D., and G. O. Poinar, Jr. 1973. Fine structure and life cycle of *Lankesteria clarki* sp. n. (Sporozoa: Eugregarinida) parastic in the mosquito *Aedes sierrensis* (Ludlow). *J. Protozool.* **20:**594–602.

Schaefer, E. 1961. Application of the cytochrome oxidase reaction to the detection of *Pseudomonas aeruginosa* in mixed cultures. *Roentgen Univ. Lab. Praxis* **14:**142–146.

Scherer, W. F., J. E. Verna, and G. W. Richter. 1968. Nodamura virus, an ether and chloroform resistant arbovirus from Japan. Physical and biological properties, with ecological observations. *Am. J. Trop. Med. Hyg.* **17:**120–128.

Scotti, P. D., A. J. Gibbs, and N. C. Wrigley. 1976. Kelp fly virus. *J. Gen. Virol.* **30:**1–9.

Sen, S. K., M. S. Jolly, and T. R. Jammy. 1970. A mycosis in the Indian tasar silkworm, *Antheraea mylitta* Drury, caused by *Penicillium citrinum* Thom. *J. Invert. Pathol.* **15:**1–5.

Shepard, M. R. N. 1974. *Arthropods as Final Hosts of Nematodes and Nematomorphs:*

An Annotated Bibliography 1900–1972. Commonwealth Agricultural Bureau, Farnham Royal, England. 248 pp.

Sikorowski, P., J. R. Broome, and G. L. Andrews. 1971. Simple methods for detection of cytoplasmic polyhedrosis virus in *Heliothis virescens*. *J. Invert. Pathol.* **17:**451–452.

Sinka, R. C. 1973. Viruses and leafhoppers. In: *Viruses and Invertebrates* (Ed. A. J. Gibbs). North-Holland, Amsterdam. 493–511.

Skou, J. P. 1972. Ascosphaerales. *Friesia* **10:**1–24.

Smith, K. M. 1967. *Insect Virology*. Academic Press, New York. 256 pp.

Smith, K. M. 1971. The viruses causing the polyhedroses and granuloses of insects. In: *Comparative Virology* (Eds. K. Maramorosch and E. Kurstak). Academic Press, New York. 479–507.

Smirnoff, W. A. 1974. Réduction de viabilité et de la fécondité de *Neodiprion swainei* (Hymenoptères: Tenthredinidae) par le flagellé *Herpetomonas swainei* sp. n. (Protozoaires). *Phytoprotection* **55:**64–66.

Speare, A. T. 1921. *Massospora cicadina* Peck, a fungus parasite of the periodical cicada. *Mycologia* **13:**72–82.

Sprague, V. 1940. Observations on *Coelosporidium periplanetae* with special reference to the development of the spore. *Trans. Am. Microsc. Soc.* **59:**460–474.

Sprague, V. 1963. Revision of genus *Haplosporidium* and restoration of genus *Minchinia* (Haplosporida, Haplosporididae). *J. Protozool.* **10:**263–266.

Sprague, V. 1977. Systematics of the Microsporidia. In: *Comparative Pathology*. Plenum Press, New York. 510 pp.

Sprague, V. 1982. Microsporida. In: *Synopsis and Classification of Living Organisms*. Vol. 1 (Ed. S. P. Parker). McGraw-Hill, New York. 589–594.

Sprague, V., R. Ormières, and J. F. Manier. 1972. Creation of a new genus and a new family in the Microsporida. *J. Invert. Pathol.* **20:**228–266.

Stairs, G. G. 1971. Use of viruses for microbial control of insects. In: *Microbial Control of Insects and Mites* (Eds. H. D. Burges and N. W. Hussey). Academic Press, New York. 97–124.

Steinhaus, E. A. 1949. *Principles of Insect Pathology*. McGraw-Hill, New York. 757 pp.

Steinhaus, E. A. 1959. *Serratia marcescens* Bizio as an insect pathogen. *Hilgardia* **28:**351–380.

Steinhaus, E. A. (Ed.). 1963. *Insect Pathology: An Advanced Treatise*. Academic Press, New York. Vol. I, 661 pp.; Vol. II, 689 pp.

Steinhaus, E. A., and M. E. Martignoni. 1970. *An Abridged Glossary of Terms Used in Invertebrate Pathology*. 2nd Ed. USDA Forest Service, Pacific Northwest Forest and Range Experiment Station. 38 pp.

Stoltz, D. B., and S. B. Vinson. 1979. Viruses and parasitism in insects. *Adv. Virus Res.* **24:**125–171.

Stone, A., C. W. Sabrosky, W. W. Wirth, R. H. Foote, and J. R. Coulson. 1965. A Catalogue of the Diptera of America North of Mexico. Agricultural Handbook No. 276. USDA, Washington, D.C. 1696 pp.

Summers, M. D. 1971. Electron microscope observations on granulosis virus entry, uncoating and replication processes during infection of the mid-gut cells of *Trichoplusia ni*. *J. Ultrastruct. Res.* **35:**606–625.

Summers, M. D. 1977. Baculoviruses (Baculoviridae). In: *The Atlas of Insect and Plant Viruses*. Vol. 8 of *Ultrastructure in Biological Systems* (Ed. K. Maramorosch). Academic Press, New York. 3–28.

Summers, M. D., R. Engler, L. A. Falcon, and P. Vail. 1975. *Baculoviruses for Insect Pest Control: Safety Considerations*. Selected papers from EPA-USDA working symposium. American Society for Microbiology, Bethesda, Maryland. 186 pp.

Sussman, A. S. 1951. Studies of an insect mycosis. I. Etiology of the disease. *Mycologia* **43:**338–350.

Sweeney, A. W. 1975. The mode of infection of the insect pathogenic fungus *Culicinomyces* in larvae of the mosquito *Culex fatigans*. *Aust. J. Zool.* **23:**49–57.

Sweeney, A. W., J. N. Couch, and C. Panter. 1982. The identity of an Australian isolate of *Culicinomyces*. *Mycologia* **74**(1)**:**162–165.

Swellengrebel, N. H. 1919. *Myiobium myzomyiae* n.g., n. sp., een parasitische Haplosporidie uit het darmkanal van eenige Anophelinen. *Meded. Burgerlijk. Geneeskund. Dienst Ned.-Indie* **10:**68–72.

Sylvester, E. S. 1977. Rhabdoviruses of insects (Sigma virus of *Drosophila*). In: *The Atlas of Insect and Plant Viruses*. Vol. 8 of *Ultrastructure in Biological Systems*. (Ed. K. Maramorosch). Academic Press, New York. 131–140.

Tanada, Y., and R. T. Hess. 1984. The cytopathology of baculovirus infections in insects. In: *Insect Ultrastructure*. Vol. 2 (Eds. R. C. King and H. Akai). Plenum Press, New York. 517–556.

Teninges, D., D. Contamine, and G. Brun. 1980. Drosophila sigma virus. In: *Rhabdoviruses*. Vol. 3 (Ed. D. H. L. Bishop). CRC Press, Boca Baton, Florida. 3–134.

Teninges, D., A. Ohanessian, C. R. Molard and D. Contamine. 1979. Isolation and properties of *Drosophila* X virus. *J. Gen. Virol.* **42:**241–254.

Tinsley, T. W. 1978. Use of insect pathogenic viruses as pesticidal agents. *Perspect. Virol.* **10:**199–209.

Tinsley, T. W. 1979. The potential in insect pathogenic viruses as pesticidal agents. *Annu. Rev. Entomol.* **24:**63–87.

Tinsely, T. W., and K. A. Harrap. 1978. Viruses of invertebrates. In: *Comprehensive Virology*. Vol. 12: *Newly Characterized Protist and Invertebrate Viruses* (Eds. H. Fraenkel-Conrat and R. R. Wagner). Plenum Press, New York. 1–101.

Tuzet, O., J. Maurand, A. Fize, R. Michel, and B. Fenwick. 1971. Proposition d'un nouveau cadre systematique pour les genres de microsporidies. *C. R. Acad. Sci.* **272:**1268–1271.

Umphlett, C. J., and C. S. Huang. 1972. Experimental infection of mosquito larvae by a species of the aquatic fungus, *Lagenidium*. *J. Invert. Pathol.* **20:**326–331.

Vance, A. M., and H. D. Smith. 1933. The larval head of parasitic Hymenoptera and nomenclature of its parts. *Ann. Entomol. Soc. Am.* **26:**86–94.

Vaughn, J. L. 1974. Virus and rickettsial diseases. In: *Insect Diseases*, Vol. I (Ed. G. E. Cantwell). Marcel Dekker, New York. 49–85.

Veen, K. H., and P. Ferron. 1966. A selective medium for the isolation of *Beauveria tenella* and of *Metarrhizium anisopliae*. *J. Invert. Pathol.* **8:**268–269.

Veremtchuk, G. V., and I. V. Issi. 1970. On the development of insect microsporidians in the entomopathogenic nematode, *Neoaplectana agriotos* (Nematodes: Steinernematidae). *Parasitologiya* **4:**3–7 [in Russian].

Verwoerd, D. W., H. Huismans, and B. J. Erasmus. 1979. Orbiviruses. *Comp. Virol.* **14:**285–346.

Vincent, M. 1927. On *Legerella hydropori* n. sp., a coccidian parasite of the malphigian tubules of *Hydroporus palustris* L. (Coleoptera). *Parasitology* **19:**394–400.

Van Zwaluwenberg, R. H. 1928. The interrelationships of insects and roundworms. Bulletin of the Experiment Station of the Hawaiian Sugar Planters' Association, Entomological Series, No. 20. 68 pp.

Wallace, F. G. 1966. The trypanosomatid parasites of insects and arachnids. *Exp. Parasitol.* **18:**124–193.

Waterhouse, G. M. 1973. Entomophthorales. In: *The Fungi: An Advanced Treatise* (Eds. G. C. Ainsworth, F. K. Sparrow, and A. S. Sussman). Academic Press, New York. 219–229.

Weinman, D. 1974. Family II Bartonellaceae. In: *Bergey's Manual of Determinative Bacteriology.* 8th Ed. (Eds. R. E. Buchanan and N. E. Gibbons). Williams & Wilkins, Baltimore. 903–906.

Weiser, J. 1955. A new classification of the Schizogregarine. *J. Protozool.* **2:**6–12.

Weiser, J. 1961. Microsporidia as parasites of insects. *Monogr. Angew.* **17:**1–149.

Weiser, J. 1963. Sporozoan infections. In: *Insect Pathology: An Advanced Treatise.* Vol. II (Ed. E. A. Steinhaus). Academic Press, New York. 291–334.

Weiser, J. 1976. The intermediary host for the fungus *Coelomomyces chironomi. J. Invert. Pathol.* **28:**273–274.

Weiser, J. 1977. Contribution to the classification of Microsporidia. *Vest. Cesk. Spol. Zool.* **41:**308–320.

Weiser, J., and J. D. Briggs. 1971. Identification of pathogens. In: *Microbial Control of Insects and Mites* (Eds. H. D. Burges and N. W. Hussey). Academic Press, New York. 13–66.

Weiss, E. 1974. Tribe III Wolbachieae, genus VII *Wolbachia,* and genus X *Rickettsiella.* In: *Bergey's Manual of Determinative Bacteriology.* 8th Ed. (Eds. R. E. Buchanan and N. E. Gibbons). Williams & Wilkins, Baltimore. 897–903.

Weiss, E., and J. W. Moulder. 1974. Genus I *Rickettsia,* genus II *Rochalimaea,* and genus III *Coxiella.* In: *Bergey's Manual of Determinative Bacteriology.* 8th Ed. (Eds. R. E. Buchanan and N. E. Gibbons). William & Wilkins, Baltimore. 883–893.

Welch, H. E. 1963. Nematode infections. In: *Insect Pathology: An Advanced Treatise* (Ed. E. A. Steinhaus). Academic Press, New York. 363–392.

Welch, H. E. 1965. Entomophilic nematodes. *Annu. Rev. Entomol.* **10:**275–302.

Whistler, H., S. L. Zebold, and J. A. Shemanchuk. 1974. Alternate host for mosquito-parasite *Coelomomyces. Nature* **251:**715–716.

Whittaker, R. H. 1969. New concepts of kingdoms of organisms. *Science* **163:**150–160.

Wille, H. 1956. *Bacillus fribourgensis,* n. sp., Erreger einer "milky disease" im Engerling von *Melolontha melolontha* L. *Mitt. Schweiz. Entomol. Ges.* **29:**271–282.

Wilson, G. G. 1981. *Nosema fumiferanae,* a natural pathogen of a forest pest: Potential for pest management. In: *Microbial Control of Pests and Plant Diseases, 1970–1980* (Ed. H. D. Burges). Academic Press, New York. 595–601.

Wirtz, R. 1908. Ein enfache Art der Sporenfärbung. *Centr. Bakteriol. I Abt. Orig.* **46:**727–728.

Woolever, P. 1966. Life history and electron microscopy of a Haplosporidian, *Nephridiophaga blatellae* (Crawley) n. comb. in the malpighian tubules of the German cockroach, *Blatella germanica* (L.). *J. Protozool.* **13:**622–642.

Wyss, C. 1974. Sporulationsversuche mit drei Varietaten von *Bacillus popilliae* Dutky. *Zentralbl. Bakteriol. Parasitenkd. Infektionskr. Hyg.* **126:**461–492.

Yarwood, E. A. 1937. The life cycle of *Adelina cryptocerci* sp. nov., a coccidian parasite of the roach *Cryptocercus punctulatus. Parasitology* **29:**370–390.

Zacharuk, R. Y., and R. D. Tinline. 1968. Pathogenicity of *Metarrhizium anisopliae,* and other fungi for five elaterids (Coleoptera) in Saskatchewan. *J. Invert. Pathol.* **12:**294–309.

INDEX